Minitab Handbook

Updated for Release 16

SIXTH EDITION

Barbara Ryan
Minitab, Inc.

Brian Joiner

Jonathan Cryer

D0071352

BROOKS/COLE
CENGAGE Learning·

Australia • Brazil • Japan • Korea • Mexico • Singapore • Spain • United Kingdom • United States

BROOKS/COLE
CENGAGE Learning

Minitab Handbook: Updated for Release 16, Sixth Edition
Barbara Ryan, Brian Joiner and Jonathan Cryer

Publisher/Executive Editor: Richard Stratton

Senior Sponsoring Editor: Molly Taylor

Assistant Editor: Shaylin Walsh Hogan

Editorial Assistant: Alexander Gontar

Media Editor: Andrew Coppola

Marketing Coordinator: Lauren Beck

Marketing Communications Manager:
 Jason LaChapelle

Rights Acquisitions Specialist:
 Shalice Shaw-Caldwell

Art and Design Direction, Production
 Management, and Composition:
 PreMediaGlobal

Manufacturing Planner: Sandee Milewski

All Material in this edition: © 2012 Cengage
 Learning
Cover Image: © leremy/iStockphoto

For product information and technology assistance, contact us at
Cengage Learning Customer & Sales Support, 1-800-354-9706
For permission to use material from this text or product,
submit all requests online at **www.cengage.com/permissions.**
Further permissions questions can be emailed to
permissionrequest@cengage.com.

Library of Congress Control Number: 2012945456

ISBN-13: 978-1-133-93944-3
ISBN-10: 1-133-93944-9

Brooks/Cole
20 Channel Center Street
Boston, MA 02210
USA

Cengage Learning is a leading provider of customized learning solutions with office locations around the globe, including Singapore, the United Kingdom, Australia, Mexico, Brazil and Japan. Locate your local office at **international.cengage.com/region**

Cengage Learning products are represented in Canada by Nelson Education, Ltd.

For your course and learning solutions, visit **www.cengage.com.**

Purchase any of our products at your local college store or at our preferred online store www.cengagebrain.com.

Instructors: Please visit **login.cengage.com** and log in to access instructor-specific resources.

Printed in the United States of America
1 2 3 4 5 6 7 16 15 14 13 12

Contents

Chapter 6: Statistical Distributions 183

Chapter 7: Simulation 215

Preface

Minitab is a powerful statistical software program that provides a wide range of basic and advanced capabilities for statistical analysis. More students have learned statistics using Minitab than with any other software package. Substantially updated and refined in Release 16, Minitab's broad, powerful capabilities and unmatched ease of use make it the ideal teaching tool. As a result, more than 4,000 colleges, universities, and high schools worldwide rely on Minitab. Developed over 40 years ago, by professors for professors and students, Minitab has become the standard for statistics education. And because Minitab is the leading package used in industry for quality and process improvement, students who learn Minitab will have the advantage of knowing how to use a real-world business tool.

The *Minitab Handbook* is designed to teach you how to use Minitab to analyze data. This book was written primarily for use as a supplementary text for first- or second-level courses in statistics (precalculus or postcalculus), as well as for researchers and practitioners new to Minitab. The *Handbook* emphasizes aspects of statistics that are particularly appropriate to computer use, such as using plots creatively, applying standard methods to real data, exploring data in depth, using simulation as a learning tool, screening data for errors, manipulating data, and performing transformations, in addition to doing multiple regression and analysis of variance.

About the Sixth Edition

The sixth edition reflects substantial changes in the Minitab software and in statistics teaching that have occurred since the fifth edition was published. Today's students are most comfortable using a menu-driven interface, so the sixth edition teaches Minitab procedures using mostly the menus. Session commands, however, are introduced early on and referred to throughout the book so that students will learn that they can accomplish some procedures more efficiently with commands. The book is based on Minitab, Release 16, first available in 2011. Release 16 is the most comprehensive release to date and includes substantial new features.

This edition makes significant use of Minitab's latest graphics capabilities and enhances several features, such as plot editing, brushing and jittering, 3D graph rotation, interaction plots, and more. Those new to Windows and Minitab will find Chapter 1 helpful in introducing the menu interface and the data management techniques available with projects. Chapter 2 introduces Session commands so that you will be prepared to use these commands later in the book when necessary. Chapter 3 introduces all Worksheet operations in order to introduce students to the procedures they will need later on to carry out such simple operations as stacking columns or sorting data. A chapter on Creating Reports is included and the new Minitab Assistant is introduced in several chapters. The core of the book—the coverage of Minitab's statistical procedures—remains the same.

In Release 16 almost everything in Minitab can be customized. You can create your own menus and toolbars, add short cuts, and so on. In this *Handbook* we assume that the default settings apply. If necessary, check the online Help pages for instructions to reset Minitab to its default values before trying the steps listed in this book.

Examples and Exercises

This *Handbook* includes numerous examples and exercises that demonstrate, step-by-step, how to use the Minitab to explore and analyze data. Many of these examples and exercises have been obtained directly from us while consulting with researchers from a wide variety of application areas. We present these examples and exercises in enough detail to give readers an idea of the scope and importance of problems amenable to statistical treatment. The chapters and sections are designed to be as independent of each other as possible. Once the main points of Chapters 1 through 5 have been covered, instructors can cover the remaining chapters in almost any order. In particular, Chapter 17, Creating a Report, may be read any time after Chapter 4—the sooner, the better. Advanced students should be able to skim much of Chapter 1, and those who don't plan to use Session commands could skip Chapter 2. However, many of the Worksheet operations in Chapter 3 are used later in the book, so early exposure to those procedures will be helpful.

Data Sets Used in This Handbook

We have incorporated over seventy data sets in this book. Most of them are distributed with the Minitab software. Several new data sets are used for illustration and exercises and nine new macros are used to reenforce various statistical concepts. The data sets and macros are available for download from the Data Library at the Web site (www.CengageBrain.com.) associated with this book.

History of Minitab

Minitab began as a student-oriented adaptation of the National Bureau of Standards "Omnitab" system, originated by Joseph Hilsenrath and developed further by David Hogben and Sally Peavy. The name "Minitab" was derived from that origin. Minitab Statistical Software was originally developed in 1972 for students in introductory statistics courses at The Pennsylvania State University. Because Minitab was developed as a portable program—that is, one that could run on many different types of computers—it was subsequently widely distributed to other universities. The first edition of the *Minitab Handbook* was published in 1976 and was the first widely used book of its kind. It helped instructors, especially those new to computers, to integrate the use of computers into teaching statistics. Since then, Minitab's use has expanded beyond introductory-level coursework, and its capabilities have been extended way beyond basic statistics.

Originally, Minitab was developed as a command-based system that used simple, English-like commands to allow the user to "speak" to the computer. However, with Release 9, Minitab, Inc. provided a full Windows interface. Commands are still available, though, for those who prefer to use them and for writing macros. For further information about the software, contact Minitab Inc., Quality Plaza, 1829 Pine Hall Road, State College, PA 16801-3008, USA, Phone: (814) 238–3280, Fax: (814) 238–4383, or visit their web site at www.minitab.com. A 30-day trial copy of the complete software may be downloaded from the Minitab web site.

Acknowledgments

We appreciate the many suggestions we have received from users of Minitab and of earlier editions of this *Handbook*. Our colleagues and students at The Pennsylvania State University and the University of Wisconsin deserve special mention. In writing and revising the *Handbook*, we are especially grateful to Thomas P. Hettmansperger, who provided extensive suggestions (particularly in the nonparametric statistics chapter); Paul Velleman, whose comments were always insightful, and thought-provoking; and Lawrence Klimko, for his careful and extensive criticisms. In the Credits, page 532, we refer to those who kindly made their data available to us. We would also like to thank The Pennsylvania State University for their partial support while we wrote the first edition of the *Handbook*. We want to acknowledge our appreciation for the thoughtful advice we received from the many manuscript reviewers for each edition. These reviewers include Sung K. Ahn, Washington State University; Jonathan D. Cryer, University of Iowa; Nicholas R. Farnum, California State University, Fullerton; Michael Martin, Stanford University; David Metcalf, University of Virginia; Richard Moreland, University of Pittsburgh; David Rogosa, Stanford University; Dale G. Sauers, York College of Pennsylvania; Bruce E. Trumbo, California State University, Hayward; David Viglienzoni, Cabrillo College; Vern C. Vincent, University of Texas–Pan American; and W. T. Woodard, University of Hawaii.

We would especially like to thank Christine Bayly, Joanne DeVoir, Linda Holderman, Cynthia Nucciarone of Minitab, Inc. and Alexander Gontar, Shaylin Walsh Hogan and Molly Taylor of Cengage Publishing for their help with this edition.

Barbara Ryan
Brian Joiner
Jonathan Cryer
June 2012

1

Introduction to Minitab

1.1 Welcome to Minitab

Modern computers have changed the way statistical analysis is performed in significant ways. They are invaluable for the storage, retrieval, manipulation, graphical display, and statistical analysis of data. The small personal computer, which may be used by itself, as a member of a local area network (LAN), or as a workstation communicating over the internet with other computers around the world, has significantly altered the way statistical analysis is accomplished.

This *Handbook* will show you how to use the statistical software Minitab® for Windows to do statistical computations, data management, and statistical graphics. Minitab is an easy-to-use, general-purpose statistical computing system. Minitab is especially easy to use on desktop or laptop computers that employ the Microsoft Windows® graphical user interface. Most of the tasks that the software can do may be selected from user-friendly menus and extensive on-screen help is always available. Minitab also incorporates a *StatGuide*™ that will help you interpret statistical procedures and an *Assistant* that will lead you through your statistical analysis. This *Handbook* is not intended as a complete reference for Minitab for Windows. Rather, it will show you how to perform statistical analysis at about the level of first, second, or third applied statistics courses.

The *Handbook* is based on Minitab for Windows, Release 16, in particular 16.2.1, and the Microsoft Windows 7 operating system, in particular, Windows Home Premium. Many of the features of Minitab illustrated will also work with earlier releases of Minitab.

Minitab allows you to enter and carry out commands in two ways: choosing them from a menu or typing them at a command prompt in the Session Window. This book introduces you to both modes of operation, but emphasizes menu commands. You can ignore Session commands altogether by skipping Chapter 2, although later chapters do sometimes employ Session commands when necessary. You might find it quicker to execute certain operations, especially common or repetitive ones, using Session commands. Many Minitab users employ a combination of methods: using menu commands for unfamiliar operations, and using Session commands for quick execution. You'll most likely find a favorite mode of working as you become familiar with the software. (You will need to learn Session commands if you wish to write Minitab macros.)

Minitab Overview

Minitab works with data in worksheets of rows and columns. It has over 200 commands to perform various manipulations and statistical analyses. Many useful graphics can be displayed easily based on data in the worksheet. Columns of the worksheets are used to store series of numbers, such as the monthly sales figures for a particular product. In addition, storage locations called constants are available to store single numbers, such as an average sales figure, and matrices to store rectangular arrays.

The procedure for starting Minitab will depend on the particular computer setup at your location. It will likely be as simple as selecting Minitab from a system menu, double-clicking the Minitab shortcut icon on your desktop (shown here at the left), or selecting Minitab from the Start ▶ All Programs list. Ask a colleague or your instructor for further information.

Minitab uses several windows: the Session Window, Data Windows, and Graph Windows. Help Window and History folders are also available. The Project Manager Window integrates most of these separate folders into a coherent form.

Here is an overall look at various Minitab Windows after data have been entered and a graph generated:

Figure 1.1   Minitab overview

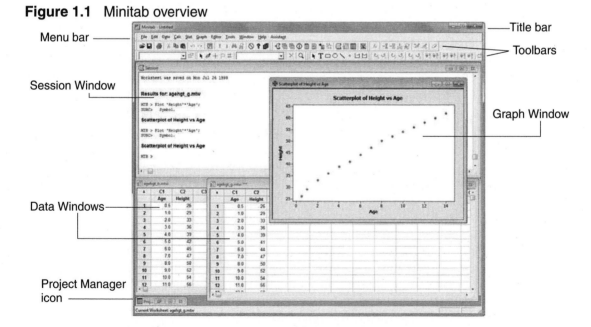

The Session Window

After the Minitab software is loaded into the computer, your screen will look something like the display below. This is the Minitab Session Window. (You may need to maximize the Session Window.)

The title bar, with the word Minitab and Age-Hgt.MPJ-[Session], identifies the window as the Session Window. Numerical results and character-style graphics are displayed here. Minitab commands may be entered in the Session Window, but are usually easier to enter using the pull-down menus. Notice the names of the menu buttons: File, Edit, Data, Calc, Stat, Graph, Editor, Tools, Window, Assistant and Help. On-screen Minitab help may be obtained by clicking on the word Help on the menu bar or by clicking on the Help icon (the question mark symbol). See "Getting Help", page 7, for more information on Minitab Help.

Here is a more detailed look at the Session Window displayed by Microsoft Windows 7 when you begin a Minitab session:

Figure 1.2 Session Window

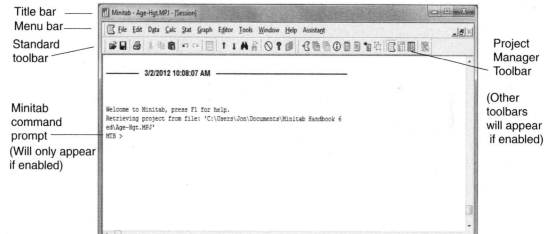

Title bar — Menu bar — Standard toolbar — Minitab command prompt (Will only appear if enabled) — Status bar — Project Manager Toolbar — (Other toolbars will appear if enabled)

Data Windows

Minitab Data Windows display data Worksheets in rows and columns. They are used to store basic data for statistical analysis but may also contain numerical results of your statistical calculations. You move from the Session Window to any of the Data Windows either by clicking on a visible portion of a Data Window, by selecting the appropriate Worksheet item from the Window menu, by pressing Ctrl+D successively until the required data Window is active, or by using the Project Manager.

In general, three asterisks, (***) show which Worksheet is "active." Data from the active Worksheet are used when you request any calculation or graph. Here is an example of a portion of a Minitab Data Window:

Figure 1.3   A Minitab Data Window

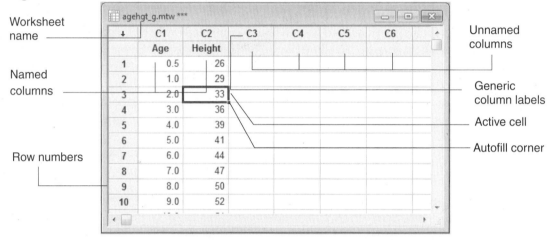

The Editor menu allows several choices for a Data Window. The Data, Session, and Graph Windows each have their own unique Editor (and Edit) menu choices. See Chapter 3 for more information on the Data Window.

When working in Minitab, we frequently move back and forth between the Session Window and a Data Window. If your screen is large enough, you may keep a large portion of several windows visible at the same time. Then you can go back and forth by simply clicking on the desired window. It is also useful to remember the keyboard shortcuts: Ctrl+M (M=Minitab) to go to the Session Window, and Ctrl+D (D=Data) to go to or cycle through the Data Windows. You may also use the Project Manager to move among the various windows.

Working with Minitab Menus

Minitab's commands are available in the menus that you open from the menu bar. To open a menu, you click the menu name (File, Edit, Data, Calc, Stat, Graph, Editor, Tools, Window, Help or Assistant). You can also open a menu by using the Alt key in combination with the underlined letter in the menu name. For example, to open the File menu, you press Alt+F, because F is the underlined letter in the word File. When you open a menu, you see a list of commands.

Note the following command conventions:

- A command followed by an ellipsis (...) opens a **dialog box**, which requests the information necessary to carry out the command. Not all commands require a dialog box.

- When a command is followed by an arrow, a **submenu** of additional commands or options opens, for example, the Other Files command under the File menu shown at the right.

- Commands that have no symbols following them, such as the Exit command on the File menu, execute immediately when you choose them.

- When a command is followed by a key combination, you can press the indicated keys to execute the command without using the menus (if, of course, you've memorized the key combination). For example Ctrl+O shown to the right opens a Minitab Project.

- All menu commands contain an underlined letter, which allows you to use the keyboard. For example, to choose Save on the File menu, you open the menu and then press S (no need to use Alt).

- Some commands appear "grayed-out," which indicates they are currently unavailable. For example, the Worksheet Description command shown above since the Worksheet was not active at the time.

- You can close a menu without executing it by clicking an area off the menu or by pressing the Esc key.

    This book abbreviates the instruction to choose a menu command like this: Choose **File ▸ Open Worksheet**. That means to click File on the menu bar and then click Open Worksheet among the choices on the drop-down menu. Choose **Stat ▸ Tables ▸ Tally Individual Variables** means to click Stat on the menu bar, click Tables on the drop-down menu, and then click Tally Individual Variables on the submenu that appears.

Working with Dialog Boxes

Minitab dialog boxes also operate according to the usual Windows conventions. Figure 1.4 shows a typical Minitab dialog box.

Figure 1.4   Minitab dialog box

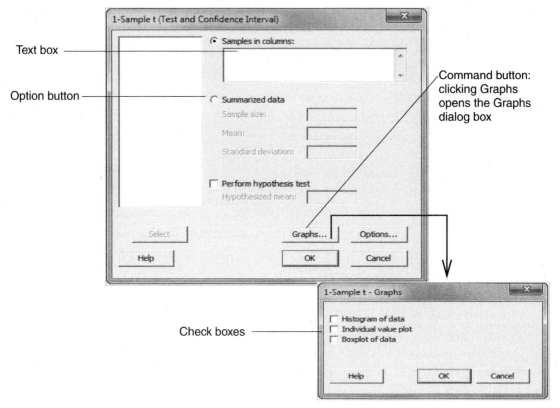

Dialog box elements function according to these conventions:

- Text boxes accept text.
- List boxes display a list of options from which you may choose one. Some lists appear only when you click a list arrow.
- Check boxes are options that can be enabled (an x appears in the box) or disabled (the box is empty). You click a check box to enable or disable it.

- Option buttons are groups of mutually exclusive options. The selected option appears with a bullet. To select an option button, you click in the circle next to the option you want.
- Command buttons open additional dialog boxes with more options.
- Click the OK button to execute the options you've selected, or click the Cancel button to cancel the operation.
- Most dialog boxes include a Help button that you can click to get context-related help relating to that dialog box.

Getting Help

An enormous amount of information about Minitab is stored with the Minitab program in the Minitab online Help system, available on the Help menu. The Help menu is shown below.

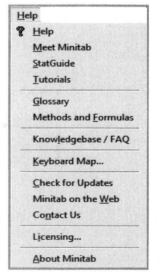

- **Help** opens the main Help system in a Microsoft Windows-style help window.
- **Meet Minitab** takes you to the Minitab Inc. site where you may download the 122 page *Meet Minitab* guide as a pdf file.
- **StatGuide** opens Minitab's StatGuide which "provides statistical guidance for interpreting statistical tables and graphs in a practical, easy-to-understand way."
- **Tutorials** opens Minitab's built-in tutorial sessions which lead you step-by-step through basic statistical analyses with the software.
- **Glossary** opens a glossary of statistical terms.
- **Methods and Formulas** opens detailed technical information about computational procedures.
- **Knowledgebase/FAQ** opens Minitab's web page of searchable frequently asked questions.
- **Keyboard Map.**
- **Check for Updates** assesses whether or not there have been updates to your Minitab software.
- **Minitab on the Web** points your browser to the Minitab Inc. Web site, which offers the most current information on Minitab products.
- **Contact Us** takes you to Minitab's web site page with contact information.
- **Licensing**.
- **About Minitab** displays information about the release of Minitab you are using and your system information.

Minitab Help is organized like most Windows Help systems, featuring both a table of contents and an index that you can search. A tree-like structure for easy navigation appears when you open Help and click the Contents tab. An example is shown at the right.

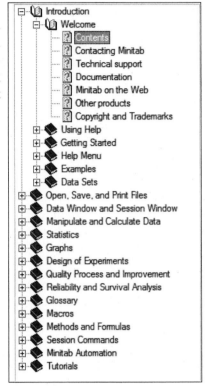

You can also get help on most Minitab dialog boxes by clicking the Help button on a specific dialog box. The Help system opens to the topic that applies to that dialog box.

The right side of the initial help screen is shown in Figure 1.5. If you click on Stat menu you are led to further choices shown on the left side of Figure 1.6. Clicking on Basic Statistics brings up the choices shown on the right side of Figure 1.6. Finally, clicking on Display Descriptive Statistics displays the information in Figure 1.7.

Figure 1.5
Minitab
Help
Overview

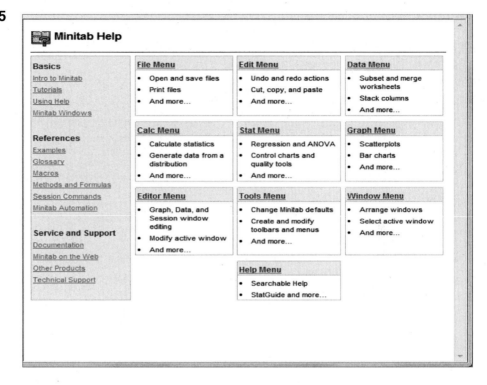

Figure 1.6
Stat Menu
Help and
Basic
Statistics

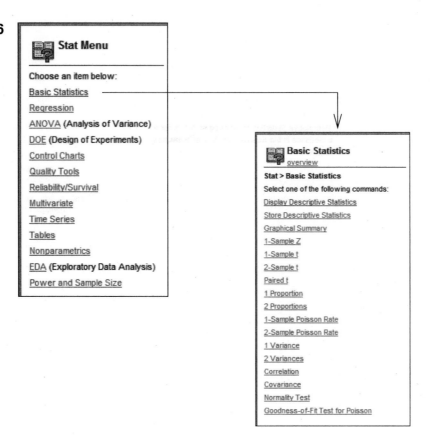

Figure 1.7

Help for
Display
Descriptive
Statistics

 Display Descriptive Statistics
overview   how to   example   data   see also

Stat > Basic Statistics > Display Descriptive Statistics

Produces descriptive statistics for each column, or for each level of a By variable.

To calculate descriptive statistics individually and store them as constants, see Column Statistics. To store many different statistics, use Store Descriptive Statistics.

Use Display Descriptive Statistics to produce statistics for each column or for subsets within a column. You can display these statistics in the Session window and optionally in a graph (see Descriptive Statistics Available for Display or Storage).

Dialog box items

Variables: Choose the columns you want to describe.

By variables (optional): Enter the column containing the by variables to display descriptive statistics separately for each value of the specified variable. Column lengths must be equal. See Graph Limits for limitations associated with using a By variable.

<Statistics>

<Graphs>

Understanding Minitab Windows and Folders

The main Minitab program Window is a container for the Minitab windows in which you do your work. In addition to the Help Window you just worked with, Minitab uses several windows and folders, described in Table 1.1. Folders appear within the Project Manager.

All open windows appear on the Window menu. You can view a window or make it active by choosing it from the Window menu or clicking any part of it. You can identify an active window by the fact that its title bar is brighter and not dimmed.

Table 1.1 Minitab Windows and Folders

Window or Folder	Description
Data Window	Displays Worksheet data in rows and columns. Each Worksheet has its own Data Window.
Session Window	Displays text output generated by your analyses. There is only one Session Window. The Session Window may also be used to enter Session commands as described in Chapter 2.

Table 1.1 Minitab Windows and Folders (Continued)

Window or Folder	Description
Graph Window	Displays up to 200 graphs, each graph in its own Graph Window.
Project Manager Window	Contains folders that permit you to view and manipulate all the parts of your project.
History Folder	Displays the sequence of Session commands executed without showing the output or results. There is only one History folder. Serves as an "audit trail" of your analysis.
Assistant screens	The Minitab Assistant displays special windows entitled Summary Report, Report Card, Diagnostic Report, Model Selection Report, and more depending on context.

When there are many windows open within Minitab, you might want to minimize some windows to reduce them to small title bars within the Minitab window. To reopen a reduced window, click the Restore button, 〔▣〕, on the reduced title bar.

1.2   Working with Data

Minitab stores data in a worksheet in three formats: columns, single-number constants, and matrices.

- **Columns** in a worksheet are designated generically as C1, C2, C3, C4.... Columns can and should be given names. Their contents can be viewed in the Data Window and also displayed in the Session Window if desired.

- **Constants** are designated generically as K1, K2, K3... and can be named. Any calculation that results in a single number can save that number in a stored constant. You can then use the stored constant in place of the number in any command. Constants cannot be viewed in the Worksheet but their values can be displayed in the Session Window if desired.

- **Matrices** are designated generically as M1, M2, M3... and can be named. Matrices store values in a rectangular block of hidden cells containing numbers. Matrices cannot be viewed in the Worksheet but can be displayed in the Session Window if desired.

The total worksheet size and number of data points available to you depends on the amount of memory installed in your computer. You usually work with data in a Data Window, which displays the columns in your worksheet, similar to a spreadsheet. Data are entered in rows, numbered 1, 2, 3, 4.... The intersection of a row and column is called a cell. A cell is identified by its column and row number. The column number, such as C1 or C2, is displayed in the column header, a cell at the top of each column. The row number, such as 1 or 2, is displayed in the row header, a cell to the left of each column. This book occasionally designates a cell by its row and column number. For example, R2C4 designates the cell in the second row, fourth column.

Missing Data

Many data sets are missing one or more observations. When you enter the data, type an asterisk (*) in place of a missing numeric value. All Minitab commands automatically take missing data into account when they do an analysis. If you ask Minitab to perform an impossible calculation, such as taking the square root of a negative number, Minitab sets the answer to *. A blank cell is used for a missing text value.

Data Types

Minitab uses three data types. Data may be either numeric, text, or date/time. All data in a given column must of the same type. **Numeric data** consists of number characters and * (the missing value code). Currency data may be read or typed into a Worksheet including dollar[†] signs and commas. Minitab operates on it as numeric format but displays the dollar signs and commas. If you are entering data in exponential notation, numeric values also contain the letter E, as in 3.2E12. If a column or constant contains any character other than numbers or *, Minitab interprets the entire column as text. **Text data** consists of keyboard characters and is often used to specify the levels of categorical variables. For example a column containing the gender of a participant in an experiment might use the text values "M" and "F" for the two levels of gender. (Note that all commands accept text designations for categorical variables.) A text value can be up to 80 characters long. **Date/time data** stores dates, times, or both. Minitab stores date/time data as numbers but displays them in whatever format you choose.

Entering Data

Table 1.2 shows reaction time data for 14 sprinters who participated in the 1996 Summer Olympics. Each sprinter's reaction time upon hearing the starting gun was recorded in the first and second rounds of the 100-meter dash.

You'll first enter these data and then use some basic Minitab commands to describe the data. When you first start Minitab, cell R1C1 is active. The **active cell** is highlighted by a dark border. See Figure 1.3, p. 4. The active cell accepts

---

[†] Minitab recognizes the currency symbols selected in the Currency tab of the Windows Control Panel Region and Language Options. It can recognize Euro (€), UK pound (£), US dollar ($), Japanese yen (¥), and the Korean currency symbol.

data you type on the keyboard. You can activate a cell by clicking it or by moving to it using navigation keys on your keyboard. Once a cell is active, you can start typing to enter data into that cell. When you type, the cursor appears. The **cursor** is a blinking vertical bar that indicates where the data will appear as you type.

Table 1.2 Sprinter data

Sprinter	Round 1	Round 2
Asahara	0.148	0.175
Bailey	0.172	0.181
Boldon	0.137	0.160
Christie	0.160	0.134
Drummond	0.140	0.116
Ezinwa	0.152	0.157
Fredericks	0.156	0.149
Green	0.175	0.147
Markoullides	0.214	0.195
Marsh	0.185	0.161
Mitchell	0.165	0.200
Nkansah	0.189	0.181
Surin	0.168	0.168
Thompson	0.184	0.172

The function of the navigation keys depends on the direction of the **data-direction arrow**, a small arrow in the upper-left corner of the Worksheet. If the arrow is pointing down, you can enter data columnwise, or one column at a time. If the arrow is pointing to the right, you can enter data rowwise, or one row at a time. Table 1.3 shows the function of the navigation keys depending on the direction of the data-direction arrow. If you typed data in a cell before using a navigation key, Minitab accepts the data as it moves. The last six operations don't depend on the direction of the data-direction arrow.

Table 1.3 Navigation keys

Key	Action when arrow points right	Action when arrow points down
Enter	Moves one cell to the right	Moves one cell down
Home	Moves to beginning of active row	Moves to top of active column
End	Moves to end of active row	Moves to bottom of active column
Ctrl+Enter	Moves to beginning of next row	Moves to beginning of next column
	Action regardless of direction arrow points	
Tab	Moves one cell to the right	
Arrow keys	Move as expected	
Ctrl+Home	Moves to R1C1 (the upper left-hand corner of the data cells)	
Ctrl+End	Moves to the last cell in the worksheet that contains data	
PageUp	Moves up through the data one screen at a time	
PageDown	Moves down through the data one screen at a time	

You can enter data in a block by highlighting the area you want to work in. You highlight an area by dragging the mouse over that area. The active cell stays within the highlighted block. To enter the sprinter data into the Worksheet in the Data Window,

1. If necessary, maximize the Data Window.
2. Make sure the data-direction arrow points to the right. If it doesn't, click it. See Figure 1.8.

Figure 1.8   Entering data in a Worksheet

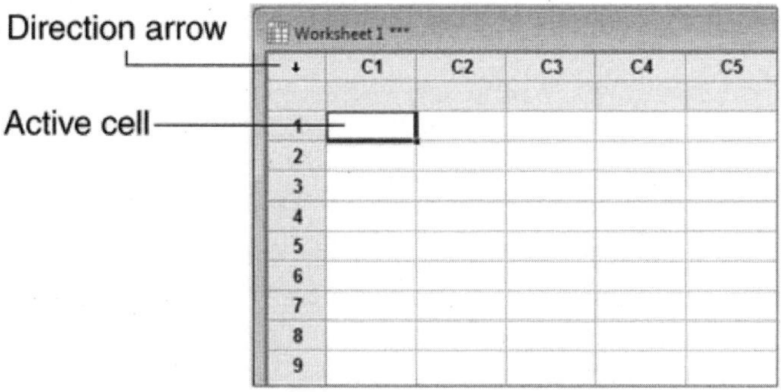

3. If necessary, click cell R1C1.

4. Type **Asahara**, the first data point in row 1. As soon as you start to type, the cursor appears.

   💡 *TIP If you make any mistakes while entering these data, don't worry. You'll learn how to correct worksheet entries shortly.*

5. Press **Enter**. The active cell moves to the right. Notice that the column header now shows C1-T. The T designates C1 as a column of text data.

6. Type **.148**, the second data point in row 1, press **Enter**, type **.175**, and then press **Enter**. The active cell continues to move to the right. However, you want to move to the beginning of the second row. (Notice that Minitab adds a leading zero to your numbers.)

7. Press **Ctrl + Enter** to move to the beginning of the second row.

8. Here is another way to simplify the data entry. Drag the mouse from cell **R2C1** to **R14C3 to** highlight the entire block rows 2-11. This will confine the active cell to this part of the Worksheet. See Figure 1.9.

Figure 1.9
Using a highlighted block to enter data

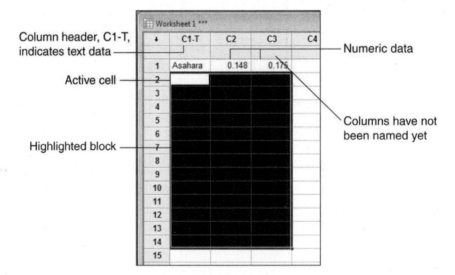

9. Type **Bailey**, the first data point in row 2, and then press **Enter**. Type **.172**, press **Enter**, type **.181**, and then press **Enter**. Notice that because you are working within a highlighted block, pressing Enter automatically moves you to the beginning of the next row.

10. Enter the remaining sprinter data in Table 1.2, using the techniques you've learned. When you reach the end of the block and press Enter the last time, the active cell returns to the first cell in the block, cell C1, row 2.

💡 TIP *You might notice as you type the entry Drummond for R5C1, Minitab automatically widens the column to accommodate the entry. Minitab widens the column even further when you type Markoullides. Minitab will continue to widen the column as necessary to accommodate the widest entry.*

11. Click cell **R1C1** to remove the highlighting from the block. The worksheet should look like Figure 1.10.

Figure 1.10
Sprinter data entered into worksheet

↓	C1-T	C2	C3	C4
	Name	Round 1	Round 2	
1	Asahara	0.148	0.175	
2	Bailey	0.172	0.181	
3	Boldon	0.137	0.160	
4	Christie	0.160	0.134	
5	Drummond	0.140	0.116	
6	Ezinwa	0.152	0.157	
7	Fredericks	0.156	0.149	
8	Green	0.175	0.147	
9	Markoullides	0.214	0.195	
10	Marsh	0.185	0.161	
11	Mitchell	0.165	0.200	
12	Nkansah	0.189	0.181	
13	Surin	0.168	0.168	
14	Thompson	0.184	0.172	
15				

Worksheet 1 ***

Editing Cell Contents

Whenever you enter data, you should proofread it carefully. If you have made a mistake, you can easily edit a cell's contents.

- To replace the entire cell contents, click the cell to make it active, type the new entry, and then press Enter. The new data replaces the old.
- To modify only certain values in a cell, double-click the cell. The cursor appears. Use the Left and Right Arrow keys on your keyboard to move the cursor to the values you want to edit. Type any values you want to insert, press Backspace to delete values left of the cursor, or press Delete to delete values to the right of the cursor. Press Enter when you are finished modifying the cell.

If you have altered a cell's contents but haven't yet pressed Enter, you can revert to the original entry by pressing the Esc key. To correct any mistakes in the Sprinter data,

1. Compare the sprinter data you just typed, shown in Figure 1.10, to the data in Table 1.2.

2. Click a cell with an error. If you made no errors, you might want to try these steps anyway so you know how to edit cell contents. When you click a cell with an error, Minitab highlights the cell. Anything you type will replace the entire entry.

3. Type the entry correctly, and then press Enter.

You can also undo any action you've taken by choosing Edit ► Undo from the menu or clicking the Undo button on the toolbar. Chapter 3 contains more information on Worksheet operations.

> *Dollar signs and commas may be typed directly into cells of a Worksheet. Minitab will display the commas and dollar signs but ignore them in any calculations.*

Naming Columns

You can refer to a column by its generic column header (such as C2), or, preferably, you can give it a name and reference it by name instead. It's usually easier to remember the name of a variable rather than the number of a column. Naming columns also makes Session Window output easier to read because output is labeled by column name.

You name a column by clicking the name cell at the top of that column (just below the column header) and then typing the column name. A name may be no longer than 31 characters. You may use any characters, but a name may not begin or end with a blank, and may not contain the symbol ' (apostrophe or single quote) or # or start with or consist entirely of the symbol *. You can refer to column names using either uppercase or lowercase; Minitab doesn't distinguish between uppercase and lowercase in column name. Thus 'Height' and 'HEIGHT' are equivalent. You can change a column name by clicking the name cell and typing a new name. To enter column names for the Sprinter data,

1. Click the name cell below the C1-T column header, shown in Figure 1.11.

Figure 1.11
Naming columns

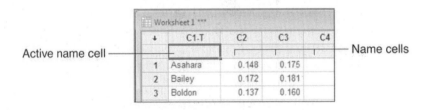

2. Type **Name** and then press **Enter**. The active cell moves to the right.

3. Type **Round 1**, press **Enter**, type **Round 2**, and press **Enter** again.

Saving Worksheets

Having taken the time to enter the Sprinter data you will want to save it in a Minitab Worksheet file so that you may open it for later analysis.

You will learn more about saving various Minitab files later. For now let's save your worksheet in MTW format to ensure you don't have to reenter the data later. To save a worksheet in MTW format,

1. Make sure the worksheet you want to save is the active window.
2. Choose **File ▸ Save Current Worksheet**.

> 💡 *TIP The button* 💾 *is the Save Project button—**not** the Save Current Worksheet button. We will use this button later.*

3. Click the **Save in:** list arrow and open the drive or folder in which you want to save the worksheet. The Windows 7 icons shown at the right may be useful for quicker navigation. You may have to navigate several levels to get to the appropriate directory or device—especially if you work on a network.

4. In the File name box, type **Sprinter** as the file name (Don't type a file extension; Minitab adds that automatically). (Save as type: This should read Minitab by default. If not, reset it with the pull-down menu.)
5. Click **Save**. Minitab saves the file with the name and location you indicated, and appends the .MTW extension to the file name.

You can now reopen the Sprinter Worksheet in Minitab any time you need it.

1.3   Working with Minitab Windows

To familiarize yourself with Minitab's windows, try performing some basic Minitab operations on the sprinter data you entered.

Choosing Variables

Many Minitab dialog boxes require you to designate the columns you want to use for a given operation. Usually you click a text box that requires a variable, and a variable list appears on the left side of the dialog box that lists the non-empty columns of the active worksheet. The variable list appears only when the cursor is in a box that accepts variables. Otherwise the list is empty or gray and unavailable. Only variables that meet the selection criteria for a given box appear for selection. For example, if the cursor is in a box that accepts numeric variables only, text variables will not appear in the variable list. When you want to enter variables in a text box, you first need to click in the text box. Then do one of three things:

- To select columns one at a time, double-click each of them in the list box.
- To select a group of columns, first highlight them. You can highlight a list of contiguous columns by dragging over them or by clicking the first one, pressing and holding down the Shift key, and then clicking the last one. You can select noncontiguous columns by pressing the Ctrl key as you click. Once the columns you want to select are highlighted, click the Select button.

You can designate consecutive columns using a hyphen shorthand. The list C2 C3 C4 C5 can be abbreviated C2-C5. If C2 is named "Before" and C5 is named "After" you can also type the range Before-After.

- Type the name or generic column name, for example C2, in the text box.

Performing Calculations in the Data Window

In Minitab, cells contain values. Cells can contain formulas that update based on other cells, such as they might in a spreadsheet program. If you have worked with spreadsheet software before, you might know that to calculate, for example, the sum of the values in two cells, you may enter a formula into a cell, and that formula recalculates the results automatically if the addends are changed. In Minitab you perform such calculations using the Calculator, and the resultant values are either static or can be updated if the calculation is Assigned as a formula. See Figure 1.12. (Graphs may also be updated automatically, manually, or not at all as you desire.)

Let's calculate how much the reaction time of each sprinter changed from Round 1 to Round 2. The Calculator command allows you to perform both simple and complex mathematical operations, such as are available on a calculator and much more. To calculate the change in reaction time and store the results in a new column,

1. Open the **Sprinter** Worksheet if it is not already open. (See page 35 if you need help.)
2. Choose **Calc ▶ Calculator**. The Calculator dialog box opens.

3. Type **Change** in the Store result in variable box. This tells Minitab to place the results in the next available empty column and name that column Change.

4. Press **Tab** to move to the Expression box where you'll enter the expression you want to calculate. You could also click in the Expression box.

5. Double-click **C3 Round 2** in the list of available columns, click the **minus** button, and then double-click **C2 Round 1**. This tells Minitab to subtract the values in C1 from those in C2, row by row. See Figure 1.12.

6. Click **OK**. Minitab performs the calculations and displays the results in C4 named Change. A portion of the results are shown in Figure 1.13.

Figure 1.12 Calculator dialog box

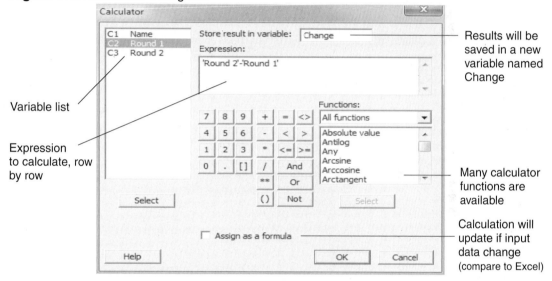

Figure 1.13
Calculating a
difference

↓	C1-T	C2	C3	C4	C
	Name	Round 1	Round 2	Change	
1	Asahara	0.148	0.175	0.027	
2	Bailey	0.172	0.181	0.009	
3	Boldon	0.137	0.160	0.023	
4	Christie	0.160	0.134	-0.026	
5	Drummond	0.140	0.116	-0.024	

If you were to select the **Assign as a formula** check box on the Calculator dialog box, then the Change column would update automatically if more data pairs were added to the Round 1 and Round 2 columns or if corrections were needed in those columns. If they are assigned, formulas work as they do in spreadsheets.

Generating Statistics in the Session Window

When you generate statistics from your data, the results appear in the Session Window. You can edit and format Session Window text with word processing tools. You can also enable the command language so that you can type commands in the Session Window rather than choose them from a menu, as you'll see in Chapter 2.

Let's calculate the average reaction time for the sprinters in Round 1 and Round 2, and the average change in reaction time. Commands that produce most statistics are located on the Stat menu. The average, also called the **mean**, is a descriptive statistic in that it describes a certain aspect of some or all of your data. Minitab displays the results in the Session Window. To obtain descriptive statistics,

1. Choose **Stat ▸ Basic Statistics ▸ Display Descriptive Statistics**. Enter the column names in the Variables box. You could double-click the variable names from the list of available columns as you did when you calculated the difference, but, if the columns are contiguous, you could also drag over them and click the Select button. You can always type the column names.

2. Drag over **C2-C4** in the Variables box and then click the **Select** button. See Figure 1.14. Minitab enters 'Round 1'-Change—the shorthand for the list 'Round 1' 'Round 2' Change.

   The variable names 'Round 1' and 'Round 2' are enclosed in single quotes since the names contain a blank space. You will need to do this manually if you type such names in any text box.

3. Click **OK**. The Session Window appears, displaying the statistics you requested. See Figure 1.15.

Figure 1.14 Descriptive statistics dialog box

Minitab will calculate and display descriptive statistics for these variables —

Remember that 'Round 1'-Change is shorthand for the list 'Round 1' 'Round 2' Change

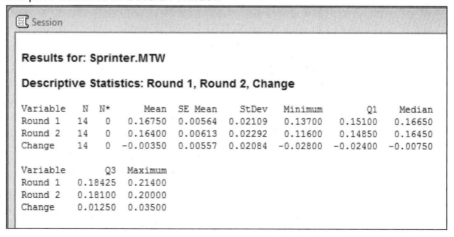

The means are located in the Mean column. The average reaction time in Round 1 is 0.16750 second and 0.16400 second in Round 2. The average change is −0.00350, so on average, racers seem to have had a faster reaction time in the second round. We will discuss the other statistics shown in Chapter 4, page 121.

Figure 1.15 Descriptive statistics in Session Window

```
Session

Results for: Sprinter.MTW

Descriptive Statistics: Round 1, Round 2, Change

Variable   N   N*      Mean   SE Mean     StDev    Minimum        Q1     Median
Round 1   14   0    0.16750   0.00564   0.02109    0.13700   0.15100    0.16650
Round 2   14   0    0.16400   0.00613   0.02292    0.11600   0.14850    0.16450
Change    14   0   -0.00350   0.00557   0.02084   -0.02800  -0.02400   -0.00750

Variable        Q3   Maximum
Round 1    0.18425   0.21400
Round 2    0.18100   0.20000
Change     0.01250   0.03500
```

When the Session Window contains more information than your screen can display, you can use the PageUp and PageDown or arrow keys on your keyboard to scroll through the information. You can also use the scroll bars.

Creating a Graph in a Graph Window

Minitab offers many graph options, ranging from simple character graphs composed of simple text characters to high-resolution graphics. Character graphs appear in the Session Window, but high-resolution graphs appear in their own Graph Windows. By default Minitab creates high-resolution graphs. Various graphs are discussed quite extensively in Chapters 4 and 5.

Let's view a dotplot of the reaction time in Round 1. Graph commands are located on the Graph menu. To create a dotplot of the Round 1 reaction time variable,

1. Choose **Graph ► Dotplot ► One Y, Simple**. Click **OK**. The dotplot dialog box appears.
2. Select **C2 Round 1** as the Graph variable.
3. Click **OK** and the dotplot appears in its own Graph Window.

Dotplots, Stem-and-Leaf displays, and Histograms will be considered more thoroughly in Chapter 4.

Printing a Window

You can print the information in any of the Minitab windows. When you choose the Print command, Minitab prints the contents of the active Window. If a Graph Window is active, the command appears as Print Graph, but if another Window were active, it might appear as Print Session Window, Print Worksheet, or Print History. Often you will want to combine Minitab Session output, graphs, and your own analysis into a report for either printing, emailing, or publishing on the Web. The Minitab ReportPad™ may be used for this purpose. See Chapter 17 for information on using the ReportPad.

Try printing the dotplot you created. To print the contents of a Window in Minitab,

1. Make sure that the Graph Window is active. (Click on it anywhere if it is visible. Otherwise use the Window menu to select the one you want.)
2. Choose **File ► Print Graph**. (You could also use Crtl+P or click the Printer button on the toolbar.)
3. Check and, if necessary, change the settings in the Print dialog box. This is the same Print dialog box that appears for any Windows application.
4. Click **OK**. Minitab prints the graph. Now try printing the Worksheet containing the Sprinter data.

5. First make sure that the Data Window containing the Sprinter data is active. (Click on it anywhere if it is visible. Otherwise use the Window menu to select the one you want.)

6. Choose **File ▸ Print Worksheet**. The Data Window Print Options dialog box appears, allowing you to specify how to print components of the Data Window, such as row and column labels, grid lines, and text justification. See Figure 1.16.

7. To add a title, click in the title box and type **Sprinter Data**.

8. Click **OK** twice and Minitab prints the Data Window.

Figure 1.16 Data Window Print Options

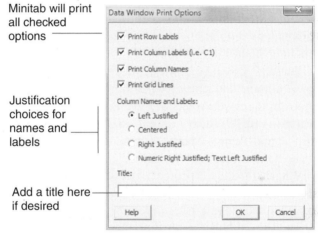

Minitab will print all checked options

Justification choices for names and labels

Add a title here if desired

1.4   Working with Minitab Projects

When you first start Minitab, a blank new project appears in the Minitab program window. A **project** is a collection of information you generate in a Minitab session including the data in your worksheets. A project includes

- the contents of each window
- the description of the project, if any, created with File ▸ Project Description
- the description of each worksheet, if any, created with Editor ▸ Worksheet ▸ Description
- the size, location, and state of each window
- the selections in each dialog box that you used

Once you have entered data into Minitab, you should save it in a computer file—at least the Worksheet if not other material that would be time consuming to reproduce. Then you will have it available for use at any later time. Minitab does not save your work automatically, so it is good practice to save your work often throughout a session to prevent data loss in the event of a computer glitch.

Saving a Project

You can save a project by using the Save Project button on the toolbar 🖫, the Save Project option on the File menu, or the Ctrl+S keyboard shortcut. The first time you use the Save Project button, Minitab opens the Save Project As dialog box, which allows you to choose a name and location for your project. After you've assigned a name and location to the project and saved it once, any time later in that session you can click the Save Project button 🖫 to save any changes you've made to your project, and Minitab will do so automatically, without opening a dialog box.

You can also save a project description, set security passwords, and set project options that specify what to include using the Description and Options buttons in the Save Project or Save Project As dialog box. By default, Minitab saves Session Window content, all Worksheets, all graphs, all dialog box settings, and Project Manager content with a project. You can choose to save only some of these. The Worksheets are always saved.

When you save a project, Minitab adds the .MPJ file extension. A **file extension** is a set of characters (usually three in Windows programs) that a program attaches to a file to identify its file type to the computer's operating system. The .MPJ file extension identifies the file as a Minitab project. Using file extensions allows your computer to classify files by file type and to associate the correct software with a file.

If you open a project, modify it, and want to retain both the old and new versions of the project, use the File ▸ Save Project As command and choose a different name for your project. Minitab saves the project with the new name without altering the contents of the original file.

Try saving your sprinter data and the work you've done during this session to a Minitab project file. To save a project,

1. Choose **File ▸ Save Project** or click the Save Project button 🖫. Notice in Figure 1.17 that the Save Project As actually opened since we have not previously saved this project.
2. Click the **Save in** list arrow and navigate to the drive and folder in which you want to save your project. Alternatively, use the Windows 7 navigation icons shown on the left side of the dialog box. See Figure 1.17. You will likely want to create a folder for your Minitab project(s).

3. In the File name box, enter **Sprinter** to give a name to your project.

4. Click the **Description** button. The first time you save a project, it is good practice to document its creator, relevant dates, and comments about the project. In subsequent saves you might want to add more information to the description.

5. Type your name in the Creator box and type the date in the Date(s) box.

6. In the Comments box type the following:

C2 and C3 contain reaction time data for 14 sprinters who participated in the 1996 Summer Olympics. Each sprinter's reaction time upon hearing the starting gun was recorded in the first and second rounds of the 100-meter dash.

Figure 1.18 shows the completed dialog box.

7. Click **OK** to close the Description dialog box and return to the Save Project As dialog box.

8. Click the **Options** button. By default Minitab saves all the information about a project, but you can deselect any of the check boxes to omit those objects from your saved project. See Figure 1.19.

9. Click **OK** to close the Options dialog box and return to the Save Project As dialog box.

10. If you want to set various password settings to protect the file, click the **Security** button. See Figure 1.20.

> *Caution: If you lose or forget your password, Minitab technical support specialists cannot help you recover it. You must keep a list of passwords and their corresponding file names in a safe place*

11. If necessary, click **OK** to close the Security dialog box and return to the Save Project As dialog box.

12. Click **Save**.

> *TIP If a message box appears informing you that the file already exists, and asking if you want to replace it, a file with the same name has already been created in the folder you specified. If you want to replace the previous file, click Yes. If you want to keep the previous file, click No, and then change the name in the File name box to a different name, such as Sprinter2. Then repeat Step 12.*

Figure 1.17 Save Project As dialog box

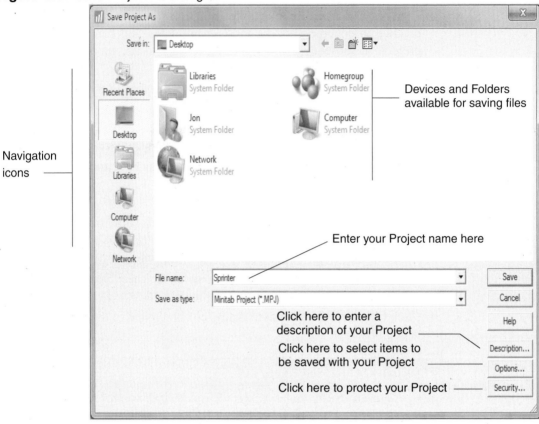

Navigation
icons

Devices and Folders
available for saving files

Enter your Project name here

Click here to enter a
description of your Project

Click here to select items to
be saved with your Project

Click here to protect your Project

Figure 1.18 Project Description dialog box

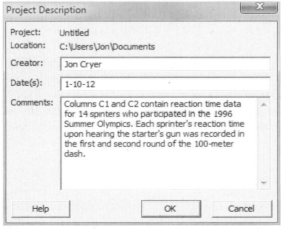

Figure 1.19 Save Project Option dialog box

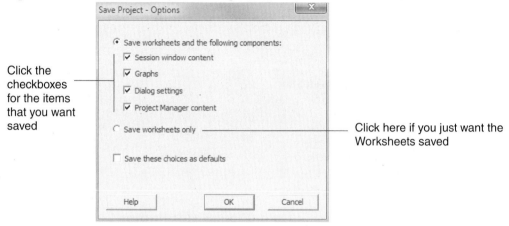

Click the
checkboxes
for the items
that you want
saved

Click here if you just want the
Worksheets saved

Figure 1.20 File Security dialog Box

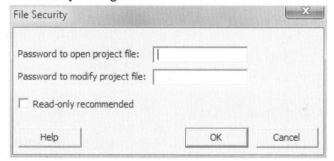

Caution: Be sure to
remember your
passwords. They
cannot be recovered
by Minitab, Inc.

Closing a Project and Exiting Minitab

You close a project by exiting Minitab, opening a new project, or opening a
different saved project. There is no "Close Project" option in Minitab: some
project must always be open. If you want to close a project but leave Minitab
running, click the Open Project button 🗁 or choose File ▸ New ▸ Minitab
Project. To close the Sprinter project and exit Minitab, click the **Close button**
🗙 on the Minitab title bar or choose File ▸ Exit. Minitab closes the project
and shuts down.

Opening a Project

When you open a project, the various Minitab Windows are repopulated with
the data, graphs, and other information that was saved with the project. If
another project is already open, Minitab prompts you to save all or part of the
current project before closing it. There are several methods of opening a
project; the method you use can depend on whether the file has been recently
opened or not.

If Minitab is open it will display the four most recently opened files at the bottom of the File menu. You can click File and then click on the file you wish to open. To close the Sprinter project and then reopen it using the File menu,

1. Choose **File ▸ New**.
2. Click **Minitab Project** and then click **OK**. The Sprinter file closes.
3. Click **File**. Notice the Sprinter Project name appears near the bottom of the File menu. See display to the right.
4. Click **Sprinter**. The Sprinter Project reopens.

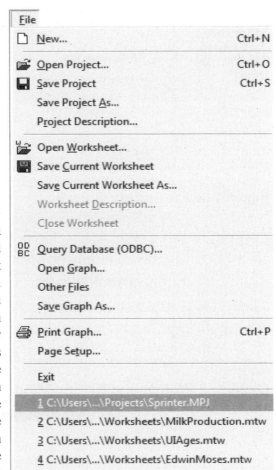

If you want to open a project that you have not used recently, click the Open Project button [icon]. When you open a project using this method, you can preview it from the Open Project dialog box. The Preview Window lists the worksheets and graphs contained in the project. This is a useful option if you aren't sure what the project contains. You'll see other uses for Preview in Chapter 3. You'll open the Sprinter project once more, but this time you'll preview it first. To preview and then open a project,

1. Click the **Open Project** button on the toolbar.
2. Click the **Look in:** list arrow and open the drive or folder that contains the Sprinter project. Alternatively, you can use the navigation icons.
3. Click the **Sprinter** project.

4. Click the **Preview** button to view project contents before opening the project. Objects in the Sprinter project appear in the box. See Figure 1.21.

Figure 1.21 Open Project listing of components

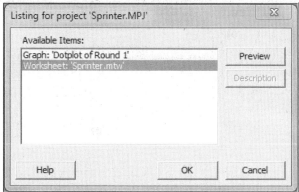

5. Click **Worksheet: 'Sprinter.mtw'** and then click the **Preview button**. A preview of the worksheet contents appears. See Figure 1.22.

6. Click **OK** twice to return to the Open Project dialog box.

7. Click the **Open** button.

If you are using Windows 7 and have pinned Minitab to the Start button, you can view the names of many of your most recently used Minitab documents by hovering the mouse pointer over the arrow as shown below. The quickest way to open a recent Minitab project or Worksheet is to click the name of the Minitab file shown to the right of the arrow. Windows starts the Minitab software and automatically opens the file you selected.

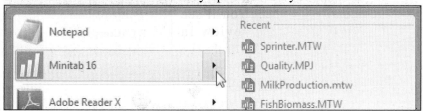

If you have pinned Minitab to the Taskbar, right-clicking the Minitab icon will also display recent file names associated with Minitab, namely, Worksheets and Projects.

 TIP The procedure described above displays only the most recently used files. If many other files have been opened since you last opened the Sprinter data, the Sprinter option might be replaced by more recently opened files. Also, if you have moved the Sprinter data to a different location than where you originally saved it, Windows won't be able to locate the file.

Figure 1.22 Open Project preview of Worksheet

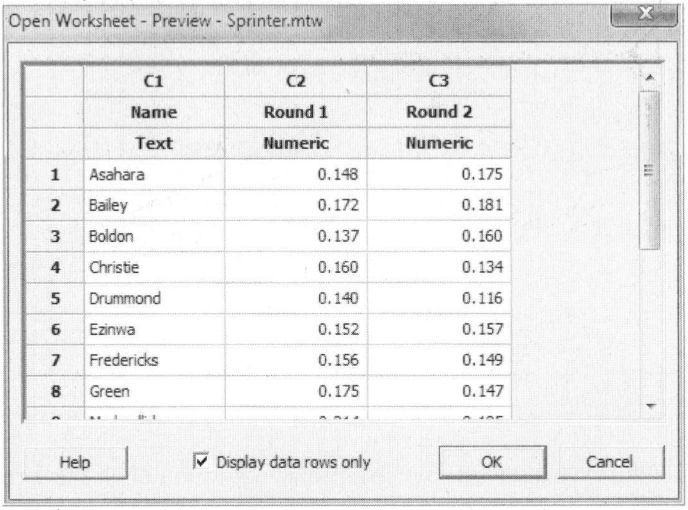

Saving Components of a Project

Minitab allows you to save parts of a Minitab session in separate files. For example you might want to save just the data in a worksheet so that you can use it in other projects. Or you might save a graph so that you can work with it in a graphics software package or insert it in a presentation file. When you save a Minitab object as a separate file, Minitab assigns a file extension to that file that identifies the file type. Table 1.4 summarizes the file types you'll use most often in Minitab.

There are other file formats that Minitab uses and recognizes; you'll learn about those in later chapters. Only Minitab can work with files in MPJ, MTW, MTP, MGF, or other specific Minitab file formats, so when you decide which format to use, make sure you first ask yourself how you will subsequently use the file. If you plan to use it only within Minitab, saving it in a Minitab file format is best. If you plan to use it in other applications, then you need to research the file formats available to you and decide which bests suits your needs.

Table 1.4 Minitab file types

Extension	File type	Description
MPJ	Minitab Project	The collection of all Minitab objects related to a statistical analysis in Minitab's proprietary format. These files can only be read or opened by the Minitab software.

Table 1.4 Minitab file types (Continued)

Extension	File type	Description
MTW	Minitab Worksheet	A file containing only worksheet data (columns, constants, and matrices) in Minitab's proprietary format. They can only be read or opened by the Minitab software.
TXT	Text	You can save the contents of the Session Window and History folder as text files that are readable by almost any application, most commonly by a text editor or word processor. Text files are not formatted (see RTF below). Text files are also known as ASCII files.
RTF	Rich Text Format	If you want to include formatting such as fonts, bolding, or italics, you can save the contents of the Session Window in an RTF file These files can be read by most Editors or word processing software.
MGF	Minitab Graphics Format	You can save the contents of a Graph Window to a Minitab Graph file, which allows you to open it in Minitab. These files can only be opened by the Minitab software.

Saving Worksheets

You can save a worksheet in MTW format (in the current or one of several previous releases of Minitab to ensure backward compatibility), MTP format, or in a format supported by other software applications, such as database and spreadsheet software, including Excel, Quattro Pro, Lotus 1-2-3, and dBase. You can also save as a Web page in .htm or .html format or in Spreadsheet XML format.

You already learned how to save a Worksheet in MTW (Minitab's proprietary data format) on page 19 This format is the most convenient format if all you want to do is reload the data into Minitab. You might want to save data in other formats for other purposes.

Saving Graphs

Using the File ▶ Save Graph As command, you can save the graph in MGF format so that Minitab can display it. Once it is displayed in Minitab it is fully editable.

For other situations you may want to save it in a standard graphics file format such as JPG (or JPEG), PNG, TIF, or BMP. These are common graphics formats that allow graphics to be displayed on the Web, used in graphics software packages, or in a word processing document. You'll save the dotplot you created as a JPG file so you can use it on a web page. You might want to review "Creating a Graph in a Graph Window", page 24. To save a graph in a stand-alone jpeg file,

1. Make sure the graph you wish to save is in the active window.

2. Choose **File ▸ Save Graph As**.

3. Click the **Save in** list arrow and select the drive or folder in which you want to save the graph.

4. In the File name box type **Sprinter Dotplot** as the file name.

5. Click the **Save as type** list arrow and select the file format you want to use, in this case, JPG. See Figure 1.23.

   💡 TIP *Some graphic file formats allow you to choose from several color palettes. The fewer number of colors you allow, the smaller the file size.*

6. Click **Save**.

Figure 1.23
Save Graph
As dialog box

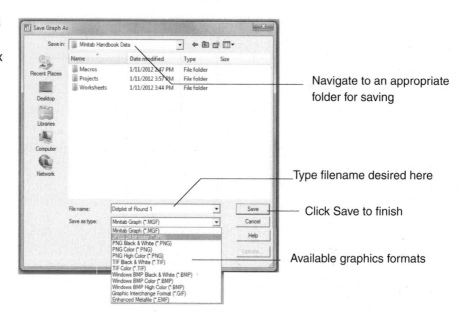

Navigate to an appropriate folder for saving

Type filename desired here

Click Save to finish

Available graphics formats

Adding an Empty Worksheet to a Project

You can use the File ▸ New ▸ Worksheet option to open a new, empty Worksheet if you want to type another set of data into your project.

To add a new, empty worksheet to your Sprinter project,

1. Choose **File ▸ New**.

2. Select **Minitab Worksheet**.

3. Click **OK**. An additional empty worksheet appears.

The worksheet is added to the current project. When a project contains more than one worksheet, each worksheet appears in its own Data Window. Any commands you execute operate on the current or active worksheet, the worksheet displayed in the active Data Window.

> TIP *Be sure that the Worksheet that you intend to work with is active. Calculations always are done with the data in the active worksheet.*

Opening a Worksheet

You can add data to a project from a Worksheet file by using **File ▶ Open Worksheet**. This option copies data from the original data file into a new Worksheet in your current open project. Because the data are copied, this procedure does not affect the original file in any way.

The Minitab software comes with many data sets stored as saved Worksheets. The examples in this book use some of those worksheets to exemplify statistical concepts. To follow along with the examples, you'll need to open and use the applicable worksheets.

Now that you have a new project open, try opening the data set named Bears that comes with the Minitab program. To add a worksheet to a project,

1. Choose **File ▶ Open Worksheet**.
2. Click the **Look in Minitab Sample Data folder** button. See Figure 1.24.
3. Scroll down the filenames and click **Bears**.
4. Click the **Open** button. Your project now contains a copy of the data stored in the Bears worksheet in a new Data Window.

> TIP *If a message appears informing you that a copy of the content of this file will be added to the current project, click OK. You can also click the check box so the message won't appear again.*

Figure 1.24 Open Worksheet dialog box

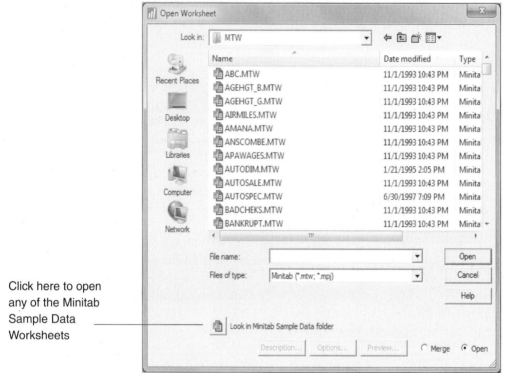

Throughout this book you will work with Minitab worksheets; whether you open them in a new project and then save your work as a worksheet (which saves just the data) or as a project (which saves everything, including your session work and any graphs) is up to you.

1.5   The Project Manager

The Minitab Project Manager permits you to work with and manage all parts of your project—the Session Window, the Data Worksheet(s), any graphs you have created, and more. In particular, it gives you access to Minitab's own mini word processor—the ReportPad. See Chapter 17 for more information on the ReportPad.

To open the Project Manager you may use any one of the following:

- Choose Window ▶ Project Manager

- Click the Project Manager icon 🖼 on the separate Project Manager toolbar

- Click the restore button on the Project Manager minimized title bar
- Use the keyboard shortcut, Ctrl+I (I for Information)

Session Folder

Figure 1.25 displays the Project Manager for a project that contains two Worksheets (cars.mtw and cars3.mtw), three graphs, and two sets of descriptive statistics. In this display the Session Window folder has been selected.

Figure 1.25 Project Manager Window with Session folder selected

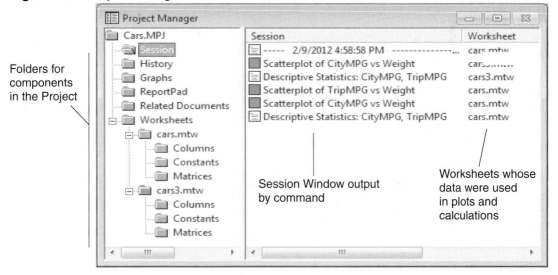

Folders for components in the Project

Session Window output by command

Worksheets whose data were used in plots and calculations

Right-clicking on either the folders or the contents of the folders gives you access to a variety of options. When in doubt, right-click an item and review the choices displayed.

The right mouse button options for the Session Window folder in the Project Manager are shown here to the right.

Bring to Front
Save Session Window
Print Session Window...

Clear Window

By Worksheet
• By Order in Session Window
By Title

A right-click on a Session command item in the right-hand pane in Figure 1.25 brings up the menu items shown at the right.

The StatGuide choice at the bottom is especially helpful to students. The StatGuide item takes you to extensive linked Help screens to assist you in interpreting statistical tables, graphs, and other statistical output in a practical and easy-to-understand way. The StatGuide is also available directly by right-clicking output in the Session Window and also by clicking the StatGuide button ![icon] on the main toolbar.

Worksheets Folder

Selecting the Columns item in the cars.mtw Worksheet folder brings up the display in Figure 1.26.

Figure 1.26 Project Manager Worksheets with Columns folder selected

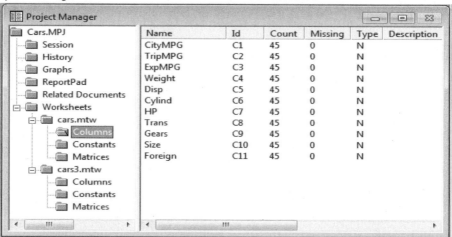

The right-click options for the Worksheets folder are shown here to the right. Right-click on each of the cars.mtw, Columns,… folders to see your choices.

Graphs Folder

Figure 1.27 Project Manager with Graphs folder selected

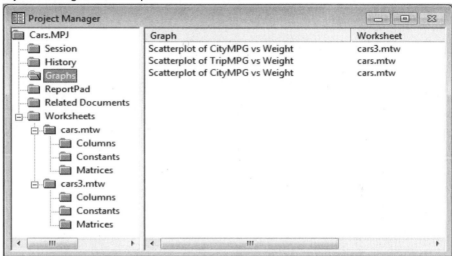

A right-click on a graph name in the right-hand pane of Figure 1.27 brings up the options shown here to the right.

Several graphs may be highlighted with Ctrl+Click and then "tiled" on one screen for easy comparisons. The Layout tool may also be used to merge several graphs into one.

A right click on the Graphs folder in the left-hand pane of Figure 1.27 displays the options shown here at the right. The bottom portion allows you to reorder the graph names in different ways.

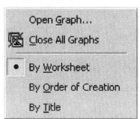

History Folder

Figure 1.28 shows a portion of the History folder for our Project. The right-hand pane displays all of the Minitab commands used in the project to open the Worksheets, produce the graphs and descriptive statistics, and so forth.

Figure 1.28 Project Manager History folder

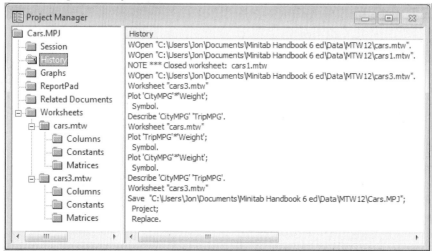

If you have trouble with a Minitab analysis and cannot determine what the problem is, it would be very helpful to print a copy of the History folder to show to a colleague or your instructor. The History folder gives an "audit trail" in command form of all of the things you have done in producing this project.

If you right-click the History folder you get choices to save, print, or clear it.

Exercises

1-1   (a) Use the Minitab Help file to explore the function of the Concatenate option on the Data menu.

(b) What example does the Help file use to show you how to use this option? (Hint: Once you have found the topic, click the How To link.)

1-2   The Descriptive Statistics output shown in Figure 1.15 printed a set of values for each row in C1, C2, and C4, such as N, Mean, Median, and so on. Try to identify what each column contains (that is, try to define N, Mean, Median, and so on). Use the Help file if necessary.

1-1 Interpret the output in Figure 1.15.
(a) On the average did the reaction time increase or decrease from the first round to the second?
(b) In Round 1 what was the slowest reaction time?
(c) The fastest?

1-3 (a) Enter the data in table below into a new Worksheet:

C1	C2	C3
15	28	30
14	30	31
16	30	34
13	27	31

(b) Calculate the sum of the values (C1 + C2 + C3) and store the sums in C4. What values does C4 contain?

1-4 Enter the data in the table below into a Worksheet.

C1	C2
32	26
34	27
32	25

Calculate the difference between C1 and C2 (C2 – C1) and store the differences in C3. What values does C3 contain?

1-5 Use Minitab to produce a temperature table in degrees Fahrenheit and the equivalent temperatures in degrees Celsius. In a blank worksheet enter temperatures in degrees Fahrenheit from 20° to 80°, inclusive, in increments of 5° in C1. To convert the values in C1 to Celsius, use the formula Celsius = (Fahrenheit – 32)*(5/9). Store the Celsius values in C2. Print your worksheet, and on the worksheet, write the expression you used in the Calculator.

1-6 Olympic regulations state that anyone who starts a race within 0.1 second before the starting gun must be "jumping the gun" and is therefore disqualified. Examine the Minimum column to see whether anyone in Round 1 or Round 2 was disqualified for jumping the gun.

1-7   **(a)** Open the Bears data from the Minitab Sample Data folder.

(b) Generate descriptive statistics for the Weight data and examine them in the Session Window.

(c) What is the weight of the heaviest bear? The lightest? What is the average weight?

1-8   Suppose C1 is named x and contains the eleven integers 0, 1, 2, ..., 10. What expression would you use with the Calculator option to calculate y with each of the following formulas? Which, if any, of the values in y will be set to the missing value code when the Calculator executes?

(a) $y = 2x + \dfrac{1}{3}$

(b) $y = x^2/5$

(c) $y = (x + 5)^2/(2x - 1)$

(d) $y = \sqrt{x^3 - 1}$

(e) $y = \dfrac{x^2 + 5}{x - 3}$

2

Session Commands

2.1 The Minitab Command Language

Command Language[†]

Minitab's functions are accessible through a command language as well as through menus. You can do your work by choosing options from menus or by typing commands, as you prefer. You saw how to enter commands using the menus in Chapter 1. In this chapter you learn how to use the command language, also called Session commands. Once you start using Minitab independently, you will probably develop preferences about when to use the menus and when to use the command language. You can carry out some procedures more quickly using the command language.

To practice working with the Minitab command language, open the worksheet named Lake, located in the Minitab Sample Data folder. The Lake worksheet contains information on lakes in northern Wisconsin. To open the Lake worksheet,

1. Start Minitab and maximize the Minitab program window.
2. Choose **File ► Open Worksheet**.
3. Click the **Look in Minitab Sample Data** button. ⬚
   (SeeFigure 1.24, p. 36.)
4. Scroll down the filenames and click **Lake**.
5. Click **Open**. The Lake data appear in the Worksheet window.

Each row contains the data for a single lake. You'll work with the first column, Area, which contains the area of each lake in acres.

Enabling Commands

The Session Window can be used to enter Minitab commands directly. However, you can enter commands only if commands are enabled. The Minitab default has commands disabled. Once you have enabled commands, the Minitab

---

[†] Since you can use nearly all of Minitab's features without learning about commands, you may want to skip this chapter and come back to it later if or when needed.

command prompt appears as MTB > in the Session Window. If the MTB > prompt appears in the Session Window, commands have already been enabled. If not:

1. Open the Session Window by clicking the Session Window button  on the main toolbar. The Session Window title bar turns dark blue to indicate it is active[†]. Notice the Session Window records the fact that you opened the Lake worksheet.

2. Check **Enable Commands** under the Editor menu.

Once commands are enabled you will see the subsequent commands that Minitab executes displayed in the Session Window even if they are executed indirectly using the Menu system. This can be a good way to learn how the commands work.

Executing a Command

To use the command language, you must know, and preferably memorize, the special words that communicate your instructions to Minitab. Most commands are simple words that are easy to remember, like Plot, Save, or Sort, so they are easy to learn. (If you don't memorize the commands and subcommands you need, it is much easier to use the menus.)

Every command starts with a command name. In most commands this name is followed by one or more arguments, also called parameters, on which the command operates. Arguments can be columns, constants, matrices, numbers, filenames, or text strings. For example C1 is the argument in the following command, which tells Minitab to describe the data in column C1:

```
DESCRIBE C1
```

After you type a command and its arguments at the prompt, you press Enter to tell Minitab to execute the command. The command is not executed until you press Enter and you can edit the command using the cursor, arrow keys, Backspace, and Delete keys in the same way that you edit text.

Try executing this command by telling Minitab to describe the data in C1, which contains the area in acres of the lakes in the Lake worksheet. To execute the DESCRIBE Session command,

---

[†] Dark blue is the default color but Minitab may be customized in numerous ways.

1. Type **DESCRIBE C1** after the MTB > prompt. The cursor moves to the right as you type.

2. Press **Enter** to tell Minitab you are done typing the command. Descriptive statistics together with a heading appear in the Session Window following the command you typed.

Command Syntax

The set of rules by which a command language operates is called the command syntax. A few commands allow fairly complicated expressions, but most are quite simple. The Minitab online Help system documents each session command, including the command's syntax, which tells you how to type the command. For example, the command syntax for the DESCRIBE command you just used is

```
DESCRIBE C...C
```

The C...C means you can type one or several columns after typing DESCRIBE. If you typed DESCRIBE C1 C2 C3, for example, Minitab would produce descriptive statistics for each of these columns. This command is the counterpart of choosing Stat ► Basic Statistics ► Display Descriptive Statistics from the Minitab menu.

Table 2.1 displays the typographical conventions this book (and the Minitab Help system) uses to describe the syntax of individual commands.

Table 2.1 Command syntax conventions

Convention	Description
C	denotes a column, such as C12 or Height
K	denotes a constant, such as 8.3 or K14
E	denotes either a constant or a column, and sometimes a matrix
M	denotes a matrix, such as M5
[ ]	encloses an optional argument
All CAPS	denotes a Minitab command or subcommand

Observe these rules when you use the command language:

1. You need type only the first four letters of the command name.[†] Thus you could write the command
```
DESCRIBE C1
```
as
```
DESC C1
```

[†] In earlier releases of Minitab you could type extra text in commands for annotation. Releases 14 and later do not permit extra text in commands.

2. Each command must start on a new line.

3. In arguments you can use variable names and variable numbers, such as C1, interchangeably but you must enclose variable names in single quotation marks. Thus if C1 is named Sales, HISTOGRAM C1 is equivalent to HISTOGRAM 'Sales'.

4. Enclose text string arguments, such as the title of a graph, in double quotation marks (").

5. You can use either uppercase or lowercase letters for command names and arguments. Thus the following are all equivalent:

```
TALLY C1
tally c1
Tally C1
```

6. You can abbreviate a list of consecutive columns or stored constants using a hyphen. To describe all columns in the Lake data, for example, you could enter

```
DESCRIBE C1-C5
```
rather than
```
DESCRIBE C1 C2 C3 C4 C5
```

7. If you type a number, don't use commas within the number. Thus 10,041 should be typed as 10041. (Commas and dollar signs may be typed directly into cells of Worksheets displayed in Data Windows. Minitab will display the commas and dollar signs but ignore them in any calculations.)

8. You can use the comment symbol (#) to add annotation on a command line. Minitab ignores everything on the line after the #. Annotation is useful when you want to take notes on your analysis plan or when you want others examining your analysis to understand what you were doing. Thus you might type:

```
# First we examine descriptive statistics on the area data
DESCRIBE C1 # C1 contains the area of each lake in acres
```

9. If a command won't fit on one line, end the first line with the continuation symbol (&) and continue the command on the next line. For example:

```
PRINT C2, C4-C20, C25, C26, C30, C33 &
C35-C40, C42, C50
```

10. To insert a blank line in the Session Window, press Ctrl+Enter.

Minitab Command Box Format

Minitab commands are described in this book in boxes like the following. These descriptions use the syntax outlined in Table 2.1. Arguments in square brackets are optional. Here is an example:

PRINT C

This command 'prints' the data in the specified column in the Session Window. (Note: it does **not** print to a printer.) To display the data in column C3 named Sales, you could type:

```
Print C3
```

or

```
Print 'Sales'
```

PRINT E...E

displays any list of columns, constants, or matrices. For example:

```
Print C1 'Sales' K1 M1
```

displays the values of the data in columns C1 and Sales, in the constant K1, and in the matrix M1. The values are all displayed in the Session Window.

Subcommands

Many commands have subcommands that allow more control over how a command works. To signal Minitab that you want to use a subcommand, end the main command line with a semicolon (;) and then press Enter. Minitab then prompts you with the subcommand prompt, SUBC >. Type your subcommands, each on its own line, ending each line with a semicolon (;) and pressing Enter after each line. End the last subcommand with a period (.) to tell Minitab you are done with the command. Minitab then executes the entire command, subcommands and all. When typing a command with no subcommand following, you do not have to type any punctuation mark after the command line. This book and the Help file indent subcommands for clarity but this is not necessary. If you forget to type a semicolon or period after a subcommand and then you press Enter, Minitab displays an error message. You can simply type the correct symbol after the MTB > prompt that appears, and Minitab will execute your commands.

To see how to enter subcommands in Minitab, try using the PLOT command, which shows the relationship between two columns. The first column is the vertical or *y*-axis variable and the second column is the horizontal or *x*-axis variable. PLOT plots one variable against another. You can use the TITLE subcommand to supply a title for your plot. The syntax looks like this:

```
PLOT C*C;
  TITLE "graph title".
```

Using the Lake data, try plotting the concentration of hydrogen ions (in C5) in the water against the pH value (in C3), which measures the acidity of the lake water. Give your plot the title Ions vs. pH. To plot C5 against C3 and give your plot a title,

1. Type **PLOT C5*C3;** and then press Enter. Make sure you type the semicolon, which signals Minitab that a subcommand will follow. The SUBC > prompt appears when you press Enter.
2. Type **TITLE "Ions vs. pH".** and press Enter. Be sure to enclose the graph title in double quotation marks ("), because it is a text string. Also, be sure to end the last subcommand with a period. (If you press Enter and fail to type the period, Minitab will assume there are more subcommands to follow and come back with the prompt SUB >. Type a single period here and Minitab will execute the commands.) The plot appears in its own graph window.

Observe these rules when you use subcommands:

1. If you want to use a subcommand of any type on the next line, end the current line with a semicolon (;). End the last subcommand with a period (.). Some subcommands have their own subcommands. The order in which you type these subcommands determines what subcommand or command they relate to.
2. Type the subcommand ABORT as the next subcommand to exit from a multiline command without executing it.

Getting Session Command Help

The Session Command Help system can help you learn command syntax for the commands you want to use. Each command has a separate Help topic. Try locating information on the DESCRIBE command. To open the Session Command Help system:

1. Click the **Help** button 🛟 on the main toolbar. The display shown in Figure 1.5, p. 9, appears.
2. On the left side of the right-hand pane, click **Session Commands**. The display shown in Figure 2.1 appears.
3. Click the **+** to the left of Basic Statistics and a long list of basic statistics command names appear.

4. Click **DESCRIBE**, the name of the Session command of interest. Figure 2.2 shows you some of the Help information for the DESCRIBE Session command. Scroll down the right-hand pane to see the extensive information available.

5. Close the Help window.

Figure 2.1   Session Command Help Window

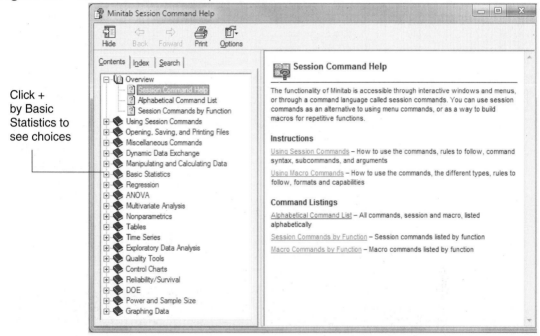

If you cannot guess which area might hold the command you are interested in, you can click either the Index tab or the Search tab to help locate the command of interest.

Figure 2.2   Session Command Help for Descriptive Statistics

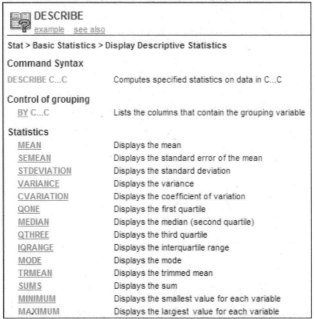

Notice that the Help topic uses the notation shown in Table 2.1. Required text appears in blue. Subcommands appear immediately after the main command. If there are additional Help topics available on the topic you chose, they are clickable and linked to more information. Subcommands, for example, always have their own Help topics. In Figure 2.2, MEAN, STDEVIATION, SEMEAN, and so forth, are all subcommands to the DESCRIBE command.

You can also access Help directly from the MTB > prompt in the Session Window. For example, you can get help on a specific command by typing HELP DESCRIBE. Of course, the down side of this method is you must know at least the first four letters of the command name.

2.2   Useful Session Commands

Upcoming chapters in this book introduce you to the session commands you are likely to find most useful, but there are a few basic commands that will help you work more efficiently in the Session Window. To practice working with these commands, you'll enter a data set and then perform a few basic operations with the data you enter.

Entering Data

The commands READ and SET let you type data into a Minitab Worksheet via the Session Window, using the following syntax:

READ C...C

This command is followed by lines of data. Each line is put into one row of the Worksheet. You may select any columns for storage. The following example puts data into columns C2 and C5:

```
READ C2 C5
3 71
5 78
4 81
END
```

You can put numbers anywhere on a line as long as they are in the correct order and are separated with at least one space, one comma, or one tab.

SET C

This command is followed by lines of data, any number per row. All the data are stacked into one column of the Worksheet. You may select any column for storage. The following example puts all 3 data points, 40, 10, 16, into column C2 in that order:

```
SET C2
40 10
16
END
```

You can put numbers anywhere on a line as long as they are in the correct order and are separated with at least one space, one comma, or one tab.

The difference between READ and SET is the way you enter the data. READ enters data one row at a time; SET enters it one column at a time. When you use one of these commands to enter data in the Session Window, Minitab uses the DATA > prompt to tell you it is ready for you to enter data. When you type END, the MTB > prompt again appears, telling you it is ready for your next command.

The READ and SET commands allow you to enter data in many different ways, but worksheet operations such as data entry and manipulation are usually quickest using the menus and the Worksheet window. See "Entering Data", page 13, and all of Chapter 3.

Table 2.2 shows data for 11 women. Each woman took her resting pulse rate, then ran in place for one minute, then took her pulse again. In addition, whether or not the woman smoked regularly was recorded, using 1 = yes and 2 = no.

We want to calculate how much the pulse rate of each woman changed after exercise, find the average pulse rates for the 11 women before and after exercising, and get the average difference between them.

Table 2.2 Pulse rate data

Pulse rate before exercise	Pulse rate after exercise	Whether the woman smokes
96	140	2
62	100	2
78	104	1
82	100	2
100	115	1
68	112	2
96	116	2
78	118	2
88	110	1
62	98	1
80	128	2

Try using the READ command to enter the pulse data one row at a time. To enter the pulse data using the READ command,

1. Choose **File ▶ New ▶ Minitab Project** and click **OK** to clear the Minitab window and start over with a new project. When asked if you want to save changes to the current project, click **No**.

2. Choose **Window ▶ Session** and then maximize the Session Window.

3. Make sure commands are enabled. If the MTB > prompt appears in the Session Window, commands are already enabled. If not, check **Enable Commands** under the Editor menu

4. Type the following commands, pressing **Enter** after each command and pressing the **Spacebar** between each data point:

```
READ C1 C2 C3
 96 140 2
 62 100 2
```

```
 78 104 1
 82 100 2
100 115 1
 68 112 2
 96 116 2
 78 118 2
 88 110 1
 62 98 1
 80 128 2
END
```

5. Compare your screen to Table 2.2 to ensure the numbers are the same.

Using SET to Enter Data from a File

The SET command is most useful for entering data from a file when the data are intended for a single column but have been saved in a rectangular block format. For example, monthly data is often laid out in a publication or Web page with months across the top and years down the rows as in Table 2.3. We might use a scanner to scan these data from a publication into a text file or you might download data in this format from the Web.

Table 2.3 Average monthly temperature, Jan 1964 - Dec 1975, Dubuque, IA

Jan	Feb	Mar	Apr	May	Jun	Jul	Aug	Sep	Oct	Nov	Dec	Year
24.7	25.7	30.6	47.5	62.9	68.5	73.7	67.9	61.1	48.5	39.6	20.0	1964
16.1	19.1	24.2	45.4	61.3	66.5	72.1	68.4	60.2	50.9	37.4	31.1	1965
10.4	21.6	37.4	44.7	53.2	68.0	73.7	68.2	60.7	50.2	37.2	24.6	1966
21.5	14.7	35.0	48.3	54.0	68.2	69.6	65.7	60.8	49.1	33.2	26.0	1967
19.1	20.6	40.2	50.0	55.3	67.7	70.7	70.3	60.6	50.7	35.8	20.7	1968
14.0	24.1	29.4	46.6	58.6	62.2	72.1	71.7	61.9	47.6	34.2	20.4	1969
8.4	19.0	31.4	48.7	61.6	68.1	72.2	70.6	62.5	52.7	36.7	23.8	1970
11.2	20.0	29.6	47.7	55.8	73.2	68.0	67.1	64.9	57.1	37.6	27.7	1971
13.4	17.2	30.8	43.7	62.3	66.4	70.2	71.6	62.1	46.0	32.7	17.3	1972
22.5	25.7	42.3	45.2	55.5	68.9	72.3	72.3	62.5	55.6	38.0	20.4	1973
17.6	20.5	34.2	49.2	54.8	63.8	74.0	67.1	57.7	50.8	36.8	25.5	1974
20.4	19.6	24.6	41.3	61.8	68.5	72.0	71.1	57.3	52.5	40.6	26.2	1975

If the file named Tempertures.dat (or Tempertures.txt) is located in a folder named Data on a USB flash drive E: we could input the data into a single column of Minitab using the Session commands:

```
SET C1;
FILE "E:Data\Temperatures".
```

You would have to manually delete the year numbers (1964, 1965,...) from various places in the column to get the final desired result. In any case, this method would be much quicker and more reliable than re-keying all of the data. If the data file is not on a flash drive, the path may be considerably longer, and you will need to replace E:Data\Temperatures by a full directory path such as C:Libraries\Documents\My Documents\August Project\Temperatures.

Using SET to Enter Patterned Data

The SET command has special features that allow you to enter data that follow a pattern. You can use a colon to abbreviate a list of numbers. For example:

```
SET C2
  5:12
END
```

enters the integers from 5 to 12 inclusive, into C2.

Use a slash to indicate an increment other than 1. For example:

```
SET C5
  1:3/.2
END
```

puts the numbers 1, 1.2, 1.4, 1.6, 1.8, 2.0, 2.2, 2.4, 2.6, 2.8, and 3 into C5.

Use parentheses to repeat a list of numbers. For example:

```
SET C2
  3(1,2) 2(9)
END
```

repeats the list 1,2 three times and then tacks on two 9s. Thus C2 contains 1,2,1,2,1,2,9,9. The number before the parentheses is called a **repeat factor**. Do not put any space between it and the open parentheses. If you use a repeat factor after the close parentheses, each number is repeated individually. For example:

```
SET C2
  (1,2)4
END
```

puts 1,1,1,1,2,2,2,2 into C2.

As a practical example, suppose we have monthly time series data which begin in July 1995 and end in February 2004. We would like to enter a 'month number' column that starts at 7 for July 1995 and goes 7, 8,..., 12, 1, 2, 3,..., 1, 2 ending with 2 for December 2004. There are 6 full years plus portions of years at the beginning and end. This would be somewhat difficult to do with menu commands but the following short Session command will do the trick quite nicely:

```
SET C1
  7:12 6(1:12) 1 2
END
```

A version of SET, namely, **DSET** permits you to enter patterns of dates or times directly. The following Session commands will put date data into C1 for the same time period as the previous example:

```
DSET C1;
DSTART "Jul-95";
DEND "Dec-04";
MONTH 1. # Increment months by 1
```

Consult the Help pages for more information on the DSET command and for using date/time variables. DSET is also available through the menu selection Calc ▶ Make Patterned Data ▶ Simple Set of Date/Time Values.

Naming Columns

You can use the NAME command to name columns, constants, and matrices. This is especially useful in Minitab macros. Otherwise, you would simply type names in the column headings of a Data Window.

You'll name the first three columns Before, After, and Change and then will name a fourth that you'll create in the next section. To give your columns names,

1. Type the following after the MTB > prompt:

   NAME C1 "Before" C2 "After" C3 "Smokes" C4 "Change"

   Be sure to put the names within quotation marks.

2. Press **Enter**. Minitab names the columns accordingly.

Displaying Data in the Session Window

When you are working with command language, you might find it cumbersome to have to move back and forth between the Session Window and Worksheet window to view your data. You can view one or more columns quickly by using the PRINT command, which displays the data in the specified columns right in the Session Window. Note that PRINT does not send your work to a printer; it simply displays it in the Session Window.

To use the PRINT command to view the data in C1,

1. Type **PRINT C1** after the MTB > prompt.
2. Press **Enter**. The data stored in C1 are displayed in the Session Window.

The PRINT command is the only way to display the values of Minitab constants (K1, K2,...) and matrices (M1, M2,...)

Performing Calculations

When you want to perform various calculations with values in columns and constants, you can use the LET command. Table 2.4 displays the symbols Minitab uses for LET command arithmetic.

Table 2.4 LET Operations

Symbol	Operation
+	Add
−	Subtract
*	Multiply
/	Divide
**	Raise to a power (exponentiation)

LET E = mathematical expression

The mathematical expression can be made up of columns, constants, arithmetic symbols (+ − * / **), parentheses, and many functions.

Functions include standard mathematical function such as LOG, ABSOLUTE, COS, SQRT, and ROUND.

LET can calculate statistical quantities such as NSCORES, RANK, SIGNS, SORT, and LAG.

LET can calculate one number answers such as MEAN, MEDIAN, STDEV, SSQ, and COUNT.

LET can calculate comparisons <, >, <=, >=, =, ~= and the Boolean AND, OR, and NOT. Answers that are true are set to 1; those that are false are set to 0.

This example shows several LET commands, first setting values for constants and then performing some arithmetic calculations.

```
LET K1=3
LET K2=5*13
LET K3=K1+K2+4
SET C1
4 6 5 2
```

```
END
LET  C2=2*C1
LET  C3=K1*C1
LET  C4=C2+1
LET  C5=C3+C4
LET  C6=C1**2
```

After these commands K1 = 3, K2 = 65, and K3 = 72. Table 2.5 shows what columns C1 through C6 contain.

Table 2.5 Results of LET command

C1	C2	C3	C4	C5	C6
4	8	12	9	21	16
6	12	18	13	31	36
5	10	15	11	26	25
2	4	6	5	11	4

You can use parentheses for grouping in an expression. For example LET C3=10*(C1+C2) tells Minitab to add the values in C1 and C2, then multiply the sum by 10, and store the result in C3. LET follows the usual precedence rules of arithmetic. Operations within parentheses are always performed first, then exponentiation, then multiplication and division, and finally addition and subtraction. If you are not sure of the sequence of a calculation, use parentheses to make sure it is done correctly.

You can also use LET to calculate many functions, including comparison operators and Boolean operators. For example LET C2 = SQRT(C1) calculates the square root of every number in C1 and stores the answer in C2.

You can also use LET to calculate single-number answers. For example LET K1=SUM(C1) calculates the sum of all the numbers in C1 and stores the answer in K1. Any calculation that results in a single number can store that number in a column or a stored constant. You can then use the stored constant in place of the number in any command. For example SUM is a command that results in a one-number answer. If C1 contains the numbers 5, 6, and 2, then SUM C1 calculates 5 + 6 + 2 = 13. Since the answer is one number, you can store it in a constant. You can use LET to do arithmetic using both columns and constants. For example:

```
READ C1
  5
  6
  2
END
SUM C1 K1
LET  K2=4
LET  K3=K1+K2
```

```
LET C2=K1+C1
```

LET can also access individual cells in a column. For example, C5(4) = 72.1 puts 72.1 into the fourth row of C5, and LET K1 = C2(5) puts the number in row 5 of C2 into K1.

LET can also use comparison operators and Boolean operators. If you want to use any of these features, use Minitab's online Help for more information.

Now you want to calculate the change in pulse rate before and after exercise. To calculate the difference between the values in C2 and C1,

1. After the MTB > prompt, type **LET C4=C2–C1**.

2. Press **Enter**. Minitab performs the calculation. You can't see the results in the Session Window without using the PRINT command. You can, however, switch to the Data window to view your worksheet contents.

3. Choose **Window ▸ Worksheet**.

4. Choose **Window ▸ Session** to return to the Session Window.

    TIP When you return to the Session Window from the Data window, the MTB > prompt might appear at the top of the screen, and your previous session commands may no longer be visible. They're still there. Simply scroll up the Session Window and you will see them again.

5. Now to get some information on the pulse rates, type **DESCRIBE C1 C2 C4** and then press **Enter**. Minitab displays descriptive statistics for the three variables in the Session Window.

Saving and Retrieving Data

You can save and retrieve data in the Session Window by using the SAVE and RETRIEVE commands. As you saw in Chapter 1, when you use the Save and Open buttons on the toolbar or the corresponding options on the File menu, you locate the data's drive and folder via the Look in list. When you want to save or retrieve data by using session commands, however, you enter the file's name if it is in the current folder, or the entire path if it is in a different folder or drive. A file's path is the drive, folder, subfolder, and filename hierarchy that describes the location of the file on one of the storage devices available to your computer, such as a local hard drive, flash drive, CD-ROM, or network drive.

You indicate the drive on which the file is stored with the drive letter, followed by a colon (:) and a backslash (\). If the file is in a folder, append the folder and subfolder names to the path, separating them with backslashes (\). Finally append the filename. Thus a file located on a flash drive in drive E: might be found as E:\Beetle. To retrieve data in other folders, you need to type the entire path name or change to that directory (CD) first.

SAVE "filename"

This command saves the current Worksheet in a file. The file contains the columns (and constants and matrices, if any) together with their names, if any. When you subsequently reopen this file in Minitab with RETREIVE (or WOPEN), all of the data will be returned to the columns, constants, and matrices from which they came.

The SAVE command has a number of subcommands including PORTABLE which permits you to move saved among Worksheets among disparate computers such as Macintosh and PCs.

RETREIVE "filename"

This command opens a Minitab saved Worksheet. After using RETREIVE (or WOPEN), all of the data will be returned to the columns, constants, and matrices from which they came. RETREIVE replaces anything currently in your Worksheet.

WOPEN "filename"
WOPEN works similarly to RETREIVE but has many more subcommands and opens many different file types such as Excel worksheets. Check the Help pages for WOPEN.

This set of steps instructs you to save the worksheet containing the Pulse data to a USB flash drive at drive E: but if you save files to a different location, type the full path name instead. To save your pulse data onto a flash drive with the name Pulse,

Type **SAVE "E:\PULSE"** at the MTB > prompt and then press **Enter**.

The SAVE command saves only your worksheet. To save the entire project, including the Session Window, Graphs, and other window elements, use the PROJECT subcommand with the SAVE command.

Getting Worksheet Information

You can view limited worksheet information by typing the INFO session command.

> **INFO [C...C]**
> Summarizes the current worksheet.
> If no columns are specified, INFO prints a list of all non-empty columns with their names and counts, all stored constants, and all matrices. If there are missing observations, a count of these is also given. If a column contains text data, the letter T is printed to the left of the column. If you list columns, information is given on just those columns.

To view information on the Pulse worksheet,

Type **INFO** and then press **Enter**. Worksheet information appears as shown in Figure 2.3.

Figure 2.3   Session Window Information from INFO Command

```
MTB > info

Information on the Worksheet

Column   Count   Name
C1          11   Before
C2          11   After
C3          11   Smokes
C4          11   Change
```

2.3   Working with the Contents of the Session Window

Editing the Session Window

You can manipulate the contents of the Session Window by using the commands on the Editor menu. When Session Window output is set to Read-Only, you can view it and copy it but cannot delete or alter it. If you want to add comments, cut and paste text or numbers, or modify Session Window text, you must first change the Session Window to Editable mode. Once you are in Editable mode, you can treat the Session Window as if it were a text file. You place the cursor where you want to edit, and start typing or erasing.

To move the cursor, click in the Session Window where you want the cursor to appear, or press the Up Arrow, Down Arrow, Left Arrow, and Right Arrow keys to move the cursor.

To move quickly through the Session Window, press Ctrl+Home to move to the top, Ctrl+End to move to the bottom, or Page Up and Page Down to move one screen at a time.

Once the cursor is in position, you can use Backspace to delete characters to the left of the cursor, Delete to delete characters to the right of the cursor, or you can type to insert characters or entire lines.

To change the Session Window to Editable mode,

1. Choose **Editor ► Output Editable**. If Output Editable is checked, you do not have to do anything. If not, click Output Editable to check it.

   💡 *TIP If Output Editable is already checked, do not change it. Leave the Session Window in Editable mode so you can practice.*

2. Now that you are in Editable mode, you can add comments about your work in the Session Window: using any of the usual Windows editing techniques.

You can change modes back and forth throughout a session. If you have the status bar enabled (Tools ► Status Bar), you will notice the message, Editable, on the status bar in the lower right-hand corner of the Minitab Window. It will display Read-Only if the Session Window is not editable.

Editing the Session Window is especially useful when others will be evaluating your session work, such as a colleague or instructor who wants to see your analysis. You can easily add your name and the date, and any comments that explain your analysis.

You can right-click objects in the Session Window to open a shortcut menu that gives you access to many options that help you work with the object you right-clicked. When you right-click text, the shortcut menu contains options that copy, cut, or delete the text, or allow you to format it with a different font.

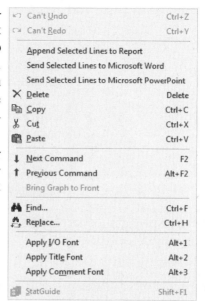

Once you have copied session commands, you can paste them at the MTB > prompt to repeat them. This is a handy trick if you want to repeat a series of commands.

 TIP You can use the Copy 🖹 *and Paste* 📋 *buttons on the toolbar instead of the Copy and Paste options on the shortcut menu to save time. The usual Windows keyboard shortcuts also work.*

Copying Commands from the History Folder

You can also copy and paste commands from the History folder, which records all the session commands you've used during a session. You can copy commands from the History folder in the same way you did the Session Window. The History folder has the advantage of listing just the commands, not the output, so it's easier to copy batches of commands. To copy commands from the History folder,

1. Choose **Window ▸ Project Manager** and click on **History** in the left pane.
2. Highlight the commands you want to execute and then click the **Copy** button 🖹.
3. Click the Session Window button.
4. Press **Ctrl+End** to move to the MTB > prompt.
5. Click the **Paste** button 📋 to paste the commands after the prompt.
6. Press **Enter** to execute the commands.

Using the Command Line Editor

The Command Line Editor is used to type, edit, and enter commands at any time. Choose Edit ► Command Line Editor to open the Command Line Editor. This works even when command language is disabled, or when the Session Window is not active. You may also use the shortcut Ctrl+L.

To execute commands using the Command Line Editor, first create the commands in the dialog box, using one of three methods:

- Highlight the commands in the Session Window or in the History folder. The highlighted commands will appear in the Command Line Editor immediately without any intervening output—no need to copy and paste. You can quickly re-execute commands using this method.
- Type the commands and subcommands directly into the dialog box.
- Paste text or data from other Minitab windows.

To execute all of the commands in the Command Line Editor, click **Submit Commands** or use the shortcut **Ctrl+Enter**. To clear all commands in the Command Line Editor, press the F3 key.

Exiting Minitab

You can use the STOP or EXIT Session commands to exit Minitab. To exit Minitab from the Session Window,

1. Type **STOP** or **EXIT** at the MTB > prompt and then press Enter.
2. If prompted to save your work, click **No** for now.

Of course, when you are doing "real" work, you will want to save your often. You can also exit Minitab using the File ► Exit menu command.

Exercises

2-1   Interpret the results of the DESCRIBE command shown in Figure 1.15, p. 23. On the average, did the pulse rates increase or decrease after exercise? Who had the greatest change? The smallest?

2-2   The following Minitab commands do some simple calculations.

```
READ C1 C2 C3
15 28 30
14 30 31
16 30 34
13 27 31
END
```

```
PRINT C1 C2 C3
LET C4=C1+C2+C3
READ C10 C11
32 26
34 27
32 25
END
LET C12=C10-C11
PRINT C4 C10 C12
```

Pretend you are Minitab and carry out the commands "by hand." Keep a worksheet as you go along. What does the worksheet contain after each command? What is printed on the screen after the last command?

2-3   The datafile Temperatures.dat is a text file that contains the data in Table 2.3, page 53, except that the heading rows and the year numbers are all preceded by the # sign so that Minitab will ignore all but the real temperature data. Use the SET command to input these data into column C2. Name the column temperatures. make sure you get all 144 values into column C2. (The datafile Temperatures.dat is available in the Data Library associated with this book at www.CengageBrain.com.) If you have a scanner available, try scanning the data into your own text file and then reading it into Minitab.

2-4   Suppose we have monthly time series data which begin in October 1997 and end in June 2004. We would like to enter a 'month number' column that starts at 10 for October 1995 and goes 10, 11, 12, 1, 2, 3,..., 1, 2,..., 6 ending with 6 for June 2004. Show how to use the SET Session command with patterned data to enter that data into column C1.

2-5   (a) Use the DSET Session command to enter monthly dates starting at May-99 and ending with Oct-04. Store the dates in a column named Time.
       (b) Use the Session command INFO to see how many months are in the Time variable.

2-6   Find the errors in each of the following. (There may be more than one error in each part).
       (a) READ C1-C3
           5 2
           6 14
           2 18
           END
       (b) READ C2 C3
           983 2
           1,102 5
           992 7
           END
```
```

```
(c) READ  C2  C5
    2 16
    4 12
    5 11
    6 12
    END
    PRNT  C2
    DESCRIBE THE 4 NUMBERS IN C2
    LET  C2=C2+C5
    LET  C10=C5+C6
    LET  C11=C1+C2+C5
(d) READ  INTO  C1  AND  C2
    5.6 23.0
    5.5 23.1
    5.4 23.3
    END
    DESCRIBE DATA IN COLUMN 1
    LET  C5=C1+C2  LET  C6=C1-C2
```

2-7 **(a)** In the following commands SET was used to enter data. Show how to use READ to do the same job.

```
SET  C1
2 3 5 7 11 15
END
SET  C2
11 18 9 16 12 10
END
SET  C3
1:6
END
SET  C4
(1 2)3
END
```

(b) In the following commands READ was used to enter data. Show how to use SET to do the same job.

```
READ  C1-C4
5 6 3 1
1 2 8 2
5 1 1 3
6 2 3 4
END
```

2-8 What values does C1 have after the following SET command is executed?

```
SET  C1
980 992 1,140 801
963 1,002
END
```

2-9 What's wrong with each of the following sets of commands?
(a) `SET C1-C2`
   ` 5  2`
   ` 13 6`
   `END`

(b) `LET C3=C1+C2 THIS IS THE TOTAL SCORE`
(c) `NAME C1 SCORE1 C2 SCORE2`
(d) `READ C1 C2`
   ` 5  3`
   ` 6  1`
   `END`
   `LET K1=C1+C2`

(e) `PRINT LENGTH`
(f) `NAME C1="length"`

2-10 Students in a small class were given three exams, with the results shown below

ID Number	Exam 1	Exam 2	Exam 3
4234	92	82	96
6457	84	84	80
5534	75	79	83
6213	98	60	72
9766	62	55	40
4538	79	72	81
4235	81	70	78

(a) Create a new Minitab project and then enter these data, using READ or SET in the Session Window. Use the NAME command to enter appropriate names for each column.
(b) Use LET to calculate the average exam score for each student.
(c) In the Session Window use the PRINT command to display a table that contains the ID number and average exam score for each student.
(d) Which student had the highest average? The lowest?
(e) Use File ▶ Print Session Window to create a hard copy of your session.

2-11 (Exercise 2-10 continued) Suppose Student with ID 6457 discovered that their third exam score had been recorded incorrectly—the 80 should have been 90.
(a) Use Session commands to change that score to the correct value.
(b) Highlight the LET and PRINT commands in the Session Window and open the Command Line Editor (with Ctrl+L, for example) and re-execute the average exam scores and the printing.

2-12   Use session commands to complete the following analysis.

(a) Create a new Minitab project and retrieve the Trees data set in the Minitab Data folder.

(b) How tall is the tallest tree? The shortest tree? Use Minitab to help answer these questions.

(c) Create three new columns that contain the data for trees that are at least 75 feet tall.

(d) How many trees are at least 75 feet tall? What is the average diameter of these trees? The average volume?

(e) Use File ▶ Print Session Window to create a hard copy of your session.

2-13   Continue to use the Trees data set from Exercise 2-12. Use session commands to complete the following analysis.

(a) How many trees have a diameter under 10 feet?

(b) Delete all trees with a diameter under 10 feet.

(c) The last tree has a diameter of 20.6 feet. Change this diameter to 21.6 feet.

(d) Add the rows of data shown below to the bottom of the current worksheet columns.

Diameter	Height	Volume
22.1	89	66.5
22.5	86	62.0

(e) Use Graph ▶ Histogram to create a histogram of Diameter.

(f) Print a copy of the graph and the Session Window contents.

3

Worksheet Operations

This chapter describes worksheet operations that you might or might not need, depending on the data you are working with. You will not need these functions for most simple data analyses, but knowing them will help you operate more efficiently in the Data Window.

3.1   Editing the Worksheet

The Edit, Data, and Editor menus offer options that help you work with data in Minitab. The Edit menu is similar to Edit menus found in other Windows applications and contains functions such as copy, paste, delete, and select. The Data menu contains data manipulation commands. The Editor menu contains worksheet navigation and formatting functions that are special to Minitab. The items on this menu change depending on which window is active. For example when the Session Window is active, the Editor menu displays session navigation and formatting options. When you work in the Data window, the Editor menu contains data editing functions. Different options appear when a Graph Window is active and depend on what is selected.

To practice using these commands, you'll enter a small data set into a new Minitab Worksheet. Columns C1-C3 will contain the gender (coded 1 = male and 0 = female), height, and weight for seven people. To enter the practice data,

1. Open a new Worksheet and maximize the Data window.

2. Type the following data into a Worksheet. (See p. 13, for a review.)

C1	C2	C3
Sex	Ht	Wt
0	66	130
1	70	155
0	64	125
0	65	115
0	63	108
1	66	145
1	69	160

3. Save the Worksheet with the name **Physical**.

Selecting Blocks of Data

Often you'll want to work with entire blocks of data, such as groups of rows or columns. Before you can work with a block of data, you need to select it.

- To select a single cell, click it.
- To select a single row, click the row number shown, for example, in Figure 1.3, p. 4. To select multiple rows, drag across multiple row numbers.
- To select a single column, click the generic column label, for example C1. To select multiple columns, drag across multiple column labels.
- To select a block of cells, drag from the upper left cell to the lower right cell of the block you want to select.

Once you've selected a block of cells, you can work with all the selected cells at once. You can access common editing operations more quickly by right-clicking a cell or block of cells and using the shortcut menu, which displays commands appropriate to the object you right-clicked. To select a cell and view its shortcut menu,

1. Click a single cell in the Worksheet to select it.
2. Right-click the selection. The shortcut menu appears as shown here to the right.
3. For now, press **Esc** to close the shortcut menu.

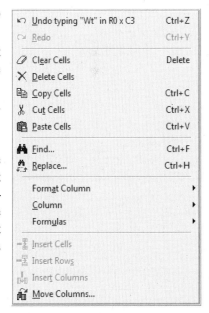

These same commands are available on the Minitab menus, but the shortcut menu provides easy access to the commands you are most likely to need for the object you right-clicked. Some of the Edit and Editor functions are also available as toolbar buttons.

Inserting Data

You can insert a cell, a row, or a column using the shortcut menu or toolbar buttons, which are usually quicker. When you insert cells, Minitab places the missing-data symbol (*) in each inserted cell. To enter new data, click the inserted cells and enter the data as you normally would.

- To use the Worksheet toolbar, make sure it is activated by inspecting the checkmark at Tools ▸ Toolbar ▸ Worksheet. Check it if necessary.
- To insert one or more cells, select the cell or cells, and then click the **Insert Cells** button shown in Figure 3.1. The selected cells move down and the new cells appear in their place. The number of inserted cells is equal to the number of cells you selected before you chose the command.
- To insert an entire row, click the **Insert Row** button on the toolbar. The row you right-clicked moves down, and the new row appears in its place.
- To insert an entire column, click anywhere in the column then click the **Insert Column** button on the toolbar. The column you selected moves to the right, and the new column appears in its place.

Figure 3.1   Data Window toolbar

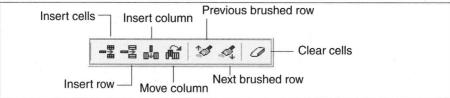

When you insert a row between other data rows, Minitab populates the new cells with *, the missing-value symbol. You then type data for the new row replacing the missing values. Try inserting a column for Age between C1 and C2. To insert a column,

1. Click any cell in **C2**.
2. Click the **Insert Columns** button on the toolbar. Minitab inserts a column to the left of the selected column.
3. Type **AGE** as the new column's name.
4. Enter the following data in C2, pressing **Enter** after each entry: **25 40 31 19 20 36 28**. For ease of entry make sure the data-entry arrow points downward.

   TIP Remember that to change the direction of the data-entry arrow, you click the arrow.

Clearing or Deleting Data

When you clear data, you erase the contents of one or more highlighted cells, without moving rows up or columns left. You select the data and then either click the **Clear Cells** button on the Worksheet toolbar or press the **Backspace** key on your keyboard. Minitab replaces the cell contents with an asterisk (*), the symbol for missing data. When you **delete** data, however, you delete not only the contents of the highlighted cells, but the cells themselves. Thus if you delete a row, all lower rows move up. Minitab lets you delete entire cells, rows, columns, or blocks of cells.

- To delete a single cell or block of cells, select the cells and then press the Delete key. You may alternatively, right-click the selected cells and click Delete Cells, or select Delete Cells on the Edit menu. Minitab moves the remaining cells up.

- To delete an entire row or column, right-click the row or column header to select the row or column and then press the Delete key. When you delete a column, Minitab moves the remaining columns to the left. When you delete a row, Minitab moves the remaining rows up.

- You can also delete entire columns, constants, or matrices by choosing Data ▸ Erase Variables, entering the variables you want to erase, and then clicking OK.

Try deleting row 4, and then clear the contents of cell R1C2. To delete data,

1. Click the row number **4** to highlight the entire row.
2. Right-click the selection and then click **Delete Cells**. Minitab deletes the row.

   TIP Once you have highlighted an entire row or column, you can also delete it by pressing the Delete key.

3. Click cell **R1C2** and then click the **Clear Cells** button. Minitab replaces the entry with the missing value symbol *.

   TIP R1C2 is a quick way of indicating the cell in row 1, column 2.

Copying and Pasting Data

You can use Copy, Cut, and Paste from the Edit menu as you can in other Windows applications. Thus you can highlight a block of rows and columns in the Data window, copy or cut them, and then paste them somewhere else in the

Data window, or in another application such as Notepad. You can copy from and paste to the Data window from a variety of sources, including cells, rows, and columns in the same Data window; cells, rows, and columns in another Data window; and documents in another application, such as a spreadsheet.

- To copy cells within Minitab, highlight the cells and click the Copy Cells button 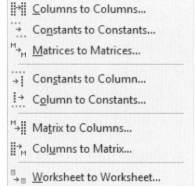. Then activate the Worksheet where you want to paste the copied cells. Click the upper-left cell of the destination location and click the Paste Cells button.

- To copy data from another application, such as a spreadsheet, highlight the cells or data, copy it in that application, then return to Minitab and paste it in the same manner. The Copy and Paste buttons you see in Minitab are standardized in most Windows applications.

- To copy columns and rows that are not contiguous, use Data ▸ Copy. You can paste columns copied this way into the same Data Worksheet or a different Worksheet. The many options are shown here to the right.

Columns to Columns...

Constants to Constants...

Matrices to Matrices...

Constants to Column...

Column to Constants...

Matrix to Columns...

Columns to Matrix...

Worksheet to Worksheet...

Try copying C3 and C4 into C6 and C7. To copy columns,

1. Drag across column headers C3 and C4 to select both entire columns.
2. Click the **Copy Cells** button.
3. Click **R1C6**. (Always select the upper left-hand corner of the block you are moving to.)
4. Click the Paste Cells button. Minitab pastes the copied cells into C6 and C7.

    TIP When you paste entire columns as you did just now, Minitab can not paste the names into the name cells because column names must be unique. Instead Minitab will name the columns Ht_1 and Wt_1, for example

Moving Columns

You can move columns from one location to another by using the Cut and Paste buttons, which work similarly to the Copy and Paste buttons, except when you click Cut, you delete the cells, not just copy them. Alternatively, you may use the Move Columns button on the toolbar or the Editor ▸ Move Columns option, both of which open a dialog box that lets you specify a destination location for the columns you have selected to move.

Try moving the columns you just pasted to a new location. To move C6 and C7 to C9 and C10,

1. Highlight **C6-C7** and then click the **Cut** button ✂.
2. Click **R1C9** and then click the **Paste** button 📋. Minitab moves the cells to the new location.
3. Delete columns C9 and C10 since we are just practicing!

Resizing Columns

You can resize a column by dragging its right border in the column header area left or right. When you are in the correct position for resizing, the pointer changes shape from ✛ to ↔. Drag to the left to shrink the column or to the right to enlarge the column. You can resize multiple columns simultaneously by highlighting the columns you want to resize and then dragging one of the borders.[†] To enlarge C2,

1. Point to the column border between C2 and C3 in the column header area. The pointer changes to ↔.
2. Drag to the right.
3. Release the mouse button when the column is the size you want.

Formatting Columns

Recall that a Minitab column will be one of three data types: numeric, text, or date/time. When you enter data in an empty column, Minitab automatically assigns a data type and format based on the characters that are entered into the first cell. Usually this works well. However, if you type the last four digits of a Social Security number in the first row of a column, Minitab identifies its type as numeric. If the digits are 0123 Minitab will drop the leading zero and enter 123—not what you wanted. One easy way to get around this difficulty is to type a letter in the first cell and press Enter. Minitab identifies the column as text. Now go back and replace the cell contents with the 'text' data 0123. If the change makes sense, you can change a column's data type using Data ▶ Change Data Type.

Minitab also assigns a specific display format to a column. You can change the format using commands on the Editor ▶ Format Columns submenu. In some cases Minitab changes the format automatically. For example if you enter

---

[†] Note that resizing columns in no way changes the contents of the cells, whether numbers or text. It only changes what you see. Since decimal numbers are stored internally in double precision, what you see is not necessarily what is in the cell.

a column of numbers with one digit after the decimal (such as 3.2, 4.3, and 5.4), but then you type a number with four digits after the decimal (as in 1.1234), Minitab changes the column format to use four decimals, and all numbers will appear in that format. For example, 3.2 will appear as 3.2000.

When you modify the format characteristics of a column, you are changing only the way that column is displayed in the Data window—you are not changing the underlying value. For example if the number in a cell is 1.2345678 and you change the format to display only two decimals, all calculations will still use 1.2345678.

Minitab stores and computes with numerical values in double precision. Double precision systems accurately represent a number that is 15 or 16 decimal digits long. If you execute the Session command LET C1=1/3, Minitab will display the result as 0.333333 but internally it will be 0.3333333333333330— close enough for most statistical work!

Try examining the format of one of the columns you've entered. To work with a column's format,

1. Click any cell in **C4** to examine its format.

2. Choose **Editor ▸ Format Column ▸ Numeric**. Notice that the other data type options are gray on the sub-menu, indicating that you can't select them. This is because Minitab has already assigned the Numeric data type to C4.

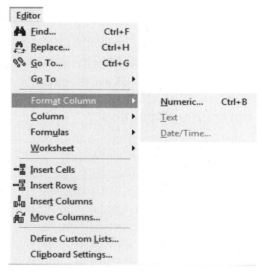

3. Click **Numeric**. Note that you can control column width and decimal places for the Numeric format.

4. Click **Cancel** to close the Numeric Column Format dialog box because you don't need to make any changes.

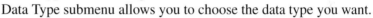

5. To change a data type, choose **Data ▸ Change Data Type**. The Change Data Type submenu allows you to choose the data type you want.

6. For now press **Esc** to close the menu without making any changes.

Date/Time columns can be formatted to display in many different ways. Figure 3.3 shows some of the possibilities. Check the Help system for more information on Date/Time data.

Figure 3.2   Format options for Date/Time data

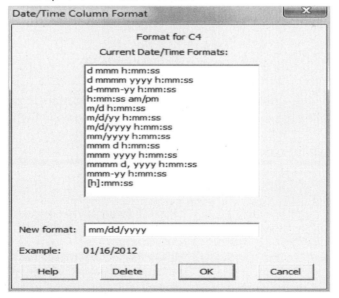

Documenting Columns

When you work with many projects and worksheets that contain many columns, it is good practice to document the definition of a given variable. This is especially true for categorical variables where the code is not obvious. For example, suppose gender is coded as 0 and 1. Are females 1 or are they 0? You can use the Editor ▶ Set Column ▶ Description command to view, edit and save text that describes the variable in a column. To document C1,

1. Click any cell in column **C1**.
2. Choose **Editor ▶ Column ▶ Description**.
3. Type the following in the Description box: **This column contains a code for the gender of each subject. 1 = male and 0 = female.** See Figure 3.3.
4. Click **OK**. The column description will be saved when you next save the Project or Worksheet and can be viewed when you reopen the Project or Worksheet.

Figure 3.3   Column Description dialog box

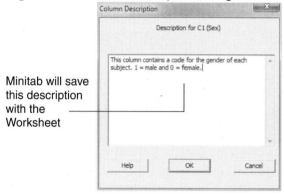

Minitab will save this description with the Worksheet

Once a description has been set up, a small red triangle is displayed in the upper right-hand corner of the column header as shown in Figure 3.4.

Figure 3.4   Red triangle for column description

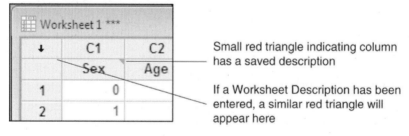

Small red triangle indicating column has a saved description

If a Worksheet Description has been entered, a similar red triangle will appear here

A quick way to read the description is to hover the pointer over the red triangle. A Tooltip description appears as shown in Figure 3.5.

Figure 3.5 Tooltip column description

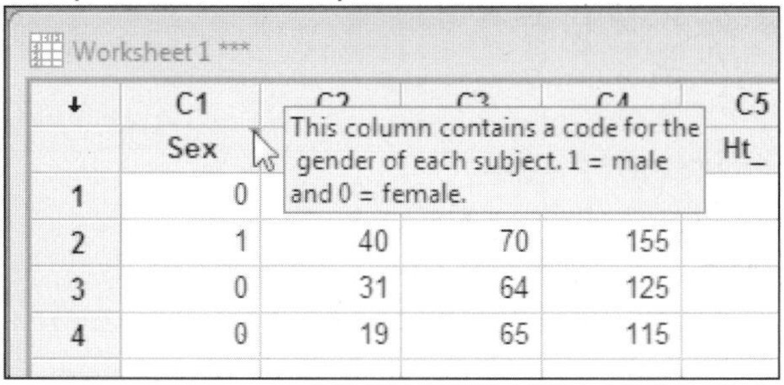

3.2 Manipulating Data

Subsetting the Worksheet

Often we need to look at just certain data values in a Worksheet. For example, we may want to analyze just the data for females or just for females under a certain age. We say we need to **subset** the Worksheet. Minitab provides several ways to do this.

The menu command Data ▸ Subset Worksheet lets you copy specified rows from the active worksheet to a new worksheet. You can specify the subset based on a condition, such as females under 25 years of age, on row numbers, or on brushed points on a graph.[†] You may also tell Minitab which data points to exclude rather than which to include.

To use the subset capabilities of Minitab to copy just the data for the females,

1. Make sure the Worksheet you want to subset is active.
2. Choose **Data ▸ Subset Worksheet** to open the Subset Worksheet dialog box.
3. Type **Females** as the Name of the New Worksheet. See Figure 3.6.
4. Make sure the option button for **Specify which rows to include** is selected.
5. Make sure the option button for **Rows that match** is selected.
6. Click the **Condition** button.
7. Type **'Sex'=0** in the condition text box. See Figure 3.7.
8. Click **OK** twice and Minitab produces a new Worksheet which contains the same variables but only for rows with Females.

[†] See page 107 for information on brushing a graph.

Figure 3.6   Subset Worksheet dialog box

Type your choice
of name for new
Worksheet
here

Click this
option button

Click this
option button

Then click
the Condition
button

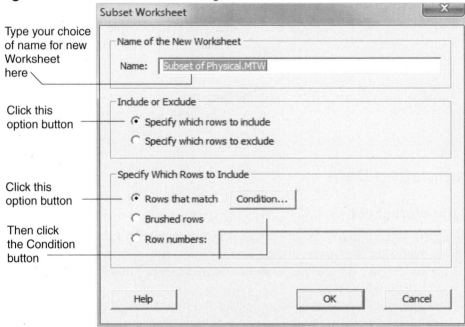

Figure 3.7   Subset Worksheet Condition dialog box

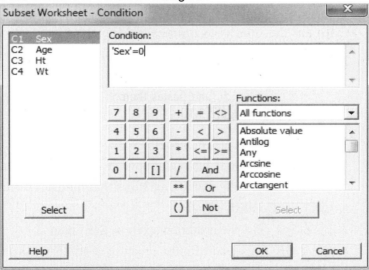

The condition text box in Figure 3.7 could be filled with more complex conditions such as (('Sex'=0) And ('Age'<25)). This would select only rows with females under the age of 25.

The Data ▸ Subset Worksheet command copies *all* columns to the subset Worksheets. You can subset just selected columns of a Worksheet using the Data ▸ Copy ▸ Columns to Columns command. For example, you can copy the height variable for just the females into a new column. To copy just the height data for the females,

1. Choose **Data ▸ Copy ▸ Columns to Columns**.
2. Enter **Ht** in the Copy from columns box. Minitab will copy data just from that column.

   💡 *TIP Remember that to enter a variable into a text box, you can type it by column number or name, double-click it from the variables list, or highlight it in the variables list and then click the Select button.*

3. Click the **Store Copied Data in Columns** list arrow and select **In current Worksheet, in Columns**.
4. Enter **'Ht.female'** into the columns box.
5. Deselect the **Name the columns containing the copied data**. (We entered our own name in the previous step.) See Figure 3.8.
6. Click the **Subset the Data** button.
7. Select the **Specify which rows to include** option. See Figure 3.9.
8. Select the **Use rows with** option.
9. Press the **Condition** button.
10. The condition dialog box is the same one we saw in Figure 3.7, p. 78. Fill it in the same way.
11. Click **OK** three times. Minitab copies the height data from C3, but only for rows where C1 = 0, and pastes them into the next empty column named Ht.female. See Figure 3.10.

Figure 3.8   Copy Columns to Columns dialog box

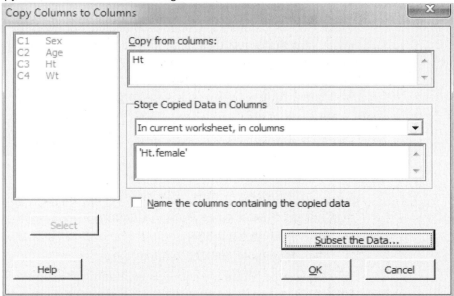

To clean up the worksheet so it's ready for your next steps, delete column **C6**. Your Worksheet should now only contain data for C1-C4.

Figure 3.9   Subset the Data dialog box

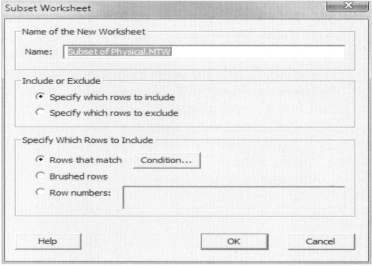

You can also subset data by specifying which rows *not* to use. If you had clicked the 'Specify which rows to exclude' option rather than 'Specify which rows to include' option, you could have told Minitab to copy all data *except* those meeting the criteria you specify.

Figure 3.10 Worksheet after copy

↓	C1	C2	C3	C4	C5
	Sex	Age	Ht	Wt	Ht.female
1	0	*	66	130	64
2	1	40	70	155	65
3	0	31	64	125	63
4	0	20	63	108	
5	1	36	66	145	
6	1	28	69	160	

Physical.MTW ***

Stacking and Unstacking Columns

Table 3.1 shows height and weight data for men and women, stored in two different forms. In the unstacked form the data for the men are in two columns, Ht.Men and Wt.Men, and the data for the women are in a second pair, Ht.Women and Wt.Women. In the stacked form the data for men and women are stacked together, in columns Ht and Wt. The column Sex indicates which rows are for men (Sex = 1) and which rows are for women (Sex = 0) and the rows in this form could be mixed up in any order. For some purposes one form is more convenient than the other.

Sometimes we need to change data from one form to the other. The commands Data ▸ Stack and Data ▸ Unstack make this relatively easy. The commands on this submenu include commands to stack or unstack one or more columns or rows.

For example to stack the columns Ht.Men, Wt.Men on the columns Ht.Women, Wt.Women and store the stacked data in columns Ht and Wt, use Data ▸ Stack ▸ Blocks of Columns. Designate the columns you want to stack, the column or columns in which you want to place the stacked data, and the column in which you want to store subscripts. The subscript column will contain either 1s in rows corresponding to the first stacked column, 2s in rows corresponding to the second stacked column, and so on or names selected from the names of the stacked columns. You can then use this subscript column for later analyses.

Similarly, to unstack the columns Ht and Wt into the columns Ht.Men, Wt.Men and Ht.Women, Wt.Women, use Data ▸ Unstack. Designate the columns you want to unstack, the column containing subscripts you want to use (which could be either numerical or text), and the destination locations for the unstacked data. The rows with the smallest subscript are stored in the first block, the rows with the second smallest subscript are stored in the second block, and so on. If you do not enter a column containing subscripts, Minitab stores each row in a separate block. Numbers in the subscript column must be integers between −10000 and +10000. They need not be in order, nor do they need to be consecutive integers.

Table 3.1 Stacked and unstacked data

Unstacked Data				Stacked Data		
Ht.Men	*Wt.Men*	*Ht.Women*	*Wt.Women*	*Ht*	*Wt*	*Sex*
70	155	66	130	70	155	1
66	145	64	125	66	145	1
69	160	65	115	69	160	1
		63	108	66	130	0
				64	125	0
				65	115	0
				63	108	0

Try unstacking the height data. To unstack Ht by Sex,

1. Choose **Data ▸ Unstack**.
2. Enter **Ht** in the Unstack the data in: box. See Figure 3.11.
3. Enter **Sex** in the Using subscripts in: box.
4. Select **After last column in use** in the Store unstacked data option area.
5. Check **Name the columns containing the unstacked data.**
6. Click **OK**. Minitab unstacks the data. C5 now contains height values for rows where Sex = 1, and C6 contains values for rows where Sex = 0. See Figure 3.12. Note the names that Minitab assigned to these columns.

Figure 3.11 Unstack dialog box

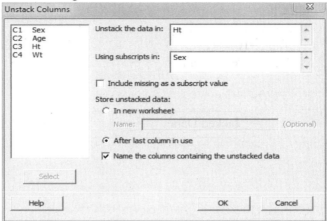

Figure 3.12 Unstacked Ht column

↓	C1	C2	C3	C4	C5	C6	
	Sex	Age	Ht	Wt	Ht_0	Ht_1	
1	0	*	66	130	66	70	
2	1	40	70	155	64	66	
3	0	31	64	125	63	69	
4	0	20	63	108			
5	1	36	66	145			
6	1	28	69	160			

To stack these columns back the way they were:

1. Choose **Data ▸ Stack ▸ Columns**.
2. Enter **Ht_0** and **Ht_1** in the **Stack the following columns:** box. See Figure 3.13.
3. Select **Column of current Worksheet:** in the **Store stacked data in:** section.
4. Enter **Height** as the name of the column to store the subscripts.
5. Check **Use variable names in subscript column**.
6. Click **OK** and Minitab stacks the data. See Figure 3.14. C7 now contains height values for both sexes and C8 tells us which is which. We could use Data ▸ Code to convert all the Ht_0 to Female and the Ht_1 to Male in C8.

Figure 3.13 Stack Columns dialog box

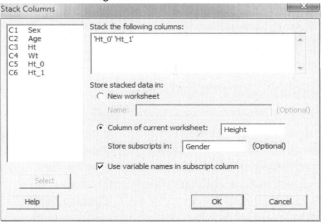

Figure 3.14 Stacked Worksheet

	C1	C2	C3	C4	C5	C6	C7	C8-T	C
	Sex	Age	Ht	Wt	Ht_0	Ht_1	Height	Gender	
1	0	*	66	130	66	70	66	Ht_0	
2	1	40	70	155	64	66	64	Ht_0	
3	0	31	64	125	63	69	63	Ht_0	
4	0	20	63	108			70	Ht_1	
5	1	36	66	145			66	Ht_1	
6	1	28	69	160			69	Ht_1	
7									

Erase columns C5-C8 to clean up the Worksheet.

Generating Patterned Data

In Chapter 2, p. 54, you learned how to use Session commands to enter data following certain patterns. Here we do similar entries with the menus. You can easily fill a column with numbers that follow a pattern, such as the numbers 1 through 100, or 5 sets of 1, 2,…, 12 to obtain 'month numbers' for 5 years of monthly data. You use the Make Patterned Data submenu to enter data that follow a pattern. This submenu contains commands for generating numeric patterns, text patterns, and date/time values that follow a pattern. You'll try entering several simple sets of equally spaced numbers. To generate patterned data,

1. Choose **Calc ▸ Make Patterned Data ▸ Simple Set of Numbers**.

2. Enter **Pattern** in the Store patterned data in: box. Minitab will store the results in the first empty column and name that column Pattern. See Figure 3.15.

3. Press **Tab** and type **1** as the first value, press **Tab** again, and type **10** as the last value. Leave everything else as is for now.

4. Click **OK**. Minitab generates the numbers 1 through 10 in C5 with column name Pattern. See Figure 3.16.

Figure 3.15 Patterned Data dialog box

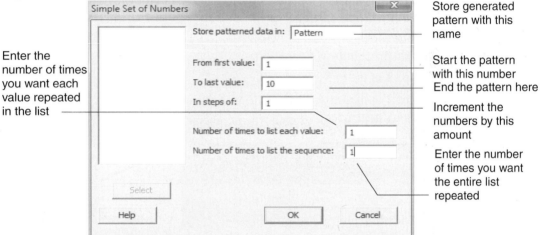

Enter the number of times you want each value repeated in the list

Store generated pattern with this name

Start the pattern with this number
End the pattern here

Increment the numbers by this amount

Enter the number of times you want the entire list repeated

Figure 3.16 Worksheet with patterned data

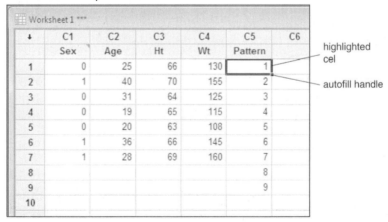

highlighted cel

autofill handle

Reusing a Dialog Box

You'll generate one more column of patterned data, but to do so, you'll use a shortcut that always allows you to open the most recently used dialog box, bypassing menus and submenus. The Edit ▸ Edit Last Dialog command opens the most recently used dialog box, with the same selections from the last time you used it. When you reopen any dialog box, you'll find that Minitab "remembers" the values you used previously in this session. This saves you from having to type information in again if you want to repeat or change some parameters in an analysis. The Edit Last Dialog command also saves you from having to search for the command you used most recently. This command is also available as a button, ▣, on the main toolbar and with the shortcut Ctrl+E.

1. Click the **Edit Last Dialog** button ▣. The Simple Set of Numbers dialog box reopens with the same entries you typed in the previous steps. Now you'll generate a sequence of numbers that increments by 2 instead of 1, and each number is repeated twice.

2. Store the results in **C6**.

3. Type **2** in the In Steps of: box.

4. Type **2** in the List each value: box. See Figure 3.17.

5. Click **OK**. C6 now contains the numbers 1, 1, 3, 3, 5, 5, 7, 7, and 9, 9.

Figure 3.17 Patterned data with steps and repeats

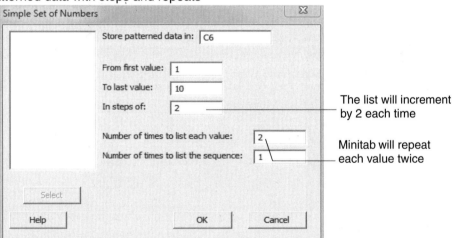

Now delete C5 and C6 for the next examples.

You can generate more complicated patterned data by using the Session commands. See "Using SET to Enter Patterned Data", p. 54, and online Help for more information.

Autofill

Minitab has an additional feature for generating patterned data called **autofill**. Consider an active Worksheet such as shown in Figure 3.16, p. 85. The highlighted cell has a darkened box around it and an **autofill handle** at the lower-right corner of the box. If you place the pointer over that corner its shape changes from the usual Worksheet cursor, ✣, to ✚, the **autofill pointer**. If you hold down the Ctrl key, the autofill pointer changes to ✚⁺. Here are some examples of using autofill:

1. Type the number **1** in the first cell of a column C6.
2. Place the cursor on the autofill handle and *hold down the Ctrl key*. The pointer should look like this ✚⁺.
3. Drag the autofill handle down the column, stopping at the tenth row.
4. The column fills with the sequence 1, 2, 3,…, 10.

1. Type the number **1** in the first cell of a column C6.
2. Place the cursor on the autofill handle. (Do **not** hold down the Ctrl key.) The pointer should look like ✚.
3. Drag the autofill handle down the column stopping at the tenth row.
4. The column fills with the sequence 1, 1, 1,…, 1.

1. Type the numbers **1**, **2**, **3**, **4** in the first four cells of a column C6.
2. Drag over all four of these cells to highlight all of them simultaneously.
3. Place the cursor on the autofill handle and **hold down the Ctrl key**. The pointer should look like ✚⁺.
4. Drag the autofill handle down the column stopping at the twelfth row.
5. The column fills with the sequence 1, 2, 3, 4, 1, 2, 3, 4, 1, 2, 3, 4.

1. Type **Jan** in the first cell of a column C6. Minitab recognizes this as the Date/Time value for January. The column header changes to C6-D.
2. Place the cursor on the autofill handle. (Do **not** hold down the Ctrl key.) The pointer should look like ✚.
3. Drag the autofill handle down the column stopping at the fourteenth row.
4. The column fills with the sequence Jan, Feb, Mar,…, Dec, Jan, Feb. Notice that the values started repeating at row 13. This will continue if we drag further down the column.

Autofill has a great many possibilities. Here we illustrated only the simplest uses. See the Help pages for more information.

Sorting Data

You can sort any number of columns, by any number of columns, and any combination of alpha and numeric data. Table 3.2 shows two examples. In the first sort C1-C4 are sorted by the values in C1 and stored in C11-C14. C1 determines the order. The smallest value in C1 is 10, so all rows in which C1 = 10 are put first. The second smallest value in C1 is 11, so all rows in which C1 = 11 are put next, and so on.

In the second SORT, C1-C4 are sorted by two columns: first C1 and then, within C1, by C2. The sorted data are stored in C21-C24.

Table 3.2 Sorting data

C1	C2	C3	C4	C11	C12	C13	C14	C21	C22	C23	C24
10	0.2	31	131	10	0.2	31	131	10	0.1	35	210
10	0.1	35	210	10	0.1	35	210	10	0.2	31	121
12	0.1	37	176	10	0.4	31	140	10	0.4	31	140
12	0.1	36	190	11	0.2	29	180	11	0.1	33	182
10	0.4	31	140	11	0.1	33	182	11	0.2	29	180
12	0.1	30	110	12	0.1	37	176	12	0.1	37	176
11	0.2	29	180	12	0.1	36	190	12	0.1	36	190
11	0.1	33	182	12	0.1	30	110	12	0.1	30	110

Hopefully, you still have the Worksheet shown in Figure 3.16, p. 85, with C5 and, possibly, C6 deleted. If not, redo that data entry now.

Try sorting the data in your Worksheet. You'll sort by C3, which contains height values, so the shortest people are listed first. To sort C1-C4 by C3,

1. Choose **Data ▸ Sort**.
2. Enter **Sex-Wt** in the Sort column(s) box.
3. Enter **HT** in the first Sort by column box. See Figure 3.18.
4. Enter **C6-C9** in the Store sorted column(s) box.
5. Click **OK**. Minitab sorts the columns by C3 and places the sorted data into C6-C9. See Figure 3.19.

If you don't indicate a column to sort by, Minitab assumes you want to sort by the first column in the Sort column(s) box.

Figure 3.18 Sort dialog box

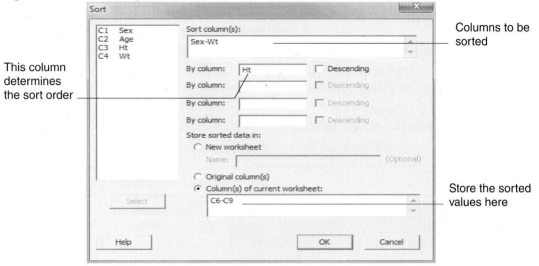

Figure 3.19 Sorted Worksheet

All the columns
have been
sorted according
to height
Rows are kept
intact, just
reordered

	C1	C2	C3	C4	C5	C6	C7	C8	C9	C1
	Sex	Age	Ht	Wt						
1	0	*	66	130		0	20	63	108	
2	1	40	70	155		0	31	64	125	
3	0	31	64	125		0	*	66	130	
4	0	20	63	108		1	36	66	145	
5	1	36	66	145		1	28	69	160	
6	1	28	69	160		1	40	70	155	

Delete the data in C6-C9 before starting the next Section.

Ranking Data

You can use the Rank command to create a new column containing the rank of
each value in the original column. The smallest value is given rank 1; the
second smallest, rank 2; the third smallest, rank 3; and so on. Note that the
Rank command does not reorder the data. Table 3.3 shows how Minitab creates
a new column, C11, containing the ranks of the numbers in C1.

The ranks for C1 are put into C11. The smallest number in C1 is 1.1, so a 1 is put into C11 in the same row as the 1.1. The second smallest number in C1 is 1.3, so a 2 is put into C11, and so on. If there are ties, the average rank is assigned to each value. For example the fourth and fifth values in C1 are both 2.0. They are both given the rank of (4 + 5)/2 = 4.5.

Try ranking the data in your Worksheet by weight. To rank the data by weight,

Table 3.3 Ranking data

C1 Data	C11 Ranks
1.4	3
1.1	1
2.0	4.5
3.1	6
2.0	4.5
1.3	2

1. Choose **Data ▸ Rank**.
2. Enter **Wt** in the Rank data in box.
3. Then type the word **Rank** in the Store ranks in box. See Figure 3.20.
4. Click **OK**. Minitab enters the ranks in the next available column and names the column Rank.

> 💡 TIP When you type a name in a text box instead of a column number, as you did with the word Rank, Minitab stores the data in the first empty column (to the right of any existing columns) and assigns the column the name you specified.

5. Delete C5 to clear unnecessary columns in the Worksheet for the next set of steps.

Figure 3.20 Rank dialog box

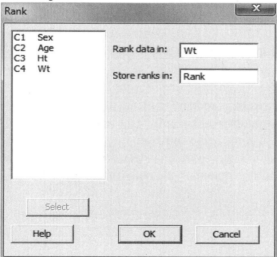

Coding Data

The CODE command searches columns for a value or set of values and replaces them with a new value. Table 3.4 shows how the Code command creates two new columns, C11 and C12, that change the values in C1-C2 that are equal to −2 or −1 into 0. All other values are left the same. The results are stored in C11 and C12, but C1 and C2 are left intact.

Table 3.4 Coding data

C1	C2	C11	C12
8	16	8	16
−2	13	0	13
6	−2	6	0
0	−2	0	0
−1	14	0	14

The Code command allows you to code numbers to new numbers, as in Table 3.4. Sometimes we need to code numbers to text. Try coding the values in the SEX column to the letters M and F. You will code 1 = M and 0 = F. The Data ▸ Code submenu contains commands for coding values from numeric to numeric, numeric to text, text to text, or text to numeric. Since you are coding numbers to letters, you will choose the Numeric to Text option. To create a column of letter codes for C1,

1. Choose **Data ▸ Code ▸ Numeric to Text**.
2. Enter **Sex** in the Code data from columns: box.
3. Press **Tab** and enter **Gender** in the Into columns: box. Now you'll tell Minitab what values you want to use for each value in the Sex column.
4. Press **Tab** and enter **1** in the first Original values box. Press **Tab** and type **M** in the New box.
5. Press **Tab** and enter **0** in the second Original values box. Press **Tab** and type **F** in the New box. See Figure 3.21.
6. Click OK.

Figure 3.21 Code Numeric to Text dialog box

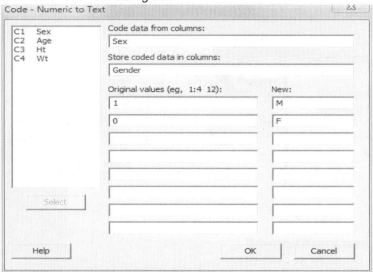

Minitab creates a new column, C5, which contains the coded data as shown in Figure 3.22.

Figure 3.22 Worksheet after coding numerical to text

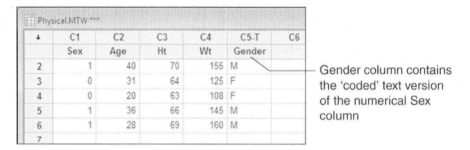

Gender column contains the 'coded' text version of the numerical Sex column

3.3   Importing and Exporting Data

Most of the time the data files you bring into your project and save from your project will be Minitab Worksheets. Those Worksheets may be stand-alone files (files with the extension MTW) or parts of a project (MPJ) file. You can exchange data with versions of Minitab on other platforms, using Minitab Portable Worksheets (MTP files). You can also open and save data files in the formats of many other applications, like Excel and Lotus 1-2-3. To exchange data with other applications, such as mainframe computer programs, you can open and save text files. To import data from database files, you can use the File ► Query Database (ODBC) command. Finally to import data with other types of applications, use File ► Other Files ►   Import Special Text and File ► Other Files ►Export Special Text.

Portable Files

Saving a Worksheet in a portable file format allows you to move the Worksheet to different versions of Minitab as well as installations of Minitab on a different type of machine, such as from a PC computer to a Macintosh. Minitab Worksheets are upwardly compatible, which means that you can always open a Worksheet saved by a version of Minitab immediately previous to the one you are using. If you want to open a Worksheet file created on a newer version of Minitab into an older release, save it as a Portable Worksheet. To save a file as a portable Worksheet,

1. Choose **File ► Save Current Worksheet As**.
2. Open the device, drive, or folder on which you want to save the Worksheet.
3. Click the **Save as type** list arrow and then click **Minitab Portable**.
4. Click **Save**. Earlier in the chapter you saved the Worksheet in the Minitab Worksheet format as Physical.mtw; now Minitab saves it as Physical.mtp, in the portable format.

   TIP When you want to open the file on any platform running Minitab, click Minitab Portable as the file type in the Open Worksheet dialog box.

Exchanging Data with Minitab-Compatible Applications

Minitab can exchange data directly with Excel, Quattro Pro, Lotus, Symphony, or dBase applications. To exchange data with other programs, you might need to use a text file, as explained in the next section. To save a Minitab Worksheet in a different file format,

1. Choose **File ► Save Current Worksheet As**.
2. Open the drive or folder on which you want to save the Worksheet.
3. Click the **Save as type** list arrow and then click the application you want.
4. Enter a name and then click **Save**.

Let's open an Excel spreadsheet data file. We'll use the file with name Forbes 2011 Top 400.xls which is available for download from the Data Library at www.CengageBrain.com associated with this book. When opening a non-Minitab file, you might want to use the Preview dialog box, not only to view the data, but also to change how Minitab will interpret the data. To open a non-Minitab file,

1. Choose **File ▸ Open Worksheet**.
2. Select the file **Forbes 2011 Top 400** and then click the **Preview** button. See Figure 3.23. Minitab assumes that the first row of the file contains the column names. See Figure 3.24.
3. If the data in the Preview dialog box does not appear as you want it to look in Minitab, close the Preview dialog box and then click the **Options** button in the Open Worksheet dialog box. The Options dialog box, shown in Figure 3.25, helps you set up the file you are opening, including column names, designating where the data begins and ends, data delimiters (items that separate data points—usually tabs, spaces, or hard returns), and how missing value codes should be converted.
4. Once you have set the options you want and have previewed the data so it appears to your satisfaction, click **Open** in the Open Worksheet dialog box.

Figure 3.23 Open Excel Worksheet dialog box

Figure 3.24 Open Worksheet Preview

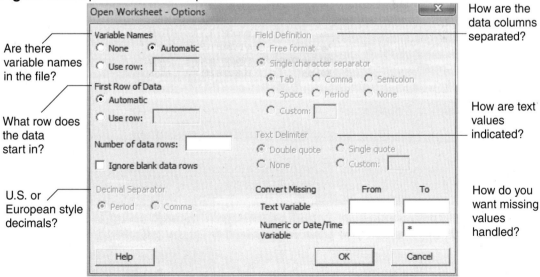

See online Help for more information on setting options so that you get the data you want. Sometimes you need do some trial-and-error with different options to get the data read in properly.

Figure 3.25 Open Worksheets Options.

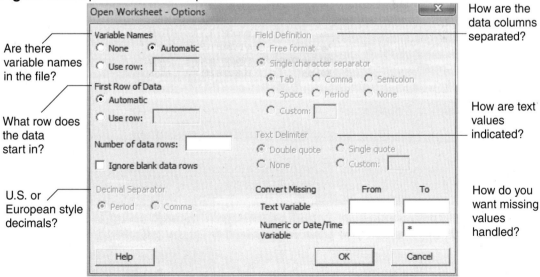

Saving Data in a Text File

When you want to open data from a file saved in an application other than Excel, Quattro Pro, Lotus, Symphony, or dBase, try opening it as a text file.

Also called an ASCII file or data file. A text file contains only the characters available on most computer keyboards, such as letters, numbers, and standard keyboard symbols. ASCII files contain no formatting, but almost any application can open and work with files that contain ASCII data. Most ASCII files you use will contain rows and columns of numbers, with at least one space, tab, or comma between numbers. However, if the file contains text data, if there are no spaces, commas, or tab characters between numbers, if blanks are used for missing values, if two or more lines of data are to go into one row of Minitab's Worksheet, or if the file has any other unusual properties, the Open Worksheet command might not be able to interpret them correctly. For example, if there is no clear delimiter that separates rows of data into columns, Minitab will place all the data into one column. In such cases you can open the file and fix it in the Data window, or you can fix it in a text editor before you open it in Minitab.

Importing Files in Custom Formats

The File ▶ Other Files ▶ Import Special Text command can be a powerful solution to importing files that are difficult to open in Minitab because you can specify a custom format. A format statement is made up of special codes that require proper syntax. The format statement allows you to define precisely how the data are organized, but you must type the format statement according to strict rules. See the Help topic "Format Statement Syntax" to learn how to write format statements that help with file import problems.

Common problems include the lack of delimiters, rows of data falling into two or more lines, and blank cells scattered randomly throughout the data file. The Help topic "When to Use the Import Special Text Command" includes a useful table of common import problems and suggests solutions.

Importing from a Database

Open DataBase Connectivity (ODBC) is a programming interface that allows one application to communicate with another application that uses a different database language. Most database programs can understand the ODBC language, which makes it possible for different programs to share data. You can import data from a database into Minitab with the File ▶ Query Database (ODBC) command. You choose the database table and fields (variables) you want to import data from. You can click the Use Rows button to specify a subset of data to import. To use ODBC in Minitab, you might have to first install ODBC software on your system. The ODBC software is available from Microsoft's Web site. Minitab's Help topic "Setting Up ODBC" can help you do this. To import data from a database,

1. Activate the Data window into which you want to import the database.

2. Choose **File ▸ Query Database (ODBC)**.

3. Select the data source you want to use. How you do this depends on the ODBC manager you have installed.

4. Click **OK**.

5. If prompted to do so, provide whatever security information is requested, and then click **OK**.

6. Select the table you want.

7. Use the arrow buttons to move items from the Available fields box to the Selected fields box. Each selected field will become a column in Minitab.

8. Click **Use rows** to set conditions for subsetting the data from a field.

9. Click **OK**.

> *TIP Importing data can take a while. Once the import is complete, save your work so you can quickly retrieve the data later.*

When you execute the query, Minitab converts the database fields and records to Minitab columns and rows, adding them to the right of any existing columns in the Data Window.

You can also use the OBDC session command, which allows you to subset data using any SQL expression from within Minitab. **Structured Query Language (SQL)** allows users to access data in databases by writing expressions, or strings, that describe conditions for data retrieval. You can also use a specific SQL string you want in the Use Rows dialog box. See online Help for more information.

Exercises

In this set of exercises you will enter a data set and work with its data. After each exercise, display the data in the Session Window using either the PRINT Session command or the Edit ▸ Display Data command. When you are finished with all the exercises the Session Window will contain a complete record of your work with the data.

3-1

(a) Create a new Minitab project and name it Classes. Enter the data shown in Table 3.5 into the Worksheet.

(b) Insert a new row for student 22 between rows 4 and 5, with Grade1=90 and Grade2=88.

(c) Delete the data for student 19, who dropped out of the course.

(d) Copy the Student ID column to C5.

(e) Save the project.

Table 3.5 Grades data

Student ID	Grade1	Grade2
38	74	81
77	93	100
19	52	40
12	78	69
63	80	82

3-2

Continue using the Classes data you entered in Exercise 3-1.

(a) Delete C5, which contains the copy of the Student ID columns.

(b) Assign a text data type to C1.

(c) Document C2 with the description "Grades for the 3/21 exam" and document C3 with the description "Grades for the 5/15 exam."

(d) Copy the grades data for students whose ID number is greater than 50 into C5-C7. (Hint: You'll need to use the Use Rows button. Can you figure out what expression to type?)

3-3

Continue to use the Classes data.

(a) Stack the two grade columns into a single column, C8, entering subscripts G1 and G2 in C9.

(b) In C10 generate the pattern 1, 2, 3, 1, 2, 3, 1, 2, 3.

3-4

Continue to use the Classes data.

(a) Delete C4-C10 from the Classes data.

(b) Sort the data in C1-C3 by Student ID and place the sorted data in C5-C7.

(c) Rank students by Grade2 and place ranks in C8. Name the new column 'Rank'.

(d) Create a new column C9 'Average' that contains the averages of Grade1 and Grade2 for each student.

(e) Code the average grades in C9 with letter grades, as follows, and store the codes in C10 'Letter Grade':
90-100 = A, 80-89 = B, 70-79 = C, 60-69 = D, 50-59 = F

3-5

Answer the following questions.

(a) When would you use the portable Worksheet format?

(b) How do you open an Excel spreadsheet file in Minitab?

(c) If you wanted to open a Minitab Worksheet in an obscure application, how would you save it?

(d) Explore the Help topics suggested in the "Importing Files in Custom Formats" section and explain what this format statement means: F3.0, 5X, F4.0.

(e) How do you open a database table created in Microsoft Access?

3-6   Use the autofill technique to enter the following patterned data into a column.

(a) twenty zeros

(b) The numbers 1 through 20

(c) The quarter symbols Q1, Q2, Q3, Q4 repeated for 12 rows, that is, for three years.

(d) The text value Male repeated in rows 1 through 10.

4

One-Variable Graphs and Summaries

4.1 Understanding Variable Types

Much can be learned from data by looking at appropriate plots and tables. Sometimes such displays are all you need to answer your questions. In other cases they will help guide you to appropriate follow-up procedures. In fact one great advantage of computers is their ability to produce a variety of data displays quickly and easily. This chapter introduces you to some of the most useful displays and some simple summary measures, such as the mean and median.

We begin with a description of basic types of variables, because the type of analysis you use often depends on the type of data you have. All numbers are not created equal. Categorical data act merely as names with, possibly, order. Quantitative data give information about numerical size.

Categorical Variables

Simple examples of categorical variables are sex, which has two values (male and female), and state in the United States, which has fifty values (Alabama, Alaska,..., Wyoming). When such data are stored in the computer, they are often entered as numbers, usually for the convenience of entry. Sex might be coded 1 = male and 2 = female, 1 = female and 2 = male, or male = 1 and female = 0. The State variable might be coded in alphabetical order, going from 1 = Alabama to 50 = Wyoming.

One problem with computers is that they will do whatever you ask—even if it's nonsense. For example a computer will gladly average categorical data that is coded as numerical, even though that average may have absolutely no meaning. Suppose, for example, you have a data set of 30 men (coded 1) and 70 women (coded 2). A computer will calculate the average sex as 1.7. There is an important exception. If two-valued categorical data are coded with zeros and ones, for example, male = 1 and female = 0, then the mean is just the proportion of ones or proportion of males in the data. This can be very convenient with variables such as sex, yes-no answers, agree- disagree answers and so on. With computers, as with other tools, it is up to you to see that they are used properly. One of the goals of this book is to help you do that.

Nominal and Ordinal Categorical Variables

Nominal variables are categorical variables where there is no natural order in the categories. Sex and state are nominal variables. Ordinal categorical variables, on the other hand, have a natural order. One example of an ordinal variable is army rank: private, corporal, sergeant, lieutenant, major, colonel, and general. We know that a general is one rank higher than a colonel and a corporal is one rank higher than a private. But is the distance from private to corporal the same as the distance from colonel to general? Does distance between army ranks really have any meaning? Probably not.

Perhaps the most common occurrence of ordinal variables is in surveys and questionnaires. For example:

"The President is doing a good job." Check one:
❏Strongly disagree ❏Disagree ❏Indifferent ❏Agree ❏Strongly agree

When entered into the computer, data on ordinal variables are often entered as numbers—for example 1 = Strongly disagree, 2 = Disagree,..., 5 = Strongly agree. Here we would know that a 4 was more favorable toward the President than a 3, but we would not have any clear idea how much more favorable. If the data are entered as text data we may have to specify the correct order to Minitab. See p. 127 for more information on this point. If our categorical data is ordinal it's usually more informative to use the order when we display the data in tables or graphs.

Quantitative Variables

These variables are usually based on measurements such as length, weight, dollars, or time. On an interval scale 4 is halfway between 3 and 5. For example, the difference between a 4-centimeter rod and a 3-centimeter rod is the same as the difference between a 5-centimeter rod and a 4-centimeter rod. Unlike categorical variables, quantitative variables occur naturally as numbers. Some investigators make an additional distinction among interval variables, ratio variables, and count variables. We don't find this distinction to be especially useful and will not pursue it here.

4.2   Exploring Graphs in Minitab

Graphs in Minitab have many options, as you'll see when you pull down the Graph menu and open a Graph dialog box. Release 14 of Minitab introduced a whole new graphics system—with advanced editing capabilities that are simple and intuitive, yet powerful. Basic graphs can be constructed with very little effort. To create more professional-looking, and perhaps, more informative graphs, however, you'll want to use some of the graph options explored in this chapter and the next.

Minitab offers graphs in four different types:

- Core graphs—high-resolution graphs that are fully customizable and editable and may be updated automatically as the data changes
- Specialty graphs—complex graphs that would require many steps to produce using core graphs; Minitab has done the work for you so that you can create complex graphs with just a few mouse clicks
- 3D graphs—plots based on three (or more) variables that are also rotatable and editable
- Character graphs—"typewriter" style graphs that appear as text in the Session Window

In Chapter 1 you learned how to create a high-resolution histogram—a core graph. In this chapter you will explore some of the many customization features available to core graphs. Most of these features appear in the Graph dialog boxes with the same layout, so once you've learned to use a customization feature for one graph type, you can apply it to any of the others. You can use online Help to learn more about these features, or you can just experiment to see what happens.

You've probably already discovered that when you reopen a dialog box in the same session of Minitab, you see that Minitab has retained the most recent settings you used. This is a nice feature when you want to do the same operation again with only some minor changes.

 TIP When you want to start an analysis 'clean,' press the F3 key after you open a dialog box. Minitab returns all the dialog box settings, even the ones currently 'hidden' under options and other buttons, to their default values.

While working through this *Handbook*, you might want to restart Minitab every time you begin a session. If Minitab is already running, use the menu sequence **File ▸ New ▸ Mintab Project** to restart quickly. This ensures that all dialog boxes are set to their default values and the Worksheet is empty. The steps given in the *Handbook* assume that you have done this.

4.3   Creating Distribution Plots

We begin with the simplest distribution display—the dotplot.

Creating a Dotplot

A **dotplot** displays a dot along a number line, one for each observation. If there are multiple occurrences of an observation, or if observations are too close together, then Minitab stacks the dots vertically. If there are too many points to fit vertically in the graph, then each dot may represent more than one point. In this case Minitab displays a message on the graph denoting the maximum number of observations that the dots represent. Sometimes a dotplot groups the data since a computer screen contains only a finite number of positions. Ideally if you had a very wide screen, wide paper, or a printer with very high resolution, a dotplot would not group the data at all.

Table 4.1 lists the 11 values of the Pulse1 variable in the Pulse1 Worksheet.

Table 4.1 Pulse rates for 11 students

Pulse1	96	62	78	82	100	68	96	78	88	62	80

To create a dotplot,

1. Open the **Pulse1** Worksheet.
2. Choose **Graph ▸ Dotplot ▸ One Y, Simple**. Click **OK**. See Figure 4.1.
3. Enter **Pulse1** in the Graph variables box.
4. Click **OK**.

Figure 4.2 shows the resulting display.

Figure 4.1
Dotplot gallery of options

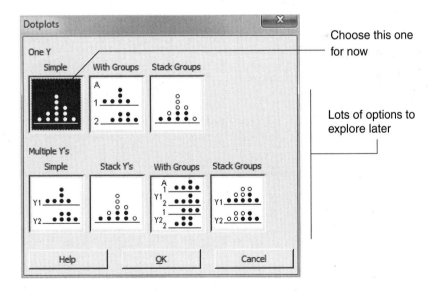

Figure 4.2
Dotplot of 11
pulse rates

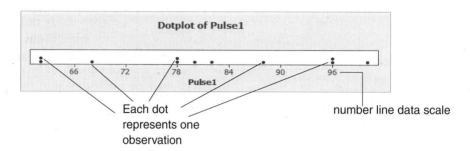

You probably noticed that you had many choices when you selected Dotplot from the Graph menu. Figure 4.1 displays the many options.

Data Tips. If you place the cursor pointer over any point on a graph, a data tips window appears on top of the graph, giving specific information about the point such as its coordinates. This works for all graph items such as individual data points and other graph items such as reference lines. Give it a try on this graph and the rest of the many graphs you will create as you work your way through this *Handbook*.

Dotplots are quite useful for comparing distributions of one or more groups. The Pulse1 Worksheet contains pulse rates for 11 women both before (Pulse1) and after (Pulse2) running in place.

To create two dotplots for comparing these distributions,

1. Choose **Graph ▸ Dotplots ▸ Multiple Y's Simple** and click **OK**. See Figure 4.1.

2. Select **Pulse1** and **Pulse2** for the Graph variables.

3. Click **OK**.

The dotplots displayed in Figure 4.3 show the substantial increase in pulse rates after the women ran in place.

Figure 4.3
Dotplots of
pulse rates
before and
after running
in place

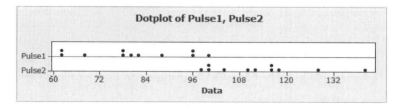

Please note that if we had selected **Graph ▸ Dotplots ▸ One Y, Simple** from the choices in Figure 4.1 and then selected **Pulse1** and **Pulse2** for the Graph variables, Minitab would have displayed *two separate dotplots*, each with its own scale. These would not be good for making comparisons.

The data in Worksheet Pulse1 is organized by putting the before and after measurements in different columns. Another way it might have been organized would be to put the pulse rates all in one column with a separate (categorical variable) column to indicate before or after for each observation. If the data had been organized that way you would use the dotplot choice **One Y, With Groups** in Figure 4.1 to produce the same two dotplots as in Figure 4.3.

To illustrate this, open the Pulse Worksheet. Here you have several variables on 92 students including their Height and Sex.

To do dotplots of Height separately by Sex,

1. Choose **Graph ▸ Dotplot ▸ One Y, With Groups** and click **OK**. See Figure 4.1.
2. Select **Height** as the Graph variable.
3. Select **Sex** as the categorical variable for grouping.
4. Click **OK**.

Figure 4.4 is the display that Minitab produces.

Figure 4.4
Dotplots of Height by Sex

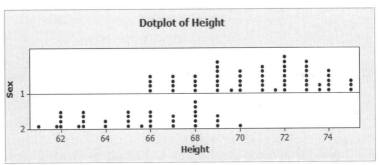

There is considerable overlap of the two distributions but the '1's are generally much taller than the '2's. Sex labelling of 1 and 2 is not at all descriptive. We could fix the graph in several ways. One would be to use the sequence **Data ▸ Code ▸ Numeric to Text** to recode the 1s and 2s in the Sex column to text data of Male and Female before creating the graph. Alternatively, we can edit the graph already produced for better labelling. That is, let's use Male and Female labels rather than 1s and 2s.

Editing a Graph

All graphs in Minitab can be edited. Make sure the graph of Figure 4.4 is active by clicking anywhere on it.

1. Double-click on either the **1** or the **2** on the Sex axis and the Edit Scale Dialog box appears.

2. Click on the **Labels** tab to obtain the display in Figure 4.5.

3. Click on the **Specified** option button.

4. Replace the **1 2** with **Male Female** (in that order) and click **OK**.

Figure 4.5   Edit Scale dialog box for vertical scale of dotplots

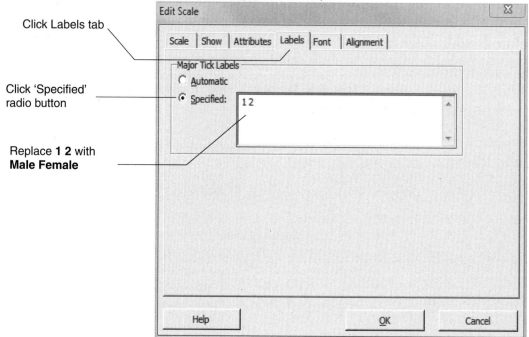

Minitab replaces the 1 and 2 on the Sex scale with Male and Female and resizes the graph a little to accommodate the new improved labels.

Much more editing is possible. Minitab put labeled tick marks every two inches of height on the graph in Figure 4.4. Perhaps you would like ticks at every inch instead. Double-click anywhere on the horizontal axis or tick marks to get the Edit Scale dialog box for the x-axis scale displayed in Figure 4.6.

To change the tick marks,

1. Double-click anywhere on the horizontal axis of the graph.

2. Make sure the **Scale** tab is 'pressed.'

3. Select the **Position of ticks** radio button.

4. Replace 62 64 66 68 70 72 75 by **61:75** and click **OK**.

Recall that 61:75 is Minitab shorthand for the list 61, 62, 63,..., 74, 75. Minitab revises the tick marks and adjusts the size of the graph as needed.

Figure 4.6   Edit Scale dialog box for horizontal scale of dotplots

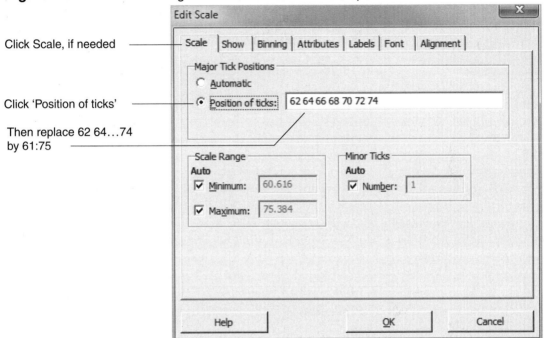

Click Scale, if needed

Click 'Position of ticks'

Then replace 62 64...74
by 61:75

Brushing a Dotplot

The purpose of a graph is to help the viewer visualize some characteristics of a set of numbers and convey quantitative information. Sometimes once we see the overall picture we are led to investigate certain individual or groups of data points.

For example, the dataset, **Forbes 2011 Top 100**, contains values of the net worth of the 100 richest Americans as listed by *FORBES* magazine in September 2011. Figure 4.7shows the dotplot of their Net Worth (in billions of dollars) of these individuals.

The distribution is quite skewed toward the large values but who are these people? Interactive **brushing** is an easy way to pinpoint individual or groups of data points.

There are several ways to initiate brushing but one simple way is to right-click the graph and select **Brush** from the choices presented. When you do so your pointer changes shape and appears as a hand with index finger pointing left ☞.

You brush a data point by either clicking or 'lassoing' the data point. When you do so that data point turns green and a small Brushing ID Window appears that contains the row number(s) for the data point(s) selected or brushed. You can also lasso groups of data points.

Figure 4.7
Dotplot of Net
Worth (in
billions of
dollars) of
richest 100
U.S.
individuals in
2011

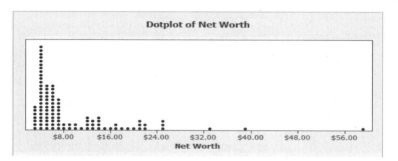

When other information is available about the same row, that is, the same individual, it is usually more informative to see that information also. For example, our Worksheet contains the names of the individuals in column C2.

To have Minitab display information from other variables of the brushed points,

1. Click the **Graph Window** to make sure it is active.
2. Click the **Editor** menu tab and select **Brush**.
3. Click the **Editor** menu tab once more and select **Set ID variables...**.
4. In the Set ID variables dialog box which appears, select **Name** for the variable. (You may leave **Include row numbers** checked or not as you wish.)
5. Click **OK** and the Brushing Window information changes to include the name of the individual(s) associated with the brushed data point(s). See Figure 4.8.

Figure 4.8 Brushing ID Window

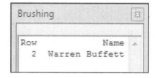

If you lasso the three data points just above 24 billion dollars in the dotplot, you get the Brushing ID Window in Figure 4.9.

Figure 4.9
Brushing ID
Window for a
group of data
points

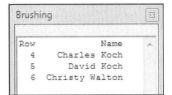

Brushing is a valuable technique for exploring data in graphs. We will return to it in investigating scatterplots in Chapter 5.

Creating a Stem-and-Leaf Plot

A stem-and-leaf display uses the digits in the data to create the distribution display. We begin with a stem-and-leaf display of the Pulse1 variable listed in Table 4.1, p. 103. To create a stem-and-leaf display,

1. Open the **Pulse1** Worksheet.
2. Choose **Graph ▸ Stem-and-Leaf**. Enter **Pulse1** in the **Variables** box.
3. Click **OK**. Minitab produces the display in the Session Window as shown in Figure 4.10.

Figure 4.10
Stem-and-
Leaf display
of Pulse1
data

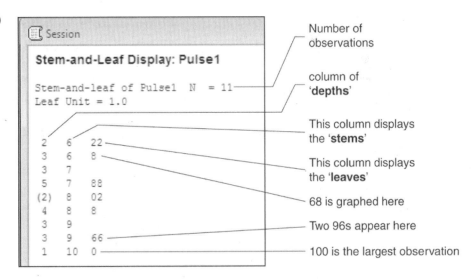

Stem-and-leaf displays are character graphs and always appear in the Session Window. The digits in the middle column of numbers are called the **stems**. The digits in the third are called the **leaves**. The display shows the digits of the actual data. The tens digits become the stems. In this case, each stem is split and displayed on two consecutive rows, with leaf digits 0, 1, 2, 3, and 4 on the first line and 5, 6, 7, 8, and 9 on the second. The ones digits are the leaves.

For example, the first observation of 96 has a stem of 9 and a leaf of 6. the second observation is a 62. Its leaf of 2 is plotted next to its stem of 6. For this data, Minitab chose to divide the stems in half so that leaves 0 through 4 are plotted on the top half and leaves 5 through 9 are plotted on the bottom half. There were no data in the 70 to 74 range and none in the 90 to 94 range.

The numbers on the far left are the **depths**, or the cumulative count of values from the top of the figure down and from the bottom of the figure up to the middle. Thus the depth of a line indicates how many leaves lie on that line or "beyond." For example the 5 on the fourth line from the top tells us there are five observations of 78 or less. The 3 on the second line from the bottom tells us there are three observations of 96 or larger.

The middle number in parentheses is the count of values in the row containing the median. If the total number of observations, N, is odd, the line with the parentheses contains the middle observation. If N is even, the line contains the middle two observations. The parentheses enclose a count of the number of leaves on this line. Note that if N is even and if the two middle observations fall on different lines, no parentheses are used in the depth column. Minitab also indicates the number of observations and the leaf unit in the display.

With a stem-and-leaf plot you can easily see the range of the data (from a low of 62 to a high of 100). There are no **outliers**. In this context outliers would be observations that are much smaller or much larger than the bulk of the data.

Looking more closely, notice that all the observations are even. Why? A reasonable conjecture, and a correct one, is that pulses were counted for 30 seconds and then doubled to get beats per minute.

In this example Minitab listed each stem on two lines. In some cases Minitab will use five lines for each stem. The number of lines per stem is always one, two, or five, and is determined by the range of the data and the number of values present.

In our example all but one of the pulse rates contained two digits, so it was easy to split each number into a stem and a leaf. When numbers contain more than two digits, Minitab drops digits that don't fit. For example the number 927 might be split as stem = 9, leaf = 2, and 7 dropped.

Decimal points are not used in a stem-and-leaf display. Therefore the numbers 260, 26, 2.6, and 2.6 would all be split into stem = 2 and leaf = 6. The heading, Leaf Unit, tells you where the decimal point belongs. For the number 260, Leaf Unit = 10; for 26, Leaf Unit = 1; for 2.6, Leaf Unit = .1; for .26, Leaf Unit = .01.

- You can control the scale of a stem-and-leaf display, or the distance from one stem to the next, by entering an increment in the Increment box of the Stem-and-Leaf dialog box. The increment must be 1, 2, or 5, with perhaps some leading or trailing zeros. Examples of allowable increments are 1, 2, 5, 10, 20, 50, 100, 200, 500, 1, 0.2, 0.5, 0.01, 0.02, 0.05. For example, setting Increment as 10 would tell Minitab to place all the numbers in the 40s on the first stem, all the numbers in the 50s in the second stem, and so on.

- You can enter a By variable to create a separate stem-and-leaf plot for groups in a column. Minitab puts all plots on the same scale.

4.4   Histograms

Histogram displays of distributions tend to be more useful with larger data sets; stem-and-leaf displays and dotplots, with smaller data sets. By appropriate grouping, histograms can be constructed to show the shape of a distribution; if there is little grouping, dotplots do not show shape well. The histogram is designed primarily for quantitative data.

Creating a Histogram

We first create a simple histogram of the Pulse1 variable from the Pulse dataset described in the Appendix, p. 529.

To create a frequency histogram,

1. Open the **Pulse** worksheet, located in the Data folder.
2. Choose **Graph ▸ Histogram ▸ Simple**. Click **OK**. See Figure 4.11.
3. Select **Pulse1** for the Graph variable. See Figure 4.12.

Figure 4.11
Histogram gallery of options

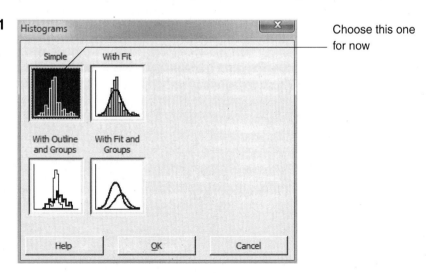

Choose this one for now

Click **OK** and Minitab produces a frequency histogram in a Graph window as shown in Figure 4.13.

Figure 4.12
Histogram
dialog box

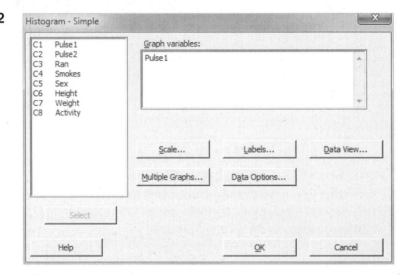

Minitab groups the pulse rates into 11 class intervals or bins, each of width 5. The first interval has a midpoint of 50, goes from 47.5 to 52.5, and contains one observation. The second interval has a midpoint of 55, goes from 52.5 to 57.5, and contains two observations. (If you let your mouse pointer hover over any one of the bars, Minitab will display a Tooltip of the count and endpoints for that bar.) The lowest pulse rate is about 50, the highest is about 100, and the most popular interval is the one at 70—one-quarter of the pulse rates are in this interval.

Figure 4.13
Frequency
histogram of
Pulse1

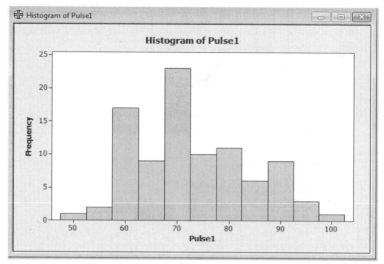

Minitab automatically chooses class intervals that are usually reasonable. If you want different intervals, you can edit the graph after you see it. For example suppose you wanted the intervals to be 40 to 50, 50 to 60, 60 to 70,..., and 100 to 110. In this example, the width, or increment, of each interval is 10 and the starting midpoint is 45. You can specify the cutpoints (the beginning or ending of an interval) or midpoints of the intervals you want to use, in order from smallest to largest. You can also use Minitab's shorthand notation that specifies the beginning and ending and the increment. Thus 10:40/5 means to use intervals from 10 to 40 in increments of 5.

Try creating a new histogram with a larger increment using 45:105/10. To specify intervals,

1. Make the Histogram of Figure 4.13 active by clicking on it anywhere.
2. Double-click anywhere on the horizontal axis. The Edit Scale dialog box appears similar to that in Figure 4.6, p. 107.
3. Click the **Binning** tab. See Figure 4.14.
4. If it is not already checked, select **Midpoint** for Interval type.
5. Select **Midpoint/Cutpoint** positions.
6. Enter the Minitab shortcut notation, **45:105/10**, into the text box.
7. Click **OK**. Minitab edits the histogram and produces Figure 4.15. Notice the increments (class intervals) are wider than those in the first histogram you created.

Since you double-clicked the graph to make it editable, there will be several small black marks along the *x*-axis indicating that you selected the horizontal axis to start the edit. To remove these marks, click Editor ▸ Select item ▸ <None> and they disappear. Alternatively, you can right-click the graph and then click on Select item ▸ <None>.

Figure 4.14
Edit Scale
dialog box for
Binning

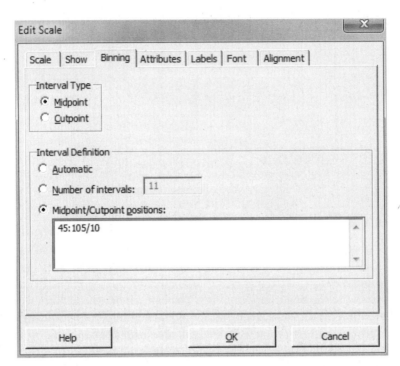

Figure 4.15
Histogram with
wider class
intervals

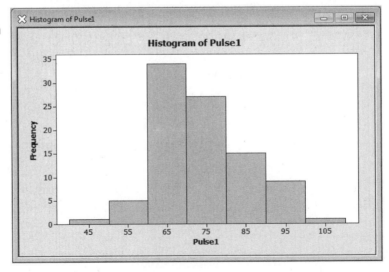

By convention, any pulse rate that falls on a boundary between two intervals is put in the higher interval. For example, a pulse of 60 would be put in the interval from 60 to 70.

Histograms from Frequency Tables

Sometimes we do not have raw data but rather data summarized in a frequency table as in Table 4.2.

Table 4.2 Chest girth in inches for 1516 soldiers of the 1885 army of the Potomac

Girth	28	29	30	31	32	33	34	35	36	37	38	39	40	41	42
Frequency	2	4	17	55	102	180	242	310	251	181	103	42	19	6	2

Minitab can produce a histogram from data in this format. To do so,

1. Enter these data into two columns named **Inches** and **Frequency**.
2. Choose **Graph ▸ Histogram ▸ Simple** and click **OK**.
3. Select **Inches** for the Graph variable.
4. Click the **Data options** button and click on the **Frequency** tab. The dialog box in Figure 4.16 appears.
5. Enter **Frequency** as the Frequency variable and click **OK** twice.

Figure 4.16
Histogram data options for frequency data

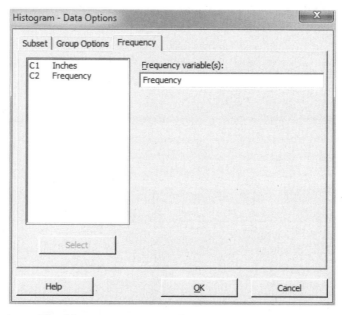

The histogram appears for the summarized data. It can be edited as any other histogram.

Relative Frequency Histograms

All the histograms we have produced so far are frequency histograms—the vertical scale is a count of the number of observations in each class interval. When we need to compare two different distributions containing different numbers of observations, we should display **relative frequency histograms**. (Density histograms also permit correct comparisons.)

You can either create the relative frequency histogram initially or edit a frequency histogram to convert it to relative frequency.

To create a relative frequency histogram,

1. Choose **Graph ▸ Histogram ▸ Simple** and click **OK**.
2. Enter the graph variable name as always. (If data are summarized, choose Data options and proceed as in Figure 4.16.)
3. Click the **Scale** button and then the **Y-Scale Type** tab. See Figure 4.17.
4. Click the **Percent** option button and click **OK** twice to obtain a relative (or percent) frequency histogram.

Figure 4.17
Histogram
scale options

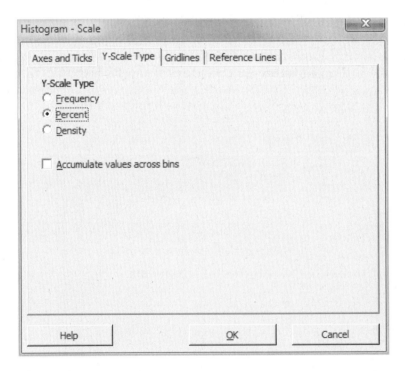

To edit a frequency histogram to change it to relative frequency:

1. Make the histogram active by clicking anywhere in the Graph Window.
2. Double-click anywhere on the vertical axis or tick marks for that axis. An Edit Scale dialog box appears.
3. Click on the **Type** tab, click the **percent** option button, and click **OK** to convert the histogram to relative frequency.

Density Histograms

Density histograms are needed when you want or need to use class intervals of different widths. A frequency or relative frequency histogram that uses class intervals of different widths is misleading and should always be avoided. But often data are available to you only in a frequency table with class intervals of different widths.

Here is an example. The data file, **UIAges**, gives the ages of all 28,311 students at the University of Iowa for the Fall of 2000. The ages are categorized by sex and also broken down as undergraduates, graduates, and professional students. The ages are given as usual *except* 17 stands for age 17 or less, 36 means ages 36 to 40, 41 means 41-45, 46 means 46-50, and 51 means age 51 or greater. Suppose we ask Minitab to display a relative frequency Histogram of these data for all 19,284 undergraduates. Figure 4.18 shows the results.

Figure 4.18
Misleading relative frequency histogram of ages

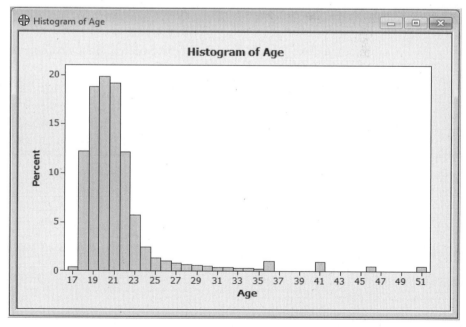

The bars at 17, 36, 41, 46, and 51 are quite misleading. These frequencies should be spread over much wider class intervals. Only a density histogram will do this correctly. In a density histogram, the height of a bar for a given class interval is given by

$$\text{Density} = \frac{\text{Relative Frequency of Class Interval}}{\text{Width of Class Interval}}$$

In our example, we do not know the ages of the youngest nor of the oldest students so we make reasonable assumptions and take 15 as the youngest and 65 as the oldest. You may experiment with other choices.

You can create a density histogram either initially using the Y-scale option button (Figure 4.17, p. 116) or by editing an existing frequency or relative frequency histogram. To edit the relative frequency histogram in Figure 4.18,

1. Make the Graph Window active by clicking on it.
2. Double-click anywhere on the vertical axis and click the **Y-Scale Type** tab of the Edit Scale dialog box. See Figure 4.17.
3. Select the **Density** option button and click **OK**. (The density histogram appears but it is still misleading since it uses the wrong class intervals.)
4. Now click anywhere on the horizontal axis and click the **Binning** tab of the Edit Scale dialog box.
5. Select the **Cutpoint** option button for Interval Type.
6. Select the **Midpoint/Cutpoint positions** radio button for Interval Definitions.
7. Enter **14.5 17.5:35.5 40.5 45.5 50.5 65.5** into the text box and click **OK**. See Figure 4.19.

Figure 4.20 displays the density histogram for the Age data. We see that the 17 year and younger ages have been spread across the interval from 14.5 to 17.5. and similarly for the ages from 36 on up. The density above 49.4 years of age is so small that you probably can't see it at all unless you maximize the graph to full screen.

Figure 4.19
Edit Scale,
Binning
dialog box for
better density
histogram

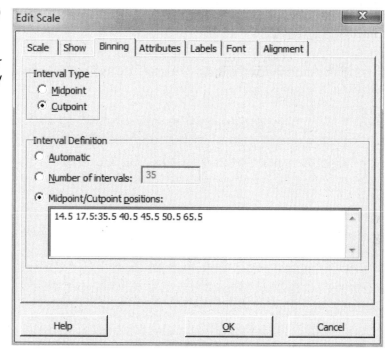

This histogram shows the age distribution correctly and without distortion.

Figure 4.20
Better density
histogram of
Age data

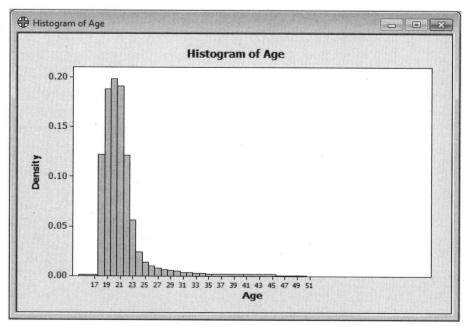

Histograms for Positive Data

Oftentimes computer generated histograms can be misleading if the data by definition must be positive but contains values close to zero.

Here is an example. Figure 4.21 gives a default histogram of the population sizes (in thousands) of the 100 counties of Iowa.

Figure 4.21
Misleading frequency histogram of positive data

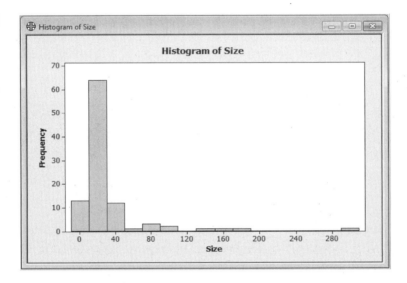

The frequency of the class interval centered at zero is misleading since it is impossible to have data below zero. An easy way to fix this is to edit the graph as follows,

1. Click anywhere on the Graph Window to make it active.
2. Double-click anywhere on the horizontal axis and the Edit Scale dialog box appears.
3. Click on the **Binning** tab and select the **Cutpoint** radio button.
4. Click **OK** and the histogram changes to the one shown in Figure 4.22.

We now have a accurate portrayal of the distribution which is quite different from the one shown in Figure 4.21.

Figure 4.22
Better
frequency
histogram of
positive data

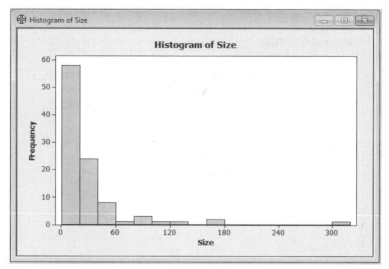

4.5 Descriptive Statistics

We often want to summarize an important feature of a set of data using a single number or a few numbers. For example you might use the mean to indicate the central tendency of the data. You could use the range, the largest value minus the smallest value, to indicate how spread out the data are.

In this section you'll learn about the Descriptive Statistics command, which prints a table of summary numbers. Then you'll see how to compute these summaries individually, first for columns of data, and then across rows.

Displaying Descriptive Statistics

The Pulse dataset contains the weights (in pounds) of 92 people. First, try creating a stem-and-leaf plot of these data, and then use the Descriptive Statistics command to describe the data. To display a stem-and-leaf plot and descriptive statistics for the Weight variable,

1. Choose **Graph ▸ Stem-and-Leaf**.
2. Enter **Weight** in the Variables box and click **OK**.
3. Now choose **Stat ▸ Basic Statistics ▸ Display Descriptive Statistics**.
4. Enter **Weight** in the Variables box and click **OK**. See Figure 4.23.
5. If necessary scroll the Session Window so you can see both the stem-and-leaf plot and the descriptive statistics. See Figure 4.24.

Figure 4.23
Descriptive
Statistics
dialog box

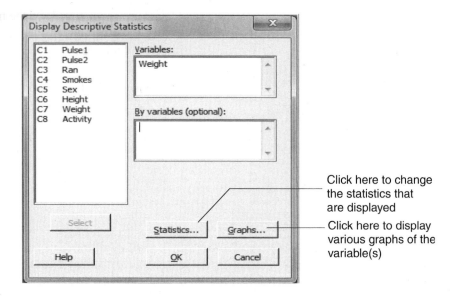

Click here to change
the statistics that
are displayed

Click here to display
various graphs of the
variable(s)

The stem-and-leaf display in Figure 4.24 shows several interesting things. For example there are numerous leaves of 0 and 5. Most people reported their weight to the nearest 5 pounds, especially the heavier people, and the heaviest person weighed 215 pounds.

The Descriptive Statistics shown in the Session Window display the following statistics:

- N = 92. This tells us there were 92 people in the study who reported their weights.

- Mean = 145.15. This is the average of all 92 weights. The mean, often written as $\bar{x}$, is the most commonly used measure of the center of a batch of numbers.

- Median = 145.00. To find the median, first order the numbers. If N, the number of values is odd, the median is the middle value. If N is even, the median is the average of the two middle values. Here $N = 92$, so the median is the average of the 46th and 47th values. These are both 145, so their average is 145. The median is another value used to indicate where the center of the data is.

- StDev = 23.74. This is the standard deviation. It is the most commonly used measure of how spread out the data are. The general formula is

$$StDev = \sqrt{\frac{\sum_{i=1}^{N} (x_i - \bar{x})^2}{N-1}}$$

For example if the data values were (1, 3, 6, 4, 6), the mean would be 4 and the standard deviation would be

$$StDev = \sqrt{\frac{(1-4)^2 + (3-4)^2 + (6-4)^2 + (4-4)^2 + (6-4)^2}{5-1}} = \sqrt{4.5} \approx 2.12$$

- SE Mean = 2.47. This is the standard error of the mean. The formula is

  $SE\ Mean = StDev/\sqrt{N}$

For the Weight variable in the Pulse dataset:

- Minimum = 95.00. The smallest value.
- Maximum = 215.00. The largest value.
- Q1 = 125.00. The first, or lower, quartile.
- Q2 = 156.50. The third, or upper, quartile.
- N*: This is the number of values recorded as "missing." Here no weights were missing, so the N* line shows a zero.

The median can be considered the second quartile, $Q2$. The three numbers $Q1$, $Q2$, and $Q3$ split the ordered data into four essentially equal parts. The concept of a quartile is very simple. However, when we try to give a formal definition, we must handle details such as how to divide ten observations into four equal parts.

There are several ways to do this, and you will find that different books define quartiles slightly differently. All, however, give answers that are very close—especially for larger datasets To examine the exact definition Minitab uses, search the Minitab Help system for quartiles and select the Methods and Formulas item.

You can select which basic statistics to have calculated when you use Stat ▶ Basic Statistics ▶ Display Descriptive Statistics. Click the **Statistics** button shown in Figure 4.23, p. 122, to see the many choices.

Figure 4.24
Descriptive
statistics and
Stem and Leaf
for Weight

```
Session

Stem-and-Leaf Display: Weight

Stem-and-leaf of Weight  N  = 92
Leaf Unit = 1.0


  1     9   5
  4    10   288
 13    11   002556688
 24    12   00012355555
 37    13   0000013555688
(11)   14   00002555558
 44    15   0000000000355555555557
 22    16   000045
 16    17   000055
 10    18   0005
  6    19   00005
  1    20
  1    21   5

MTB > Describe 'Weight'.

Descriptive Statistics: Weight

Variable   N  N*    Mean  SE Mean  StDev  Minimum      Q1  Median      Q3
Weight    92   0  145.15     2.48  23.74    95.00  125.00  145.00  156.50

Variable  Maximum
Weight     215.00
```

Calculating Column and Row Statistics

The Display Descriptive Statistics command prints a collection of summary statistics in the Session Window. You can also calculate and store each statistic separately, using the Calc ► Column Statistics command.

You can store a calculated value in a constant by designating a storage location. If you enter a name, Minitab automatically names the next available constant accordingly.

You can also calculate these statistics by using Calc ► Calculator. The main difference is that Calculator stores the result in an empty column; Column Statistics displays the results in the Session Window if no storage location is specified. Similar capabilities exist when using Calc ► Row Statistics but results are always stored in worksheet columns.

4.6   Creating a Boxplot

Boxplots, also called box-and-whisker plots, display the main features of a batch of data. They are especially useful for comparing two or more groups of data, as we will do in later chapters. To display a boxplot of the weights of the 92 people in the Pulse data set,

1. Choose **Graph ▸ Boxplot ▸ One Y, Simple** and click **OK**. See Figure 4.25.
2. Enter **Weight** as the Graph variable.
3. Click **OK**. See Figure 4.26.

Figure 4.25 Boxplot gallery

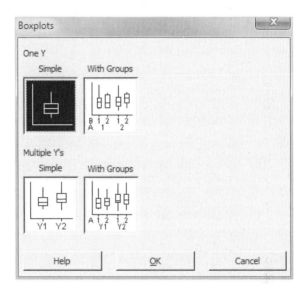

(The arrows and dashed lines shown here are **not** part of the boxplot.)

Figure 4.26
Boxplot of
Weight

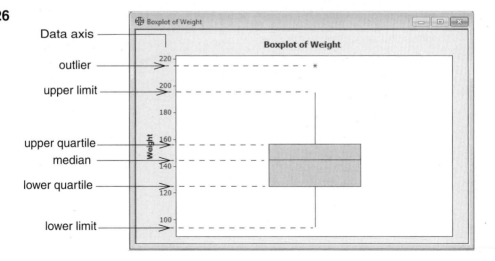

The box represents the middle half of the data. The line in the middle of the box is the median of the data. The upper box edge is the upper hinge, HU, essentially the upper quartile. The lower box edge is the lower hinge, HL, essentially the lower quartile. The whiskers are the lines that extend from the ends of the box to the adjacent values. The adjacent values are the lowest and highest observations that are still within the following limits:

Lower limit $= Q_1 - 1.5\,(Q_3 - Q_1)$

Upper limit $= Q_3 + 1.5\,(Q_3 - Q_1)$

Observations outside these limits are plotted with asterisks. These points are far from the mass of the data, and are often called **outliers**. There is one outlier in Figure 4.25, a weight of over 200 lbs.

> TIP To read precise information for Q_1, median, Q_3, interquartile range, N, or an outlier from the graph, hover your cursor over any part of the boxplot or outlier asterisk symbol and Minitab will display precise information in a Tooltip. These Tooltips work on all Minitab graphs.

Now you'll display boxplots for weight, separately for men and women. This time you'll use some of the graphic options to customize your graph. To create separate side-by-side boxplots,

1. Choose **Graph ▸ Boxplot ▸ One Y with Groups** and click **OK**. See Figure 4.25.
2. Select **Weight** as the Graph variable and **Sex** as the Categorical variable for grouping.
3. Click **OK** to obtain the boxplots shown in Figure 4.27.

Figure 4.27
Boxplots for
Weight by Sex

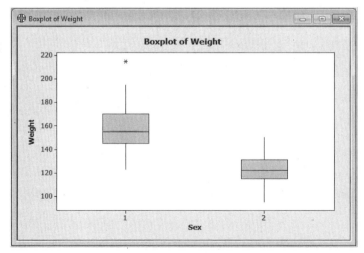

Of course, we should now edit the labels on the Sex axis and replace 1 by Male and 2 by Female to obtain our final display. To do this double-click anywhere on the Sex axis, click the Labels tab, select the Specified radio button, replace 1 2 in the textbox by Male Female, and click OK.

4.7 Summarizing and Displaying Categorical Data

The preceding sections described techniques that are designed primarily for quantitative data. This section introduces the Tally command, which is designed especially for categorical data. It gives four statistics: count, cumulative count, percent, and cumulative percent.

The Pulse data set contains several categorical variables. When working with categorical data coded as numbers, it is much more convenient to recode the data into meaningful text. Recall from "Coding Data", p. 91, how to do this. Briefly, use Data ▶ Code ▶ Numeric to Text and code 1 to low, 2 to moderate, and 3 to high. Store back in the Activity column. With ordinal text data, such as Activity, we need to inform Minitab of the proper order for the text values.

Value Order for Text Data

To set the value order for a text column,

1. **Right-click** the text column bringing up a variety of choices.
2. Select **Column ▶ Value Order.**
3. Select the option button for **User-specified Order:** (**NEW ORDER** should be selected under Choose an order:)
4. The text values moderate, high, low appear in a text box. This is the order that they happened to be in the Activity column. By deleting and typing or by cutting and pasting enter the values into the correct order, one per line. Namely, **low, moderate, high,** press **Add Order** and press **OK.**

Now Minitab will use this order in graphs and tables involving this variable. This information will also be saved within a Minitab Project or Worksheet.

Tallies

Tallies—lists of counts and percents—are the simplest way to describe the data in one categorical variable.

Try tallying the Sex and Activity variables. To tally variables,

1. Choose **Stat ▶ Tables ▶ Tally Individual Variables**.

2. Enter **Sex** and **Activity** in the Variables box.

3. Make sure just the **Counts** check box is selected. See Figure 4.28.

4. Click **OK**. The tally appears in the Session Window as shown in Figure 4.29.

Figure 4.28
Tally dialog
box

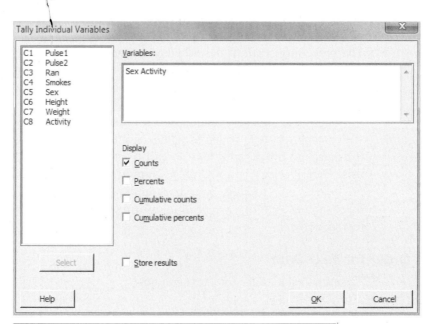

Figure 4.29
Session
Window tally
for Sex and
Activity
variables

Tally for Discrete Variables: Sex, Activity

Sex	Count	Activity	Count
Female	35	moderate	61
Male	57	high	21
N=	92	low	9
		N=	91
		*=	1

Sex was coded as 1 = Male and 2 = Female. Activity was recoded as low, moderate, and high. The first Count column, for Sex, tells us there are 57 men and 35 women, for a total of $N = 92$. The table is much more readable since we recoded the Sex variable to text values before we made the tally.

The second Count column, for Activity, has a curious feature: one person has a blank activity level and so we get the *= 1 indication. This is the person, in row 54, who originally had a zero recorded for activity level. When we recoded these numbers to text, the zero was replaced with a blank for missing text. Does this mean no activity? If you cannot find the correct value, you should change the 0 to * (Minitab's missing-value code) before doing further analysis.

Bar Charts

A bar chart is sometimes useful to display the "distribution" of a categorical variable—especially ordinal variables.

To construct a bar chart of raw categorical data,

1. Choose **Graph ▸ Bar Chart ▸ Simple** and click **OK**. See Figure 4.30.
2. Enter **Activity** as the Categorical variable and click **OK** to obtain the bar chart shown in Figure 4.31.

Figure 4.30 Bar charts gallery when bars represent unique values

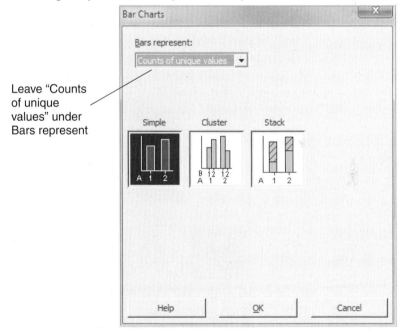

If you prefer to have percents on the vertical axis you can either redo the chart and select **Show Y as Percent** on the **Bar Chart options** page or edit the present chart by right-clicking the vertical axis, selecting **Graph Options**, choosing **Show Y as Percent**, and clicking **OK**.

The bar for the one missing value appears alerting us to the problem with the data in row 54 who reported a 0 for Activity level.

> *TIP If you let the mouse pointer hover over any one of the bars, Minitab will display the exact height of the corresponding bar.*

Figure 4.31
Bar chart of
Activity
variable

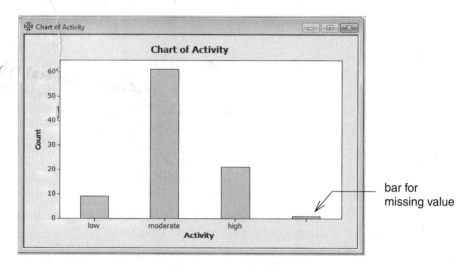

bar for
missing value

If you have categorical data that are already summarized in a table, you need to select the pull-down list to set the "Bars represent" to "Values from a table." See Figure 4.32. Then proceed as before.

Figure 4.32   Bar charts gallery when bars represent values from a table

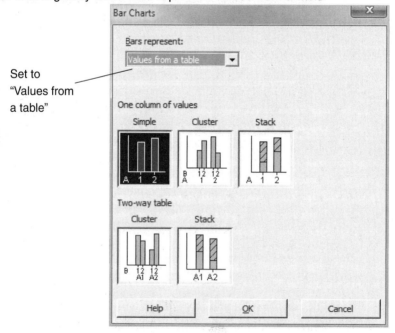

Set to
"Values from
a table"

Two-Way Tables

Minitab's Cross Tabulation command provides two general types of capabilities. The first makes tables of counts and percents, which is a generalization of the Tally command you learned earlier. The second provides summary statistics, such as means and standard deviations, for related variables. Statistical tests of association also can be done with the Cross Tabulation command but they will be deferred until Chapter 12. The input to the cross tabulation command can be either raw data, frequency data, contingency table data, or indicator variables data. We will illustrate the use of raw data here. Consult the Minitab Help System for information on the other possibilities.[†]

Because tables arise naturally in sample survey work, we illustrate the Cross Tabulation command using data from an actual survey, the Wisconsin Restaurant Survey, conducted by the University of Wisconsin Small Business Development Center. This survey was done primarily to "allow educators, researchers, and public policy makers to evaluate the status of Wisconsin's restaurant sector and to identify particular problems that it is encountering." A second purpose was to develop data that would "be useful to small business counselors in advising managers as to how to effectively plan and operate their small restaurants."

Nineteen of Wisconsin's counties were selected for study. Lists of restaurants were drawn up from telephone directories and these were sampled in proportion to the population of each county. A sample of 1000 restaurants yielded 279 usable responses.

In a full analysis of these data we would be concerned about possible biases resulting from the sampling method and from the 28% response rate. Here, however, we are interested only in providing summaries of the returned questionnaires—not in making inferences for all restaurants.

The database thus consists of 279 cases, one for each restaurant with usable data. In this book we will use only a few of the many variables in the survey. The data set is described in the Appendix, p. 530, and is saved in the worksheet called Restrnt. Each of the 279 respondents failed to answer at least one question. Thus there are a number of missing values, each of which is coded as an asterisk (*).

Tables of Counts and Percents

As a first example we will classify the restaurants by type of ownership and size. The variable Owner takes on three values: 1 = sole proprietorship; 2 = partnership; and 3 = corporation. Size also takes on three values: 1 = under 10

[†] Strictly speaking cross-tabulation involves more than one variable. However, we feel this is the best chapter in which to place this discussion.

employees; 2 = from 10 to 20 employees; and 3 = over 20 employees. The tables will be much more readable if we take these numerical values and recode them into meaningful text. Use the Data ▸ Code ▸ Numerical to text menu sequence to code the values according to Table 4.3.

Table 4.3 Text values for four categorical variables

Variable	Text values					
Owner	sole proprietorship	partnership	corporation			
Size	under 10	10 to 20	over 20			
TypeFood	fast food	supper club	other			
Outlook	very unfavorable	unfavorable	mildly unfavorable	mildly favorable	favorable	very favorable

Here is the Code Numeric to Text dialog box filled in for one of these conversions:

Figure 4.33
Code Numeric
to Text

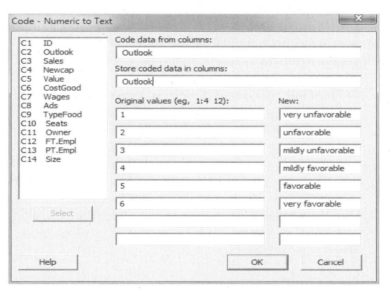

The value order for each of these new text variables needs to be set as we did on p. 127. Once you have made these conversions and set the value order for each of these variables, you should save the Worksheet under the new name Restaurant.mtw, so that you can use it in the Exercises and in later chapters.

You can use the Stat ▸ Tables ▸ Cross Tabulation and Chi-Square command to display a table of counts for these variables. Cross Tabulation displays one-way, two-way, and multiway tables of counts. Counts are printed by default, but you can also display various percentages.

You can control what is displayed in each cell by checking the appropriate boxes in the Cross Tabulation and Chi-Square dialog box. The Count check box displays a count of the number of observations in each cell. The other check boxes produce percents defined as follows:

$$\text{Row percents} = \frac{\text{Number of observations in the cell}}{\text{Number of observations in the row}} \times 100$$

$$\text{Column percents} = \frac{\text{Number of observations in the cell}}{\text{Number of observations in the column}} \times 100$$

$$\text{Total percents} = \frac{\text{Number of observations in the cell}}{\text{Number of observations in the table}} \times 100$$

To display a table of the counts for Owner and Size,

1. Open the **Restaurant** worksheet. (recoded and value ordered)
2. Choose **Stat ▸ Tables ▸ Cross Tabulation and Chi-Square**.
3. Enter **Owner** for Rows and **Size** for Columns as the classification variables as shown in Figure 4.34.
4. Make sure the Display **Counts** box is checked and then click **OK**. Minitab produces the table in Figure 4.34 and displays it in the Session Window.

Figure 4.34
Cross
Tabulation
and
chi-square
dialog box

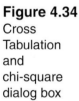

Figure 4.35
Number of restaurants classified by ownership and size

```
Tabulated statistics: Owner, Size

Rows: Owner    Columns: Size

                    under  10 to  over
                      10     20    20   Missing   All

sole proprietorship   83     18     2        3    103
partnership           16      6     4        1     26
corporation           40     42    50        4    132
Missing                1      0     1        8      *
All                  139     66    56        *    261

Cell Contents:        Count
```

The first column of this table gives counts for the restaurants with under 10 employees. There were 83 sole proprietorship restaurants with under 10 employees. Similarly there were 16 partnerships and 40 corporations with under 10 employees. Altogether there were 139 restaurants with under 10 employees. The grand total, 261, is the total number of restaurants represented in this table. This is less than the 279 restaurants that are in the database because only 261 restaurants had usable data for *both* Owner and Size. The other 18 had missing data for either Owner or Size or both.

The last row and last column are called the **margins** of the table, and the counts there are called **marginal statistics**. The other counts form the main body of the table.

It is often easier to interpret a table if we convert counts to various kinds of percents. The Cross Tabulation dialog box includes three check boxes that allow you to do this. To calculate these percentages,

1. Click the **Edit Last Dialog** button (shown at the right) ▣.
2. Deselect the **Counts** check box and select the **Row percents**, **Column percents**, and **Total percents** check boxes.
3. Click **OK**. Figure 4.36 shows all the percentages. (We used the **options** page to stop the display of missing value information.)

The first number in each cell is the row percent. To calculate these, Minitab divided the count in each cell by the row total, then multiplied by 100. (Refer to Figure 4.35.) There are 103 restaurants in row one, sole proprietorships. Of these, 83 have under 10 employees. Therefore (83/103)100 = 80.58% of the sole proprietorships had under 10 employees. This is the row percent shown in the first cell of Figure 4.36. Let's look at the rest of this row. Only 1.94% of the sole proprietorships had more than 20 employees. The remaining 17.48% had between 10 and 20 employees. What about the third row, corporations? They are split fairly evenly among the three sizes: 30.30% have under 10 employees, 31.82% have from 10 to 20 employees, and 37.88% have over 20 employees.

The second number in each cell is the column percent. To calculate these, Minitab divided the count of each cell by the column total, then multiplied by 100. Notice that among restaurants with under 10 employees, the majority, 59.71%, are sole proprietorships, whereas among restaurants with over 20 employees, the overwhelming majority, 89.29%, are corporations.

To calculate the total percent in a cell, Minitab divided the count in that cell by the grand total, 261, then multiplied by 100. For example, (83/261)100 = 31.80% of the restaurants were sole proprietorships with fewer than 10 employees. Thus almost one third of the restaurants in the study were small: one owner and under 10 employees. The last row, the row margin, tells us that over half the restaurants, 53.26%, have under 10 employees. And the last column, the column margin, tells us that just over half the restaurants, 50.57%, are owned by corporations.

Figure 4.36
Various percentages for restaurants

```
Tabulated statistics: Owner, Size

Rows: Owner    Columns: Size

                      under 10  10 to 20  over 20     All

sole proprietorship      80.58     17.48     1.94  100.00
                         59.71     27.27     3.57   39.46
                         31.80      6.90     0.77   39.46

partnership              61.54     23.08    15.38  100.00
                         11.51      9.09     7.14    9.96
                          6.13      2.30     1.53    9.96

corporation              30.30     31.82    37.88  100.00
                         28.78     63.64    89.29   50.57
                         15.33     16.09    19.16   50.57

All                      53.26     25.29    21.46  100.00
                        100.00    100.00   100.00  100.00
                         53.26     25.29    21.46  100.00

Cell Contents:      % of Row
                    % of Column
                    % of Total
```

Tables of Summary Statistics

So far we have discussed tables whose cells contain counts or percentages. We can also create tables whose cells contain summary statistics on related variables.

Any number of summary statistics, on any number of variables, can be put in a table. These options are available from the Stat ▸ Tables ▸ Descriptive Statistics menu sequence. Try creating a table where the cells contain three summary statistics: median sales, minimum sales, and maximum sales. To create a table of related variables,

1. Select **Stat ▸ Tables ▸ Descriptive Statistics** button again to open the Table of Descriptive Statistics dialog box.
2. Enter **Owner** and **Size** as the categorical variables for rows and columns.
3. Click the **Associated Variables** button to open the Descriptive Statistics— Associated Variables dialog box.
4. Enter **Sales** as the associated variable.
5. Select the check boxes for **Medians**, **Minimums**, and **Maximums**.
6. Click **OK** twice. See Figure 4.37.

Figure 4.37 shows three statistics in each cell: the median sales, the minimum sales, and the maximum sales.

This table merits careful study. First, look at the cell with sole proprietorship and under 10 employees. The minimum sales figure is 0. Thus one restaurant reported no sales for 1979. Suppose we look at the data set. This sales figure is for the restaurant in row 203. This restaurant also has its market value listed as 0. All other responses, however, seem reasonable. Perhaps the restaurant owner forgot to answer these two questions, or gave unreadable answers, and the person who entered the data into a computer file mistakenly typed a zero instead of an asterisk. Or perhaps the owner answered zero when he meant to answer, "I don't know." Unfortunately the original questionnaire had been destroyed by the time these problems were discovered, so the best we can do now is replace these two zeros by *, the missing-data code.

The problem of a zero value for Sales was very easy to find. If we carefully look at the patterns in the table, we can find another unusual value. Within each Owner category, minimum sales increase as the number of employees increases. The medians follow the same pattern. Now look at the maximums. In the first two Owner categories, maximum sales increase as we go across the row. In the last Owner category, however, this pattern does not hold: Corporations with under 10 employees have a much larger maximum than corporations with 10 to

20 employees. In fact, the sales figure of 3450 (that's $3,450,000) is suspiciously large!

Again if we go back to the data not shown here, this sales figure is in row 111. Look at the other information for this restaurant. The owner said he was very pessimistic about the future, had made only $10,000 in capital improvements, and valued his business at just $100,000. These responses do not make much sense for a restaurant that had gross sales of $3,450,000. This sales figure is almost certainly in error. We would replace the 3450 value in Row 111 of Column 3 by *.

Figure 4.37
Sales for restaurants (in thousands of dollars)

```
Tabulated statistics: Owner, Size

Rows: Owner    Columns: Size

                      under 10   10 to 20   over 20    All

sole proprietorship     98.0       220.0     579.0    110.0
                           0          39       425        0
                         435         507       733      733

partnership             76.0       267.5     750.0    125.5
                           2         200       480        2
                         250         550       800      800

corporation            133.5       274.0     565.0    290.0
                           2         100       225        2
                        3450         720      8064     8064

All                    100.0       250.0     600.0    200.0
                           0          39       225        0
                        3450         720      8064     8064

Cell Contents:    Sales  :  Median
                  Sales  :  Minimum
                  Sales  :  Maximum
```

The simple table in Figure 4.37 brought two errors to our attention. There are probably many more still in the data set. Most data sets initially contain many errors, and one of the first steps in any analysis is to try to find them. Simple displays and summaries and the Cross Tabulation command can help. Use the Tally command on each categorical variable to discover values out of range. Use Display Descriptive Statistics and perhaps Stem-and-Leaf on all quantitative variables to find values out of range and unusual patterns. Both Descriptive Statistics and Stem-and-Leaf would have uncovered the zero on Sales. Neither, however, would have revealed the restaurant with $3,450,000 in Sales. In itself $3,450,000 is not unusual, but in the context of Owner and Size it is. This value is called a **multivariate outlier**; it is unusual in the context of several variables. The plotting commands of Chapter 5 and the Cross Tabulation command display several variables at once and can help find multivariate outliers.

Exercises

4-1   For each of the following, indicate whether the variable are best considered as nominal, ordinal, or quantitative and justify your choice.
  (a) The response of a patient to treatment: none, some improvement, complete recovery
  (b) The style of a house: split-level, one-story, two-story, other
  (c) Income in dollars
  (d) Temperature of a liquid
  (e) Area of a parcel of land
  (f) Highest political office held by a candidate
  (g) Grade of meat: prime, choice, good, or utility
  (h) Political party

4-2   The Appendix, p. 521, contains a collection of data sets. Look through these data sets and select at least two examples of the three basic data types: nominal, ordinal, or quantitative.

4-3   Use the Pulse dataset. Display dotplots of both first and second pulse rates for all the students in the same graph. (Similar to Figure 4.3, p. 104.) Interpret the output.

4-4   **(a)** The Lake dataset (described in the Appendix, p. 527) provides measurements for 71 lakes in northern Wisconsin. Display a dotplot of the areas of these lakes. What are the striking features of this plot?
  (b) How many lakes are under 2,000 acres? How many are under 1,000 acres?
  (c) Use brushing to discover the row number for the largest lake. Remove that row from the dataset. Display a dotplot of the areas of the remaining lakes. How does this plot compare to the one in part (a)?

4-5   Consider the stem-and-leaf plot shown in Section 4.10, page 109.
  (a) What were the lowest and highest pulse rates?
  (b) How many people had a pulse rate of 58?
  (c) How many had a pulse rate in the eighties?
  (d) How many had a pulse rate of 64 or lower?

4-6   (a) Do a stem-and-leaf display of the following numbers by hand: 36, 43, 82, 84, 81, 84, 45, 60, 64, 71, 81, 78, 79, 43, 79.

(b) Now use Minitab to do a stem-and-leaf display of the numbers in part (a). Compare it to your hand-drawn display.

(c) Multiply the numbers in part (a) by 10. Then use Minitab to get a stem-and-leaf display. Explain the differences between the displays in (b) and (c).

(d) Multiply the numbers in part (a) by 2. Use Minitab to get a display of these numbers. How does this display compare to the one in part (b)?

(e) Multiply the numbers in part (a) by 5 and get a stem-and-leaf display. Compare this display to the one in part (a).

4-7   (a) Use Minitab to get a stem-and-leaf display of the variable Height from the Pulse data set.

(b) Convert Height from inches to centimeters (to do this, multiply by 2.54) and then get a stem-and-leaf display. Comment on the "unusual" appearance of this display. Compare it to the display in part (a), where height was measured in inches.

4-8   (a) Make a stem-and-leaf display of the variable Weight from the Pulse data set. Do you see any special pattern in the leaf digits of the display?

(b) What increment did Minitab choose for this display? What is the next smaller increment? Use this value in the Increment box to make a new display. Compare this display to the one in part (a). Now use the next larger increment to make a display. Compare this display to the one in part (a).

4-9   Consider the histogram in Figure 4.13, p. 113.

(a) How many pulse rates fell into the interval with midpoint 65?

(b) Are there more pulse rates above 75 or below it? Can you tell from this histogram?

(c) Were there any outliers (that is, any pulse rates that were much smaller or much larger than the others) in this data set?

4-10   (a) Make a histogram of the following numbers: 36, 43, 82, 84, 81, 84, 45, 60, 64, 71, 81, 78, 79, 43, 79.

(b) Make a histogram for numbers that are 10 times the numbers in (a). Compare this display to the one in part (a). Is the overall impression the same?

(c) Repeat (b) with numbers 5 times those in part (a). Compare this display to those in parts (a) and (b).

(d) Make a histogram of the numbers in part (a), using the class intervals 37.5:87.5/5. The intervals will be five units wide. The first interval will contain values from 35 through 39, the second from 40 through 44, and so on. Compare the overall shape of this display to the one in part (a).

4-11   The dataset, **USRichest2002**, contains values of the wealth of the 400 richest Americans as listed by *FORBES* magazine in 2002. Additional information is also given in the file. (As far as we know there are only two statisticians on this list.)

(a) Display a histogram of the wealth data. What are the striking features of this distribution?

(b) Do you see any difficulties with the histogram display such as those considered in "Histograms for Positive Data", p. 120?

(c) Edit the histogram of part (a) to obtain a better display for the distribution of these data.

4-12   Use the Pulse data set.

(a) Display the Stem-and Leaf for the Weight variable. Use this display to answer the following questions:

(b) How many people are exactly 150 pounds? Over 150 pounds?

(c) The mean weight was 145.15. How many people weigh less than the mean?

(d) The range of a set of numbers is defined as (maximum value) − (minimum value). Find the range for the weight data.

(e) There is a connection between the range and the standard deviation of a data set. In many data sets the range is approximately four times the standard deviation. Is this true for the weight data?

(f) Calculate the two values (Mean minus StDev) and (Mean plus StDev). In many data sets, approximately two-thirds of the observations fall between these two values. Is this true for the weight data?

4-13   Consider the following 11 numbers: 5, 3, 3, 8, 9, 6, 9, 9, 10, 5, 10.

(a) Use hand calculations to obtain each of the 8 statistics printed by Minitab's Descriptive Statistics.

(b) Now use Minitab to check your answers.

4-14   Suppose Minitab did not have the command StDev. Show how to use the Calculator and the formula for StDev to calculate the standard deviation of the 11 observations in Exercise 4-13.

4-15   The median is said to be "resistant" to the effects of a few outlying points in a data set. That is, the median will not be very different even if there are a few unusually large values or abnormally small values in the data set. However, the mean is not resistant to outliers.

(a) Enter the observations in Exercise 4-13 into C1. Use Stat ▸ Display Descriptive Statistics to find the mean and median.

(b) Change the 5 in row 1 to a 25. Now find the mean and median. How have they changed?

(c) Now change the first observation to 100. Find the mean and median. How have they changed?

4-16 (a) Use Minitab to compute the standard deviation of the numbers 6, 8, 4, 10, 12, 3, 4, 10.

(b) Add 29 to each number. Now compute the standard deviation. How does your answer compare with that in (a)?

(c) Multiply the data in (a) by 16 and compute the standard deviation. How does your answer compare with that in (a)? If you are not sure, divide the standard deviation in (c) by the one in (a).

4-17 Use Boxplot to summarize each variable in the Pulse data set. When is Boxplot most useful? What does the display show you in each case?

4-18 Use Boxplot to study two variables, Area and Depth, in the Lake data. What do the plots tell you about the data?

4-19 Use the Pulse data set. Display two Boxplots in the same graph to compare Pulse1 for those who smoke and those who do not.

4-20 Table 4.4 gives the number of "hits" experienced by the ten most popular Web search engines as reported by the *Financial Times* in 2001. Hit are in millions per day in year 2000.

Table 4.4 Search engine data

Search Engine	Hits
AltaVista	9.01
AskJeeves	11.33
Atomz	3.15
Best20sites	2.50
Clickheretofind	2.78
Directhit	3.63
Dogpile	3.86
Google	11.56
GoTo	11.20
Looksmart	7.81

(a) Use Minitab to create a chart of Hits versus Search Engine with Search Engines listed alphabetically.

(b) Use Minitab to create a chart of Hits versus Search Engine with Search Engines listed from most hits to least hits.

4-21   The Furnace data set (described in the Appendix, p. 526) contains several categorical variables. These include the type of furnace in a house and the shape of the chimney. Use Tally to help you answer the following questions.

(a) What was the most popular chimney type? The least popular?

(b) What percent of the houses had a forced-water furnace?

(c) Why does chimney shape have $N = 89$ and chimney type have $N = 90$?

(d) What percent of the houses had a round chimney?

4-22   (a) Two more variables in the Pulse data set are best suited for Tally. Which are they?

(b) Make a tally of each. Look carefully at the results. Are there any things that seem strange?

4-23   Use Tally to help you answer the following questions concerning the Cartoon data set (described in the Appendix, p. 524).

(a) How many people got a perfect score on the immediate cartoon test? On the delayed cartoon test?

(b) How many people failed to take the delayed cartoon test?

(c) What percentage of people got a score of 5 or less on the immediate cartoon test? Of 7 or higher?

4-24   The following questions refer to Figure 4.35, p. 134.

(a) How many restaurants had corporate ownership? How many were partnerships?

(b) How many were partnerships with 10 to 20 employees? How many were partnerships with over 20 employees?

(c) How many restaurants had 10 or more employees? Twenty or fewer employees?

(d) How many had 10 or more employees and were owned by a corporation? How many had 10 or more employees and were not owned by a corporation?

4-25   Refer to Figure 4.36, p. 135. In each part indicate what percent of the restaurants had the specified characteristics.

(a) What percent had fewer than 10 employees? More than 20 employees? Ten or more employees?

(b) What percent were partnerships with fewer than 10 employees? Corporations with 10 to 20 employees?

(c) Of the partnerships what percent had 10 to 20 employees? More than 20 employees?

(d) Of the restaurants with fewer than 10 employees what percent were partnerships?

(e) Of the partnerships what percent had 10 or more employees?

4-26

(a) Make a table of the Pulse data (described in the Appendix, p. 529) in which the two classification variables are Sex and Ran. Calculate row and column percentages.

(b) What percent of the females ran in place? Males? Do either or both of these results seem surprising to you in light of the data description in the Appendix? Discuss.

4-27

Data on all 58 persons committed voluntarily to the acute psychiatric unit of a health care center in Wisconsin during the first six months of a year are given in the table below and stored in the file HCC. There are three variables: (1) **Reason** for discharge, with 1 = normal and 2 = other (against medical advice, court ordered, absent without leave, and so on); (2) **Month** of admission, with 1 = January, 2 = February,…, 6 = June; and (3) **Length** of stay, in number of days. See Table 4.5. (The data file is set up as 58 rows with three columns for the three variables.)

Table 4.5 Health care center data

Reason	1	1	1	1	1	1	1	1	1	1	1	1	1	1	1	1	1	1	1	1	1	1	1	1	1	1	1	1	1
Month	1	1	1	1	1	1	1	1	1	2	2	2	2	2	2	2	2	3	3	3	3	3	3	3	3	4	4	4	4
Length	1	2	5	8	8	9	10	13	25	0	1	7	7	1	8	11	11	0	1	2	4	5	13	1	25	2	4	5	12
Reason	1	1	1	1	1	1	1	1	1	1	1	1	1	1	2	2	2	2	2	2	2	2	2	2	2	2	2	2	2
Month	4	4	4	5	5	5	5	6	6	6	6	6	6	1	2	3	4	4	3	4	6	6	6	6	5	3	4	4	4
Length	25	35	45	1	11	18	19	1	1	3	15	35	75	0	9	1	2	3	1	5	2	5	6	14	1	4	6	19	25

(a) First code the categorical numerical data to appropriate text data, set the value order for ordinal text data, and save in a new file.

(b) Make a table of the data in which the two classification variables are Reason and Month.

(c) What month had the highest number of discharges? The lowest?

(d) What percent of all discharges are normal?

(e) How many discharges are there in the winter months January through March? In the spring months April through June?

(f) Compare the length of stay, month by month, for the two types of discharge. Is one always longer than the other?

4-28   Refer to Figure 4.37, p. 137. If the desired results are not available from the figure, please say so.

(a) What was the median sales volume of partnerships with under 10 employees? Of all partnerships? Of all restaurants?

(b) With regard to ownership, which type of restaurant had the highest median sales volume? The lowest?

4-29   Figure 4.37, p. 137, gives data on Sales classified by Owner and Size. Use the Cross Tabulation command to determine how many of the restaurants were used in this table. Compare this to the total number of restaurants in Figure 4.35, p. 134. Why is there a difference?

4-30   The following are based on the Pulse data (described in the Appendix, p. 529).

(a) Find the mean of Pulse1 classified by Sex and Activity. Discuss the output. How do men and women compare on pulse rate?

(b) Find the mean of Pulse2 classified by Sex and Ran. Discuss the output.

4-31   The Stat ▶ Tables ▶ Descriptive Statistics command can be used with just one classification variable. Use the Restaurant data to answer the following:

(a) Find the mean and median number of seats for each type of food (C9 TypeFood). Discuss the results.

(b) Find the mean and median amount of sales per seat for each different type of food. Discuss the results.

4-32   Refer to the HCC data in Exercise 4-27.

(a) Obtain a table of the average length of stay in the acute psychiatric unit by reason and month.

(b) Does the average length of stay in April seem to be the same for both types of discharge?

(c) Repeat parts (a) and (b) for median length of stay.

(d) Does average length of stay seem to be constant over time for normal discharges? For other discharges? If any pattern is apparent, is it similar for both types of discharge?

4-33   The Cross Tabulation command can be used with three classification variables. Use the Restaurant data with the variables Owner, Size, and TypeFood for Rows, Columns, and Layers. (Hopefully, you have already converted the categorical data coded as numerical to appropriate text data and set the value order for the text data. If so, use that file here.)

(a) One cell contains the count 32. What does this mean?

(b) One cell contains the count 42. What does this mean? One cell contains the count 58. What does this mean?

(c) One cell contains the count 108. What does this mean?

5

Two- and Three-Variable Graphs

5.1 Creating Scatterplots

The plots we have used so far—dotplot, stem-and-leaf, boxplot, and histogram—have involved only one variable. However, often we are interested in the relationships between two or more variables, such as the relationship between height and weight, between smoking and lung cancer, or between temperature and the yield of a chemical process. Scatterplots allow us to investigate such relationships. A **scatterplot** is so-named because typically its points do not lie on a simple, smooth mathematical curve. Rather, it shows the variation in the data—possibly around some mathematical ideal.

Furnace Example

If both variables are quantitative, the most useful display is the familiar scatterplot. Table 5.1 presents data from a study of the effectiveness of an energy-saving furnace modification. The data are for one house over an 11-week period. The first column gives the week. The second column gives the average amount of natural gas used per hour (per square foot of house area). The third column gives the average daily temperature during the week. The variable, Furnace Modification, will be explained later.

Table 5.1 Furnace data

Week	Gas Used (BTU/hour)	Average Temperature (°F)	Furnace Modification
1	2.10	45	Out
2	2.55	39	In
3	2.77	42	Out
4	2.40	37	In
5	3.31	33	Out
6	2.97	34	In
7	3.32	30	Out
8	3.29	30	In
9	3.62	20	Out
10	3.67	15	In
11	4.07	19	Out

Now try plotting the amount of gas used versus the outside temperature. To create a scatterplot,

1. Open a new Minitab Project to reset all components to their default values.
2. Open the **Furnace1** worksheet. (Make sure you open Furnace1 and not Furnace.)
3. Choose **Graph ▸ Scatterplots ▸ Simple** and click **OK**. (Notice the variety of options in Figure 5.1 that Minitab permits with scatterplots.)
4. Enter **BTU/Hr** as the first Y variable and **Temp** as the first X variable. See Figure 5.2.

   💡 *TIP If you open a dialog box and its settings don't match those shown in the figure, or your results are different, the dialog box may be using previous settings. You may always press F3 to reset a dialog box to its defaults.*

5. Click **OK**. Minitab produces the graph in its own Graph Window. See Figure 5.3. Maximize the Graph Window to get the best resolution of the plotted points.

Figure 5.1   Scatterplots gallery

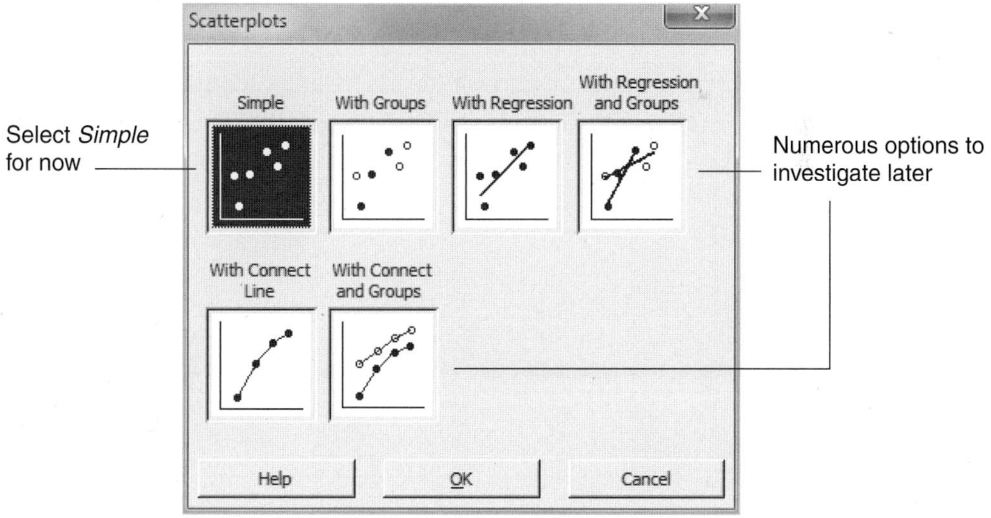

Figure 5.2   Scatterplots (Simple) dialog box

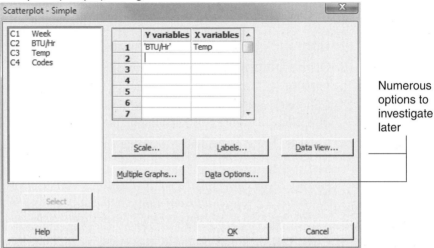

Notice that the amount of natural gas consumed by this house decreased more or less in as a straight-line as the average temperature varied from 15°F to 45°F.

Figure 5.3   Simple scatterplot of BTU/Hr versus Temperature

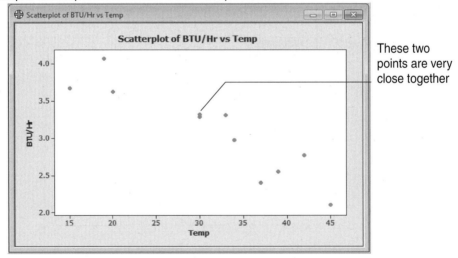

TIP You can read precise values from a graph using the crosshairs pointer. Choose Editor ▶ Crosshairs. (or use Tools ▶ Toolbars ▶ Graph Editing to activate the Graph Editing toolbar and click the crosshairs button: ⊞.) Now your pointer displays large crosshairs on the graph. The x, y coordinates of anywhere you point on a graph are displayed in a small window to the upper left of the graph.

Jittering Points

The two points very close to each other on the plot mean that there were two weeks when the average temperature was about 30°F and the gas consumption was about 3.3. Reviewing the data, we see that this happened in weeks 7 and 8.[†] It is possible that other points lie on top of one another and we just can't see it in the plot. When points are on top of each other or very close to each other, it is good practice to apply **jittering** to your plot. Jittering offsets plotted points in either the vertical and/or horizontal directions a small random amount so you can better see points that are overlapping. Try jittering the plot you just produced. Jittering is added to a plot by editing the plot. To jitter the plot,

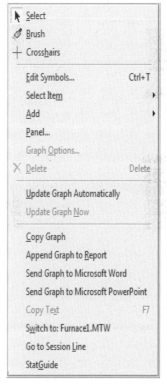

1. **Click** on any of the points in the graph. All the plotted points are now selected. (You cannot jitter single points.)
2. **Right-click** any one of the highlighted points and the menu shown here at the right appears.
3. Select the **Edit Symbols** and the Edit Symbols dialog box appears.
4. Click the **Jitter** tab to obtain the dialog box shown in Figure 5.4.
5. Check the option **Add jitter to direction**. Leave the default X and Y direction jitter settings as is, and click **OK**. Figure 5.5 shows the resulting jittered plot. Yours will be different as the actual amount of jittering comes from random numbers.

 TIP While the data points are still highlighted it is useful to successively click the Undo ↶ and Redo ↷ buttons to "animate" the changes from the original plot to the jittered plot. Doing this several times allows you to "see" the effect of the jittering and look for more points that may have been covered up in the original plot.

[†] We could also use **brushing** to find out which points these were. Go to page 107 to review the procedure for brushing a graph.

Figure 5.4   Edit Symbols with jitter dialog box

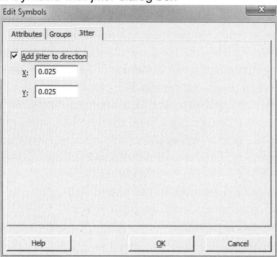

Note that both points are now visible, having been moved a random amount based on the jittering numbers in the dialog box. Use jittering with care. While the resulting plot will show all the data points more visibly, the actual data values are being misrepresented if only slightly.

Figure 5.5   Jittered scatterplot of Furnace data

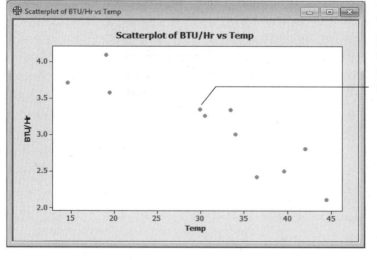

Using The Graph Editing Tools

A graph may also be jittered using the **Graph Toolbar**. To jitter with the graph toolbar,

1. First make sure the Graph Editing Toolbar is Enabled. Select **Tools ▶ Toolbars** and check the status of Graph Editing. Click on **Graph Editing** if it is not already checked. The Graph Editing toolbar appears.

graph editing button

2. Select **Symbols** from the drop-down list and all of the plotted points highlight.

3. Click the **Graph Editing** button .

4. The Edit Symbols dialog box appears.

5. Click **jitter** and proceed from there as before.

See p. 160, for more on graph editing.

Cartoon Example

The Cartoon data (in the Appendix, p. 524) were collected in an experiment to assess the effectiveness of different sorts of visual presentations for teaching. In one part of the experiment participants were given a lecture in which cartoons were used to illustrate the material. Immediately after the lecture a short quiz was given. In addition each participant was given an Otis test, a test that attempts to measure general intellectual ability.

We first create a **marginal plot** of Cartoon1 versus Otis. Marginal plots allow you to see the relationship between two variables as any scatterplot does, but also display the **marginal distribution** of each of the two variables. The marginal distributions may be shown as either histograms, boxplots, or dotplots. To make a marginal plot,

1. Choose **Graph ▶ Marginal Plot** and select **With Dotplots**.

2. Click **OK** to obtain the Marginal Plots dialog box shown in Figure 5.7.

3. Enter **Cartoon1** for the Y variable and **Otis** for the X variable and click **OK**.

Figure 5.6   Marginal Plots dialog box

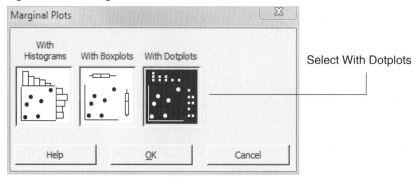

Minitab produces the graph shown in Figure 5.8. the dotplots clearly show that the distribution of Cartoon1 scores is highly skewed and piles up at the higher values. Otis scores are much more "normal."

Figure 5.7   Marginal plots dialog box

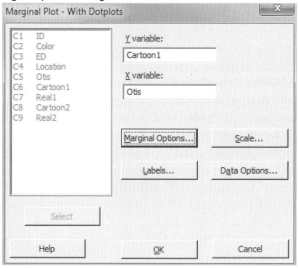

Figure 5.8   Marginal plot of Cartoon data

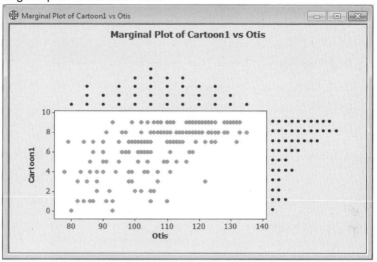

There are a large number of data points (179 people) and the Cartoon1 variable is quite **granular**, that is, it can assume only a small number of values: 0, 1, 2,…,9. In this situation there is a very good chance that points will plot on top of one another. This should be another example where jittering is a valuable tool. To plot Cartoon1 vs. Otis score with jittering,

1. Open the **Cartoon** worksheet if necessary.
2. Choose **Graph ▸ Scatterplot ▸ Simple** and click **OK**.
3. Enter **Cartoon1** as the first Y variable and **Otis** as the first X variable.
4. Click **OK**. (This is the same as Figure 5.8 but without the marginal distributions.)
5. Now double-click the points to jitter this plot **but only in the Y direction**. See Figure 5.9. This will preserve the correct Otis scores but alter the Cartoon1 scores slightly. We get Figure 5.10. (Yours will be similar but not identical to ours.)

Figure 5.10 shows several interesting facts. People with higher Otis scores tend to get higher cartoon scores, as you might expect. Note also that the highest cartoon score possible is 9, so that a person with very high ability cannot really demonstrate the full extent of his/her knowledge. The best a person can get is a 9. In this plot the scores of people who had Otis scores below 100 show no evidence that they were held down to the maximum score. But the scores of people with Otis scores over 100 do show the effect of this limitation.

Figure 5.9 Edit Symbols dialog box

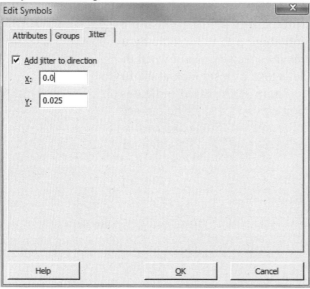

This effect is sometimes called **truncation**; it results in some loss of information. If the experiment were to be done over again, it might be better to develop a slightly longer or harder test so that everyone would have a chance to demonstrate his or her full ability.

Figure 5.10 Jittered plot of Cartoon1 versus Otis

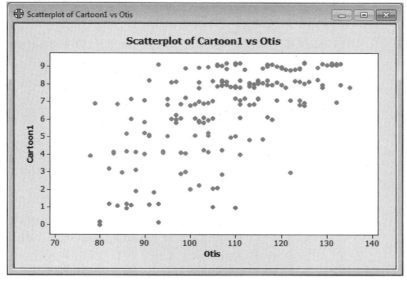

5.2   Creating Plots with Groups

A scatterplot shows the relationship between two variables. We can add information about a third variable by using different symbols for different points. This is especially useful when the third variable is categorical. For an example let's take a closer look at the furnace modification data in Table 5.1. The last column in the table indicates whether or not the energy-saving modification was installed in the furnace that week. In the file this variable is named Codes and is coded numerically as 1 for In and 2 for Out. To obtain more meaningful labelling, let's first recode this variable to text values. Call the new variable Modification and code 1 to In and 2 to Out. Review p. 88 for the procedure to do this. To create a plot marking groups with different symbols and colors,

1. Return to the **Furnace1** worksheet.
2. Choose **Graph ▸ Scatterplot ▸ With Groups** and click **OK**. (See Figure 5.1, p. 147.)
3. Enter **BTU/Hr** as the first Y variable and **Temp** as the first X variable.
4. Enter **Modification** as the Categorical Variables for Grouping. Figure 5.11 shows the dialog box with the correct entries.
5. Click **OK**. Figure 5.12 shows the resulting plot.

Figure 5.11 Scatterplot with Groups dialog box

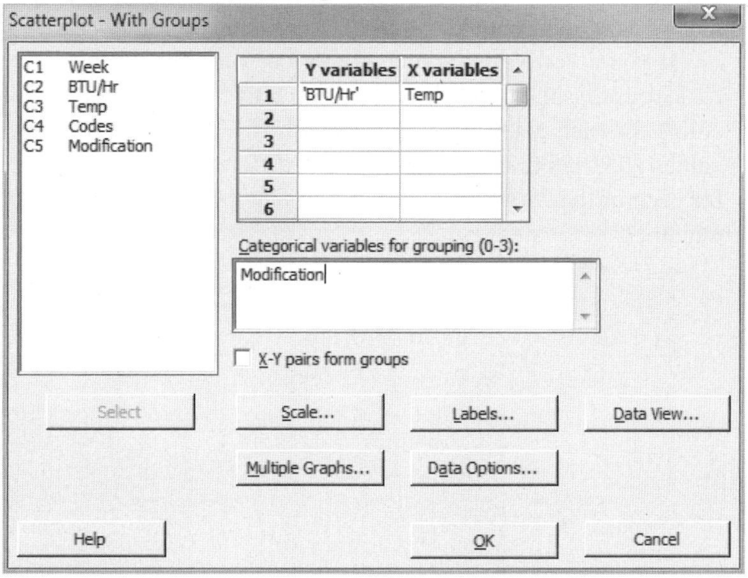

Figure 5.12 Scatterplot of Furnace data with groups indicated

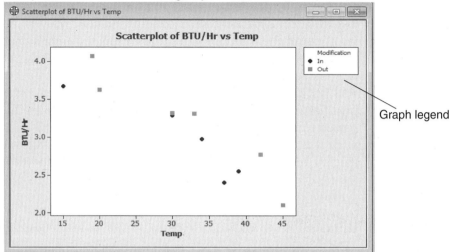

        Minitab plotted each observation having the furnace modification with a
(black) solid dot and each observation not having the modification with a (red)
square symbol. In general, Minitab will use different colors and symbol shapes
for each different value of the categorical variable. Looking closely at this plot,
we see some indication that the furnace modification might produce a slight
savings in gas consumption. Overall, for a given temperature, the In's seem to
be just a bit lower than the Out's, but the effect is not dramatic. (As you might
expect, you may edit the symbol shapes and colors.)

        If you enter two or more categorical variables in the Categorical Variables
for Grouping in Figure 5.11, Minitab will use different symbols and different
colors for each combination of categories. For example, if one variable has two
values and the other has three values, different symbols and colors will be used
for each of the $2 \times 3 = 6$ combinations. The plot legend will inform you which is
which.

5.3   Creating Plots with Several Variables

Table 5.2 gives winning times (in seconds) for three Men's Olympic races. There were no games in 1916 because of World War I and none in 1940 and 1944 because of World War II.

Table 5.2 Winning times in olympic track races

Year	100-Meter Race	200-Meter Race	400-Meter Race
1900	10.8	22.2	49.4
1904	10.8	21.6	49.2
1908	11.0	22.4	50.0
1912	10.8	21.7	48.2
1920	10.8	22.0	49.6
1924	10.6	21.6	47.6
1928	10.8	21.8	47.8
1932	10.3	21.2	46.2
1936	10.3	20.7	46.5
1948	10.3	21.1	46.2
1952	10.4	20.7	45.9
1956	10.5	20.6	46.7
1960	10.2	20.5	44.9
1964	10.0	20.3	45.1
1968	9.9	19.8	43.8
1972	10.1	20.0	44.7
1976	10.1	20.2	44.3
1980	10.3	20.2	44.6
1984	10.0	19.8	44.3
1988	9.9	19.8	43.9

A quick look tells us that the winners have gotten faster over the years. Suppose we want to compare the "progress" made in the different races. We could plot these winning times, but because the races are of different lengths, probably the average speeds are easier to compare. So let's plot speed versus year for each of the three races. Recall that speed = distance/time. The speeds will be in meters per second. We'll put all three plots on the same set of axes to make comparisons easier. We will also connect the points over time to visualize the changes better. To enter the data and then plot multiple variables,

1. Open a New Project to set everything to their defaults.
2. Open the **Trackm** worksheet.

3. Name empty columns C5-C7: **100 meters**, **200 meters**, and **400 meters**, respectively.

4. Use **Calc ▸ Calculator** to calculate the speeds as **'100 meters'=100/ '100m'**, **'200 meters'=200/'200m'**, and **'400 meters'=400/'400m'**. (Of course you can enter all of these column names by clicking rather than typing.)

5. Choose **Graph ▸ Scatterplot ▸ With Connect Lines** and click **OK**.

6. Enter **100 meters** as the first Y variable and **Year** as the first X variable. Enter **200 meters** as the second Y variable and **Year** as the second X variable. Enter **400 meters** as the third Y variable and **Year** as the third X variable. See Figure 5.13. (If you were to click OK at this point Minitab would produce three separate graphs in three separate Graph Windows— each with its own scale and so forth.)

7. Click the **Multiple Graphs** option button. The subcommand dialog box appears as shown in Figure 5.14.

8. Select the **Overlaid on the same graph** and click **OK**.

9. Click **OK** once more and Minitab produces the graph shown in Figure 5.15. (Since three variables are plotted, Minitab put the generic label "Y-Data" on the vertical axis. We will edit that label to something more appropriate in the next section.)

Figure 5.13 Scatterplot dialog box with several variables and connect lines

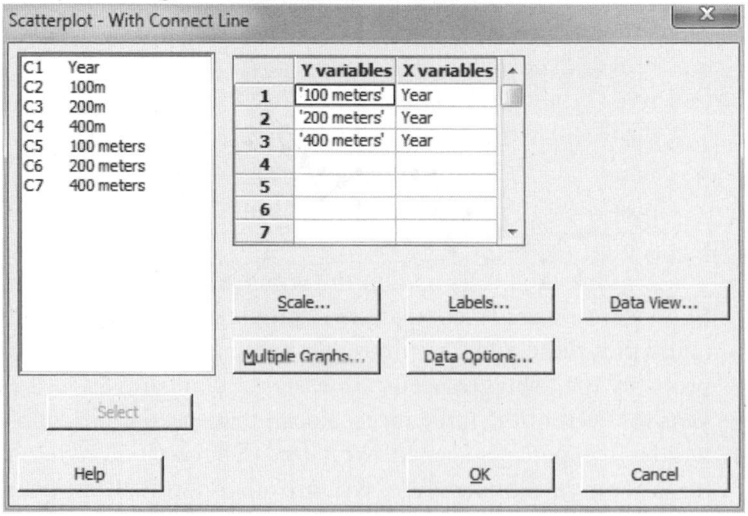

Figure 5.14 Scatterplots Multiple Graphs dialog box

Select 'Overlaid on the same graph'

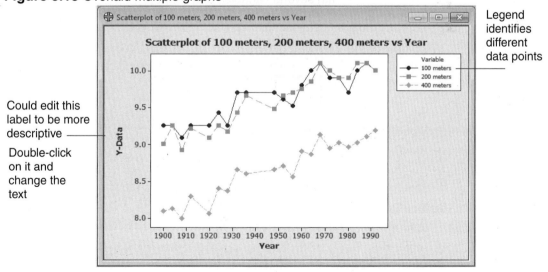

Figure 5.15 Overlaid multiple graphs

Legend identifies different data points

Could edit this label to be more descriptive

Double-click on it and change the text

By looking at this plot, we can get a very good idea of what happened. On the average the speeds of all three races increased over the years. Notice that the 100-meter and 200-meter races are run at about the same speed. Some years the 100-meter was faster; some years the 200-meter was faster. The 400-meter race, however, has always been run at a much slower pace. We can also spot several very "good" years, when progress was made in all three races—1924, 1932, 1960, and 1968. (Brushing with Years as ID variable would let you pinpoint these years with ease.) We can also see that essentially no progress was made in the three Olympics following World War II. Creative plotting can reveal a lot that is not apparent in a simple table of numbers.

5.4   Graph Editing and Annotation

Graphs are easily edited and annotated within Minitab. This has become especially easy with Release 14 and later. You have already done some simple editing of a dotplot, p. 105, and adding jitter, pages 149 and 151. Here we will explore graph editing further.

Graph Editing

You can use graph editing to change the attributes (color, size, shape, text, font,...) of current graph elements, alter a graph after updating its data or changing the variables used, and change the view or layout of graphs.[†]

To enter edit mode, make the Graph Window active and do *one* of the following:

- Double-click the graph element to edit
- Select the graph element, right-click, and choose Edit
- Select the graph element and choose Editor ▸ Edit
- If the Graph Editing toolbar is enabled, you can choose the graph element from the drop-down menu on the graph toolbar and then click the Graph Edit button 🖻 (See p. 151)

Titles and axis labels can be edited to change their text, font, size, color, and orientation. Let's edit the axis label on the *y*-axis of the graph in Figure 5.15. To edit the label,

1. Make sure the graph is active. (Click anywhere on the graph.)
2. **Double-click** the *y*-axis label, "Y-Data." The Edit Axis Label dialog box appears. See Figure 5.16. (Be sure the Font tab is selected.)
3. In the Text: box near the bottom, replace Y-data by **Speed** and click **OK**.
   Minitab relabels the axis and resizes the graph if necessary.

[†] Consult the Minitab online Help system for information on automatic or manual updating of graphs. You may also find the graph layout information of interest.

Figure 5.16 Edit Axis Label dialog box

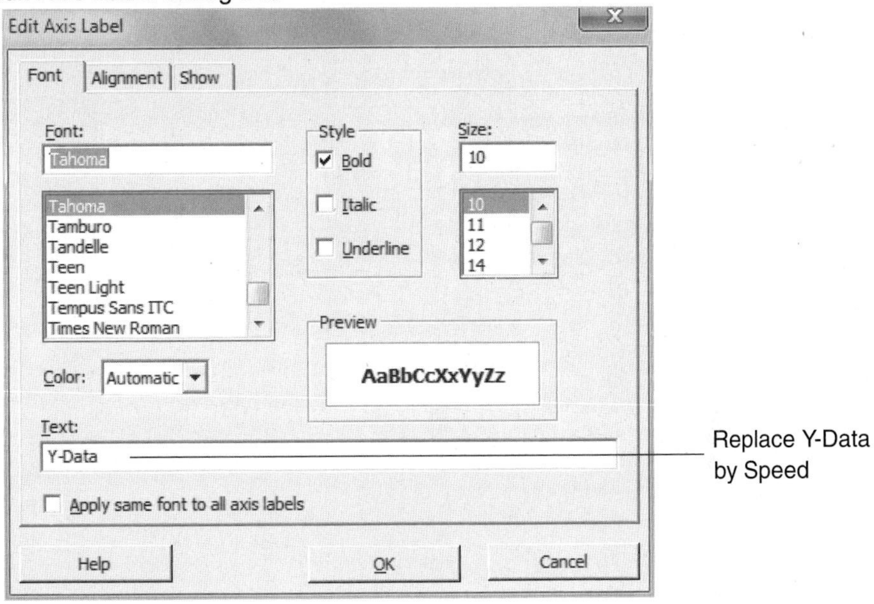

Replace Y-Data
by Speed

Selecting Points to Edit

You can select all points on a graph, a set of points belonging to a group, or an individual point. You can also select connecting lines, bars on histograms, boxes on boxplots, and other data representations.

If you click any point on a graph once, all points are selected. Click the point a second time and that individual point is selected—unless the plot is based on groups.

If the graph is based on groups, click the same point a second time and only the points in that group are selected. Click a point a third time to select that individual point. Similar rules apply to selecting connecting lines.

Let's edit the connecting lines in Figure 5.15 and make them all solid lines. To do so,

1. Make sure the graph is active. (Click anywhere on the graph to activate.)
2. **Double-click** on any of the connecting lines (and *not* on any of the plotted points). The Edit Connecting Lines dialog box appears. See Figure 5.17.
3. Be sure the **Attributes** tab is selected and choose **Custom** for Lines.
4. Select the **solid line** from the drop-down list of Types. Leave Color empty so that the three connecting lines will remain their current distinct colors. Alternatively, you could make them all black, for example.
5. Click **OK** and Minitab makes all of the connecting lines in Figure 5.15 solid.

Figure 5.17 Edit Connect Lines dialog box

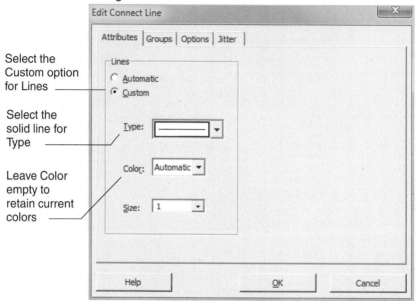

Select the Custom option for Lines

Select the solid line for Type

Leave Color empty to retain current colors

Many more possibilities are available for editing graph items. Investigate them by selecting anything on a graph and right-clicking to see the options available. You should also check the Help screens for more on graph editing.

Annotating Graphs

You can use the **graph annotation tools** to add text and objects such as markers, lines, circles, and rectangles to a graph. First make sure the Graph Annotation Toolbar is enabled. Select **Tools ► Toolbars** and check the status of Graph Annotation Tools. Click on Graph Annotation Tools if it is not already checked. Figure 5.18 shows the Graph Annotation Toolbar. It's elements function much like the tools in many other drawing programs with which you may be already familiar. Check the Help screens for more information.

Figure 5.18 Graph annotation toolbar

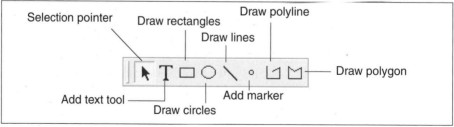

Draw polyline

Selection pointer   Draw rectangles

Draw lines

Draw polygon

Add text tool   Add marker

Draw circles

5.5 Graphs with Three or More Variables

Using graphs to display relationships among three or more variables is rather difficult. Some progress can be made with three-dimensional plots and real time rotation of such plots. As an example consider the data in Table 5.3. These are the first ten values of a residential property data set. The variables are square footage of the house, assessed and market values, and a categorical grade variable with values: −1 = low grade, 0 = medium grade, 1 = high grade. The grade variable has also been converted to three indicator variables in columns Low, Med, and High. There arc 60 rows (cases or houses) in the full data set. Your goal is to understand or predict the response variable, Market, from the others.

Table 5.3 First ten cases of property data

ID	Sq.ft	Grade	Assessed	Market	Low	Med	High
1	521	-1	7.8	26.0	1	0	0
2	538	-1	28.2	19.4	1	0	0
3	544	0	23.2	25.2	0	1	0
4	577	-1	22.2	26.2	1	0	0
5	661	-1	23.8	31.0	1	0	0
6	662	0	19.6	34.6	0	1	0
7	677	0	22.8	36.4	0	1	0
8	691	-1	22.6	33.0	1	0	0
9	694	0	28.0	37.4	0	1	0
10	712	0	21.2	42.4	0	1	0

Matrix Plot

Matrix plots allow you to assess relationships among several pairs of variables at once by displaying an array of small scatterplots on a single screen. Let's graph all pairs of scatterplots among Market, Sq.ft, and Assessed while indicating Grade membership with different plotting symbols. To do so,

1. **Open** the Worksheet named **Property** (available from the Data Library at www.CengageBrain.com associated with this book).
2. **Code** the numeric Grade column to the values Low, Medium, and High.
3. Set the **Value Order** for the new text-valued Grade variable.
4. Choose **Graph ▸ Matrix Plot** to obtain the matrix Plot gallery shown in Figure 5.19.
5. Select **With Groups** and click **OK**.

Figure 5.19 Matrix plot gallery

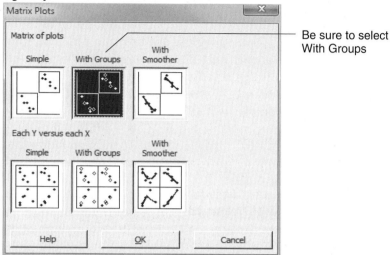

6. Enter **Market Sq.ft Assessed** in the Graph variables text box. See Figure 5.20. (Entering the response variable, Market, first has the advantage of displaying graphs of response versus each predictor across the top of the matrix plot.)

7. Enter **Grade** in the Categorical variables for grouping text box.

8. Click the **Matrix options** button and select **upper right** for the Matrix Display. (This will reduce the number of graphs displayed but the ones missing are redundant.)

9. Click OK twice and the matrix plot shown in Figure 5.21 is displayed.[†]

[†] Actually, in Figure 5.21 shown here, we edited the symbols in the matrix plot replacing solid circles, squares, and diamonds with open circles, open squares, and open diamonds. This helps see the many overlapping points somewhat better. Of course, the symbols are three different colors on your computer screen and this helps distinguish the three grades.

Figure 5.20 Matrix Plot dialog box (with groups)

Figure 5.21 Matrix plot of Market, Sq.ft and Assessed with Grade indicated

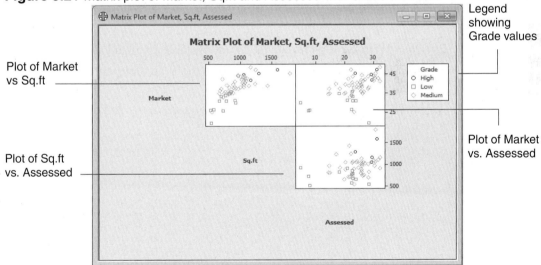

Three small scatterplots are shown in Figure 5.21. The top two show the relationships, if any, between the response variable, Market, and each of the predicator variables, Sq.ft and Assessed. There is a strong linear relationship between Market and Sq.ft but the relationship is weaker between Market and Assessed. You can also see the effect of the categorical variable, Grade. The bottom graph shows that the two predictor variables, Sq.ft and Assessed, are also somewhat related and you may have to be concerned about multicollinearity in a regression model of these variables. See Chapter 15.

3D Scatterplots

Three-dimensional scatterplots allow you to graph three variables simultaneously to see more complicated relationships among them. Try graphing Market versus Sq.ft and Assessed in a 3D scatterplot. To do so,

1. Choose **Graph ► 3D Scatterplot** to see the 3D Scatterplot gallery shown in Figure 5.22.

Figure 5.22 Three-dimensional scatterplot gallery

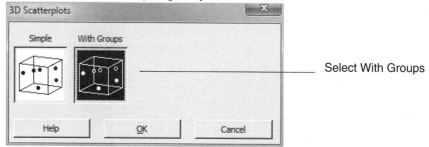

Select With Groups

2. Select **With Groups** (to see the Grade variable also) and click **OK**.
3. Enter **Market** as the Z variable, **Sq.ft** as the Y variable, and **Assessed** as the X variable. See Figure 5.23.
4. Enter **Grade** as the Categorical variable for grouping and click **OK**.
5. Figure 5.24 shows the 3D scatterplot.[†]

[†] In Figure 5.24 shown here, we again edited the symbols in the plot replacing solid circles, squares, and diamonds with open circles, open squares, and open diamonds. On your computer screen the symbols are three different colors and this helps distinguish the three grades. You may also want to maximize the graph to full screen to see more detail.

Figure 5.23 Three-dimensional scatterplot dialog box (with groups)

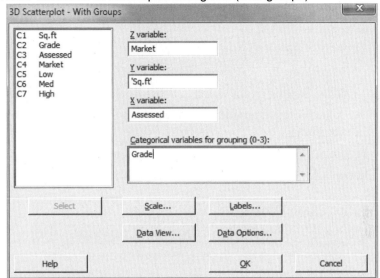

Figure 5.24 Three-dimensional scatterplot of Market vs. Sq.ft and Assessed Value

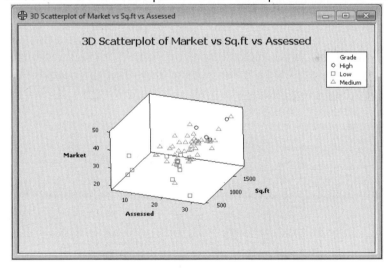

Figure 5.24 shows the general increase in Market as both Sq.ft and Assessed increase but our viewpoint is fixed. Perhaps the relationships are easier to see if we look from other vantage points—that is, we would like to rotate the graph and change the viewpoint.

Rotating a 3D Graph

To rotate a three-dimensional scatterplot,

1. Make sure the Graph Window is active.
2. Make sure the 3D Graph Rotation Toolbar is enabled. See Figure 5.25.
3. If the Toolbar is not visible, choose **Tools ▸ Toolbars ▸ 3D Graph Tools.**
4. You can rotate around any of the three axes. Try the Z axis by clicking the ⤵ button.

Figure 5.26 shows the view where we stopped the rotation. Press the Home button 🏠 to return to the initial or home view.

Figure 5.25 3D graph rotation toolbar

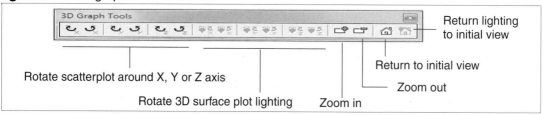

We might see a somewhat linear relationship in this rotated plot. This would, of course, mean a plane might fit these data reasonably well. Market would be modeled as a linear function of the two variables Sq.ft and Assessed. We return to this topic in Chapter 15

Figure 5.26 Another view of three-dimensional scatterplot after rotation

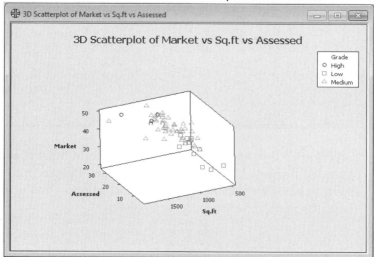

5.6   Creating Time Series Plots

A sequence of observations taken over time is called a **time series**. Examples include the milk output of a cow recorded each week, the average temperature in Chicago recorded daily, the consumer price index recorded each month, and the annual number of homicides in the United States. Minitab's Time Series command is designed to plot time series data with equally spaced time values.

Table 5.4 gives the number of people employed in the food and kindred-products industry in Wisconsin from January 1970 to December 1974 (in units of 1000 employees).

Table 5.4 Number of employees in the food products industry

Year	Jan	Feb	Mar	Apr	May	Jun	Jul	Aug	Sep	Oct	Nov	Dec
1970	53.5	53.0	53.2	52.5	53.4	56.5	65.3	70.7	66.9	58.2	55.3	53.4
1971	52.1	51.5	51.5	52.4	53.3	55.5	64.2	69.6	69.3	58.5	55.3	53.6
1972	52.3	51.5	51.7	51.5	52.2	57.1	63.6	68.8	68.9	60.1	55.6	53.9
1973	53.3	53.1	53.5	53.5	53.9	57.1	64.7	69.4	70.3	62.6	57.9	55.8
1974	54.8	54.2	54.6	54.3	54.8	58.1	68.1	73.3	75.5	66.4	60.5	57.7

These data are entered in column C2 of the Employ worksheet. To create a time series plot of these data,

1. Open the **Employ** worksheet.
2. Choose **Graph ▸ Time Series Plot ▸ Simple** and click **OK**. See Figure 5.27.

Figure 5.27 Time series plots gallery

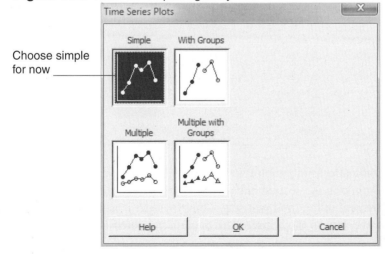

Choose simple for now

3. Enter **Food** as the Series variable. See Figure 5.28.

Figure 5.28 Time Series Plot dialog box

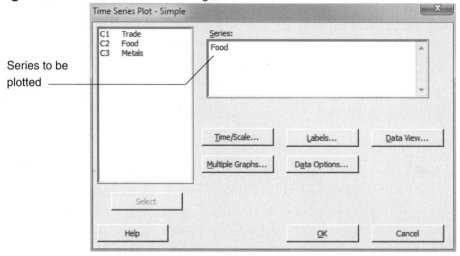

4. Click **OK and** Minitab displays the plot shown here.

Figure 5.29 Time Series Plot of Employees

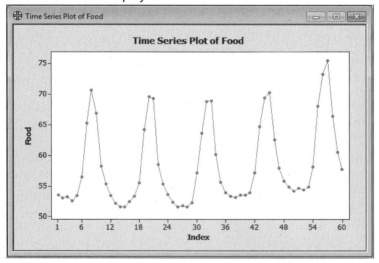

There are two patterns in the plot: a very strong repeating cycle and perhaps a slight upward trend in the second half of the series. The cycle comes from the seasonal employment in the food industry—employment is naturally quite high in the summer and low in the winter. The cycle is repeated every 12 months.

Many time series contain repeating patterns. For example if you record hourly temperature for 10 days, you will probably see a repeating cycle every 24 hours. Temperatures generally tend to be higher in the daytime and lower in the night.

Indicating a Time or Date Scale

The default time series plot shows the row number or sequence number on the time axis. You can change the plot to show real time instead of an index of integers. To indicate a real time scale,

1. Click the **Edit Last Dialog** button ⊞ .
2. Click on the **Time/Scale** button. See Figure 5.28.
3. Be sure the Time tab is selected. Select the **Calendar** option then click the list arrow and choose **Month Year**. See Figure 5.30.
4. Enter **1** for the Start Value for Month and **1970** for the Start Value for Year.
5. Click **OK** twice. Minitab now labels the time axis with months and years. See Figure 5.31.

Figure 5.30 Time/scale options for time series plots

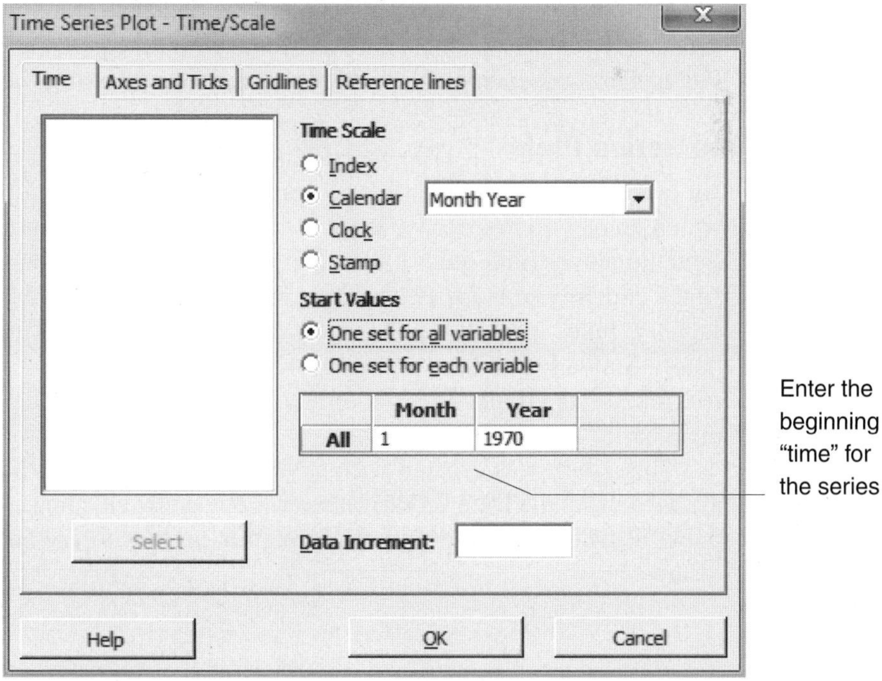

Figure 5.31 Time series plot of employees with time labelling on time axis

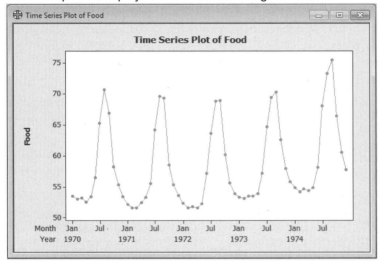

Now you can identify the seasonality of the pattern a little more clearly. The highest months each year are late summer, and the lowest are the winter months after harvest and before planting.

Note that you can also use the Graph ➤ Plot option to display time series data, but if you do, you must have a column with time in it. Graph ➤ Time Series Plot is most convenient when you choose not to enter a time variable.

Seasonal Time Series Plots

An even better way to identify seasonality is to use special plotting symbols which convey time information within the plot itself. Figure 5.32 shows the same Employee time series plotted with plotting symbol J for all Januaries (and Junes and Julys), F for Februaries, M for Marches (and Mays), and so on. January, June, and July can be easily distinguished by looking at neighboring points. Similarly for April and August and March and May. It would be wise to maximize this Graph Window on your screen so that the details can be read more easily.

This graph was created with a Minitab macro named Monthly which is available from the Data Library at www.CengageBrain.com associated with this book. See Chapter 18 for more information on how to use and create Minitab macros.

Figure 5.32 Seasonal time series plot of employee data

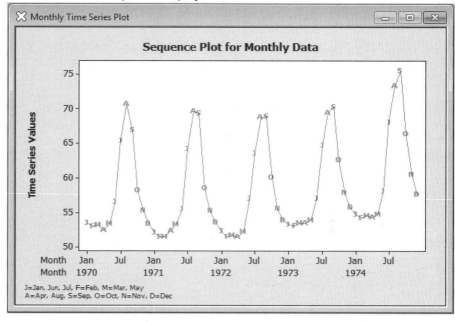

Using Time Series Plots to Find Problems

Nearly all data must be collected either across time or across space or both. For example chemical and industrial measurements are often made one after the other; agricultural data are often collected on various areas scattered across a field. Plotting the data in ways that reflect these connections often reveals unusual features or problems.

For example, track-etch devices were under consideration as an inexpensive way to measure low-level radiation. An experiment was done to study the usefulness of one such device, called an open cup. A total of 25 open cups were hung along a string that ran the length of a chamber. The chamber was designed to have a uniform level of radiation. The level of radiation was measured by each of the 25 cups. Here "time" is really position along the chamber. To generate a time series plot of the radiation levels,

1. Enter the following radiation levels into C1: 6.6, 4.8, 5.0, 6.3, 5.9, 6.5, 6.1, 5.8, 6.4, 5.3, 5.2, 5.2, 4.8, 4.9, 5.5, 5.1, 5.0, 5.7, 4.8, 5.3, 4.2, 5.3, 4.5, 4.5, 4.2.
2. Name the column Radiation Level.
3. Choose **Graph ▸ Time Series Plot ▸ Simple** and click **OK**.
4. Enter **Radiation Level** as the Series variable.
5. Click **OK**.

Figure 5.33 Time series plot of track-etch radiation data

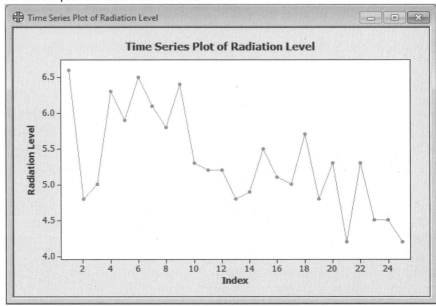

This plot shows a fairly systematic pattern: radiation levels at the left of the plot are generally larger than those at the right.

Why this pattern? These 25 observations are plotted in the order in which the cups had been hung along the string. One conjecture is that the radiation level in the chamber was not, in fact, constant, but changed gradually from one end of the chamber to the other. Further checking uncovered another possibility. The cups had been put in the chamber in the same order as their serial numbers. Thus a change in sensitivity of the devices during the manufacturing process could account for the observed trend. If the cups had not been placed in the chamber in the same order as their serial numbers, but instead had been placed in a random order, it might have been possible to pinpoint the real source of the problem.

Exercises

5-1   The following questions refer to the plot in Figure 5.8, p. 153.
   (a) How many people got a perfect score on the cartoon test?
   (b) How many people with an Otis score of 90 or less got a perfect score on the cartoon test?
   (c) Approximately what is the lowest Otis score? The highest?

5-2   **(a)** Figure 5.3, p. 148, shows a plot of gas consumption versus temperature. Suppose we want some idea of how much gas would be consumed at a temperature of 25°F. Use this plot to make an estimate. Explain how you arrived at your estimate.
(b) Repeat part (a) for 10°F. How much faith do you have in your estimate?

5-3   Table 5.1, page 146, shows data on gas consumption over an 11-week period. Plot weekly 'average temperature' versus week. How did temperature change during this period?

5-4   In this problem we will examine the Trees data set (described in the Appendix, p. 531) with an eye to developing a way to predict volume from measurements that are easier to make.
(a) Plot volume versus diameter.
(b) Plot volume versus height. Compare this plot to the one in part (a). Which seems to be a better predictor of volume: height or diameter? Which do you think would be easier to measure in a forest?
(c) Use Calc ▶ Calculator to create a new variable, V, defined as $V = \text{height} \times \text{diameter}^2$.
(d) Plot volume versus V. Does V seem to predict volume better than either diameter or height alone? Can you think of a mathematical explanation for this? A matrix plot listing volume as the first variable will make these comparisons easier.

5-5   Table 5.5 gives the marriage rate and divorce rate per 1,000 from 1920 to 1990. Let's see how these rates have changed over the years.

Table 5.5 Marriage data

Year	Marriage	Divorce	Year	Marriage	Divorce
1920	12.0	1.6	1960	8.5	2.2
1925	10.3	1.5	1965	9.3	2.5
1930	9.2	1.6	1970	10.6	3.5
1935	10.4	1.7	1975	10.0	4.8
1940	12.1	2.0	1980	10.6	5.2
1945	12.2	3.5	1985	10.1	5.0
1950	11.1	2.6	1990	9.8	4.7
1955	9.3	2.3			

(a) Open the Marriage worksheet and plot marriage rate versus year. What patterns do you see? (Use Graph ▶ Time Series ▶ Simple. Choose Calendar and Year for Time/Scale and 5 for Increment.)

(b) Plot divorce rate versus year. What patterns do you see?

5-6   Table 5.6 contains track records as of September 5, 1992.

Table 5.6 Track records

Distance (meters)	Women's Track		Men's Track	
	Minutes	Seconds	Minutes	Seconds
100	0	10.49	0	9.86
200	0	21.34	0	19.72
400	0	47.60	0	43.29
800	1	53.28	1	41.73
1500	3	52.47	3	28.86
2000	5	28.69	4	50.81
3000	8	22.62	7	28.96
5000	14	37.33	12	58.39
10000	30	13.74	27	8.23

First we will look at women's track. Open the Trackmw worksheet and calculate the speed at which each race was run in miles per hour. (Note: There are 1609.4 meters in a mile.) What is the fastest speed? The slowest?
(a) Plot the record speed versus distance in meters for women's track. What is the pattern?
(b) Repeat parts (a) and (b) for men's track.
(c) How do men's and women's track compare?

5-7   **(a)** Use the Pulse data set and plot weight versus height.
(b) Use Graph ▸ Scatterplot ▸ With Groups to make the same plot using one symbol for men and another for women. Interpret the plot.

5-8   The following questions refer to the Trackm data in Figure 5.2, p. 157.
(a) Were any records set in these three races in the 1992 Olympics? That is, were any of these three races run in the least time ever?
(b) In what year was the record set for the 100-meter race? For the 200-meter race? For the 400-meter race?
(c) The best time for the 100-meter race was 9.9 seconds. How fast is this in miles per hour? (Note: There are 1,609.4 meters in a mile.)
(d) Consider the 100-meter race. In which years was a new Olympic record set?

5-9   Table 5.7 lists the monthly average temperatures over many years for five U.S. cities. The data are stored in the file named Cities. Plot temperature versus month for the five cities all on the same graph. (Use Graph ▸ Time Series Plot ▸ Multiple and set the Time/Scale to Calendar Month with Start Value 1.) Write a paragraph comparing the temperature patterns in these five cities.

Table 5.7 City temperatures

Month	Atlanta	Bismarck, ND	New York	San Diego	Phoenix
Jan	42	8	32	56	51
Feb	45	14	33	60	58
Mar	51	25	41	58	57
Apr	61	43	52	62	67
May	69	54	62	63	81
Jun	76	64	72	68	88
Jul	78	71	77	69	94
Aug	78	69	75	71	93
Sep	72	58	68	69	85
Oct	62	47	58	67	74
Nov	51	29	47	61	61
Dec	44	16	35	58	55

5-10   During the winter months you have probably noticed that on days when the wind is blowing it seems particularly cold (at least in colder climates). The explanation lies in the fact that wind creates a slight lowering of atmospheric pressure on exposed flesh, thereby enhancing evaporation, which has a cooling effect. Thus the effective temperature (how cold it seems to you) depends on two factors: thermometer temperature and wind speed. Table 5.8 was prepared by the U.S. Army. It gives the effective temperature corresponding to a given thermometer temperature and wind speed.

Table 5.8 Wind Chill index

	Thermometer Temperature ($°F$)			
Wind Speed	20	10	0	−10
10	4	−9	−21	−33
15	5	−18	−36	−45
20	−10	−25	−39	−53

(a) There is an error in this table. Can you find it? Plotting the data will help. Open the Wind worksheet and plot the Wind Chill Index versus wind speed for the four temperatures. Do you see any points that do not seem to follow the overall pattern in the plot?

(b) Try to correct the data point that is in error. What value would you substitute? Give reasons for your choice. Use both the data table and the plot to help.

5-11   Table 5.9 shows an alternative way to set up the wind chill data shown in Table 5.8.

Table 5.9 Wind chill index (alternative form)

Wind Chill	Wind Speed	Temperature
4	10	20
−5	15	20
−10	20	20
−9	10	10
−18	15	10
−25	20	10
−21	10	0
−36	15	0
−39	20	0
−33	10	−10
−45	15	−10
−53	20	−10

(a) Enter the data into a Minitab worksheet in this format. (The data in this form is available as a file with name WindChill on the Data Library at www.CengageBrain.com associated with this book.)

(b) Use the Matrix Plot command to view plots of each of these three variables versus each of the others.

(c) Inspect these plots using brushing to highlight any unusual data points. Which points appear unusual?

5-12   Again use the Wind Chill data as arranged in Table 5.9.

(a) Use the 3D Scatterplot command to display these data in three dimensions. Plot Wind Chill as the Z variable and Wind Speed and Temperature as X and Y.

(b) Rotate the graph in various ways and see if you can spot the erroneous data point.

5-13   How do automobile accident rates vary with the age and sex of the driver? Table 5.10 gives the numbers of accidents per 100 million miles of exposure for drivers of private vehicles.

Table 5.10 Accidents

Age of Driver	Accidents that involved a casualty		Accidents that did not involve a casualty	
	Male	Female	Male	Female
Under 20	1436	718	2794	1510
20-24	782	374	1939	851
25-29	327	219	949	641
30-39	232	150	721	422
40-49	181	228	574	602
50-59	157	225	422	618
60 and over	158	225	436	443

(a) Open the Auto worksheet. Note that the Age column contains only one representative age for each category in the table. For example 18 stands for "Under 20," 22 stands for "20-24," and so on.

(b) Plot the casualty accident rate for males versus age and the casualty accident rate for females versus age on the same graph. Describe in a sentence or two how these accident rates depend on the age and sex of the driver.

(c) Repeat (b) for the noncasualty accidents. How well do the conclusions in (b) and (c) agree?

(d) Compute overall accident rates for male and female drivers by adding up the casualty and noncasualty rates separately for each sex. For example use C11 = C1+ C3 and C12 = C2 + C4. Repeat (b) for these overall rates.

(e) For each age and sex group compute the proportion of accidents that involve a casualty. Repeat part (b) for these proportions.

(f) Suppose we want to look at the total casualty accident rate (that is, the number of casualty accidents per 100 million miles of exposure) for each age category. Would (C1 + C2) provide us with appropriate data? Can we get the appropriate figures from the data we have? Explain.

5-14   In this chapter you plotted employment data for the food industry, using the Employ data set, Table 5.4, page 169. Make a similar time series plot of the data for the wholesale and retail trade industries, stored in the variable named Trade. Interpret this plot. Did employment tend to increase over this period? Is there a seasonal pattern to employment in wholesale and retail trade? Can you give any reasons for the patterns in the data? (Use the Monthly macro to make the time series plots.)

5-15   Use the Employ data set. Make a time series plot to display the data for employ-ees in fabricated metals from the Metals variable. Describe the pattern of employment in this industry. Did employment tend to increase over this period? Is there a seasonal pattern? (Use the Monthly macro to make the time series plots.)

5-16   Just by looking at the time series plot in Figure 5.33, p. 174, can you figure out the following?
(a) Which cup gave the highest reading? Approximately what value did the highest reading have?
(b) Two cups seem to be tied for lowest. Which are they? Approximately what value did they have?
(c) How many cups were there in all?

5-17   Figure 5.34 shows a portion of the Gallop Poll Organizations Web page that was displayed on October 13, 2003.
(a) Discuss the small graph shown in the display from the point of view of showing changes over time in the percentage of families experiencing trouble from drug abuse. Notice that the four time points are not equally spaced—not even approximately.
(b) Use Minitab to produce a time series plot that would show the changes more correctly—that is, use a proper time scale.

Figure 5.34 From the Gallop Poll Organizations Web Page, October 13, 2003

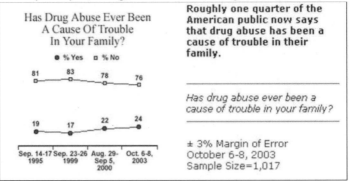

5-18   Table 5.11 displays measurements from a diffusion porometer, an instrument for measuring changes in resistance across a leaf's surface. Any treatment that closes the stomates of a leaf causes the resistance to go up; when the stomates are open, the resistance goes down. Four treatments were used with six observations of each treatment.

Table 5.11 Leaf treatment

							Treatment							
			1							2				
Resistance	7.6	8.9	8.4	14.5	15.8	12.8		8.2	10.5	6.8	14.3	13.4	13.2	
			3							4				
Resistance	6.5	9.9	8.9	12.6	13.5	14.6		11.4	8.6	12.2	10.9	9.9	15.3	

The observations were taken in the order in which they are listed across the rows. Open the Leaf worksheet and create a time series plot to plot these data in this order. Write a concise summary of these data, including an opinion as to whether the treatments differ very much. Also discuss any problems you see with the data.

5-19   Edwin Moses was a fabulous hurdler. From September 2, 1977 to May 29, 1987, he won 122 consecutive races in the 400-meter high hurdles. Table 5.12 gives his winning times to the nearest hundreth of a second. His first winning time was 47.58 seconds and the winning time of his final race was 48.90 seconds. (The data are available in the file named EdwinMoses in the Data Library at www.CengageBrain.com associated with this book. In the file the times are in a single column.)

(a) Display these data in a time series plot.
(b) Inspect the plot for specific behavior. Did he "quit while he was ahead"? Did he wait too late to quit? Are there any discernible trends?

Table 5.12 Edwin Moses winning times in 400-meter high hurdles (read across rows)

47.58	48.66	48.50	48.85	49.70	48.62	48.20	49.90	48.60	49.11	48.55	48.34
48.72	48.85	47.94	48.51	48.76	49.71	48.50	49.15	47.69	48.98	49.70	48.75
47.89	49.76	49.51	48.72	48.58	48.43	48.51	49.59	48.58	47.67	48.15	49.05
47.53	49.36	48.91	49.00	48.28	50.11	49.41	49.23	48.22	47.90	49.10	47.13
49.00	48.62	48.53	48.36	48.65	48.51	47.17	48.53	48.74	47.81	49.89	48.67
48.87	48.86	50.19	49.50	48.61	48.65	48.29	50.18	47.59	48.35	47.99	47.14
47.64	47.27	47.37	48.69	49.28	49.21	49.02	48.43	49.47	47.84	48.46	47.98
49.53	49.00	48.50	49.54	48.11	47.50	48.48	47.37	47.43	47.02	48.74	47.93
48.71	49.61	48.25	48.83	47.58	47.56	49.33	48.51	47.75	48.49	47.95	47.32
48.01	48.89	47.94	48.21	47.66	48.21	47.76	47.53	48.28	47.38	48.73	48.89
49.19	48.90										

5-20   Table 5.13 lists monthly milk production data over several years. The data set is available in a file named MilkProduction from the Data Library at www.CengageBrain.com associated with this book. In the file the time series data are given in a single column. Several date columns are given also.

(a) Plot the time series using a basic time series plot versus time index, 1, 2, 3,…. Can you see any patterns in the series?

(b) Now plot the series versus "real time", that is, Jan-76 through Jul-93. Are the patterns any clearer now?

(c) Use the Monthly Minitab macro to display a seasonal time series plot with special plotting symbols for the months. Is the seasonal pattern now quite clear?

(d) Finally, use Graph ► Time Series Plot (With Groups) and use Month as the Grouping variable. Set the Time/Scale option to Calender Month Year with Start Values 1 and 1976. What patterns are shown in this plot?

Table 5.13 Milk production per cow per month, January-76 through July-93

Year	Jan	Feb	Mar	Apr	May	Jun	Jul	Aug	Sep	Oct	Nov	Dec
1976	867	840	931	949	1007	978	946	918	870	872	836	880
1977	899	845	964	979	1037	1003	976	948	897	900	860	899
1978	915	851	968	982	1034	1005	975	945	897	906	864	901
1979	923	862	981	989	1049	1023	997	971	929	936	895	935
1980	958	926	1016	1023	1084	1050	1025	997	955	964	926	966
1981	994	934	1062	1061	1110	1064	1044	1018	974	982	945	990
1982	1008	943	1065	1059	1112	1074	1055	1031	992	1001	963	1007
1983	1050	980	1099	1091	1154	1122	1103	1067	1023	1037	997	1034
1984	1046	1008	1097	1094	1151	1104	1082	1055	1012	1024	986	1031
1985	1057	981	1109	1113	1187	1147	1148	1129	1073	1089	1045	1080
1986	1104	1022	1144	1147	1215	1168	1158	1133	1076	1086	1050	1094
1987	1123	1052	1190	1191	1259	1208	1200	1173	1124	1148	1107	1158
1988	1178	1135	1237	1230	1282	1227	1222	1196	1158	1179	1140	1193
1989	1172	1130	1233	1222	1276	1225	1218	1186	1148	1165	1130	1186
1990	1218	1141	1294	1274	1325	1263	1259	1232	1172	1202	1172	1225
1991	1250	1169	1308	1296	1337	1261	1253	1238	1189	1226	1194	1256
1992	1291	1238	1344	1314	1367	1321	1322	1295	1246	1278	1237	1292
1993	1310	1216	1356	1344	1404	1354	1351					

6

Statistical Distributions

6.1   The Normal Distribution

Probability Density Function

The normal distribution is unequivocally the most important distribution in statistics. Many populations in nature are well approximated by a normal distribution. But more importantly, the sampling distributions of many important statistics are normal to a useful approximation. An example of a naturally occurring 'normal' population is given by the data in Table 6.1.

Table 6.1 Chest girth in inches for 1516 soldiers of the 1885 army of the Potomac

Girth	28	29	30	31	32	33	34	35	36	37	38	39	40	41	42
Frequency	2	4	17	55	102	180	242	310	251	181	103	42	19	6	2

Since these data are already summarized in a frequency table, we could display their distribution with either a bar chart or a regular histogram. The picture will be the same in either case. Figure 6.1 shows the density histogram which Minitab produces from summarized data. (See Chapter 4,"Histograms from Frequency Tables", p. 115 and "Density Histograms", p. 117 to review the procedures for carrying this out.)

Figure 6.1
Histogram of
chest girths

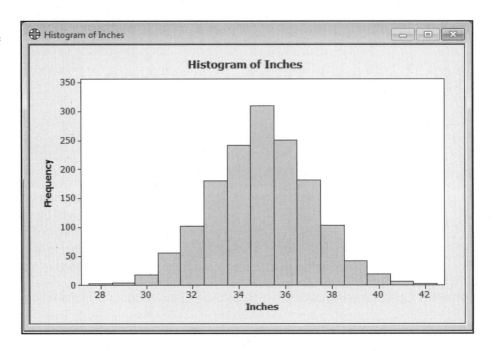

Figure 6.2
Histogram of
chest girths
with a normal
curve overlaid

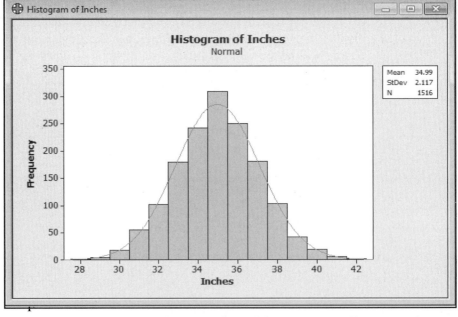

igure 6.2 shows the same histogram but with a corresponding normal curve
superimposed. The technical term for this normal curve is the **probability
density function** (pdf) of the normal distribution. The equation for this curve is

given in most statistics textbooks and in Section 6.5, p. 204. This formula contains two parameters, the mean, μ, and the standard deviation, σ. The center of the normal curve is μ and the spread is specified by σ. Large values of σ give curves that are spread out. Small values of σ give curves that are much more compact. The smooth curve in Figure 6.2 is the pdf of a normal distribution with μ = 34.99 and σ = 2.117—the same as the sample mean and standard deviation of the data.

Minitab will produce the values of the pdf for any normal distribution (through the menu sequence Calc ▸ Probability Distributions ▸ Normal) but we rarely need such values for statistical analysis. Rather, we usually need values for the areas under the normal curve below, above, or between various data values. These areas can be viewed as probabilities.

Cumulative Distribution Functions

A cumulative distribution function (cdf) gives the cumulative area or probability associated with a pdf. In particular a cdf gives the area under the pdf, from the left, up to a value you specify. Most statistics textbooks have tables that give a range of values for the cdf of a standard normal distribution. With Minitab you can calculate the cdf for any values you want, and you can calculate it directly for a normal distribution with any mean and standard deviation.

Figure 6.3 gives some examples of areas under a normal curve with μ = 34.99 and σ = 2.117. Minitab's CDF command can be used to compute cumulative probabilities related to these graphs.

Figure 6.3   Certain areas under the normal curve with μ = 34.99 and σ = 2.117

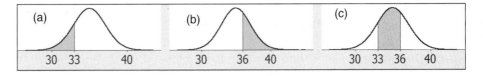

In panel (a) we show the area under the normal curve from the left up to the value 33. You can calculate this value by using the Cumulative probability option. To calculate the cdf for 33,

1. Open the **Normal Distribution** dialog box with the menu sequence **Calc ▸ Probability Distributions ▸ Normal**.
2. Make sure the **Cumulative probability** option button is selected.
3. Input 33.99 for the **Mean** and 2.117 for the **Standard Deviation**.
4. Click the **Input constant** option button and type **33** in the adjacent text box. See Figure 6.4.

5. Click **OK**. Minitab displays the cdf value in the Session Window as 0.173606.

6. Repeat these steps to calculate the cdf for 36.

Figure 6.4   Calculating the Normal CDF at 33

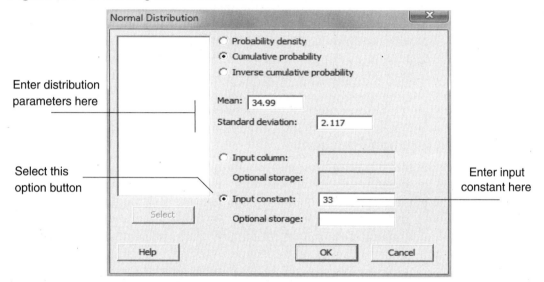

The CDF command calculates the area under the normal curve up to the value 33 as 0.173606. Panel (b) displays the area above 36. The CDF command gives only areas below a value. Therefore we find the area below 36. This is 0.683351. The area under the entire normal curve is 1. Thus $1 - 0.683351 = 0.316649$ is the area above 36. In panel (c) we show the area between 33 and 36. This is given by (area below 36) − (area below 33) = $0.683351 - 0.173606 = 0.509745$.

Visual displays with numerous examples will often help you learn more about the properties of normal curves and other distributions. To facilitate this Minitab from Release 15 on has commands that automate the work. You need only input the mean and standard deviation of the normal curve and the data values between which you want areas. To display various areas under the normal curve,

1. Select **Graph ▸ Probability Distribution Plot**.
2. Click on **View Probability** as shown in Figure 6.5.

Figure 6.5   Probability distribution plots gallery

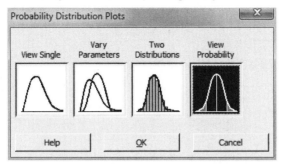

3. Click on **OK** and the View Probability dialog box opens as in Figure 6.6.
4. Select the proper distribution (**Normal** is the default) and enter the required parameters.

Figure 6.6   View Probability

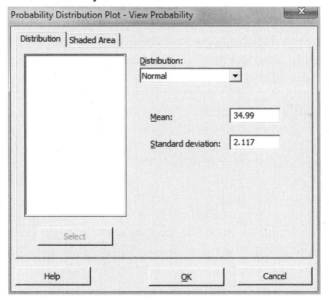

5. Now click on the **Shaded Area** tab and the dialog box shown in Figure 6.7 appears

Figure 6.7   Shaded Area dialog box

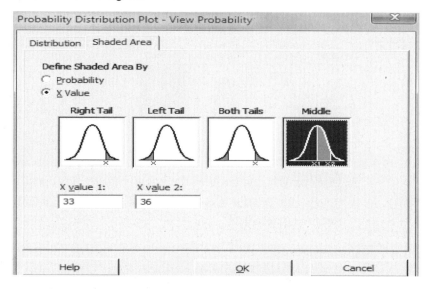

6. Click on **X Value**, **Middle**, and enter the two X values **33** and **36**.

7. Click OK twice and the graph shown in Figure 6.8 is displayed.

Figure 6.8
Normal
distribution
area display

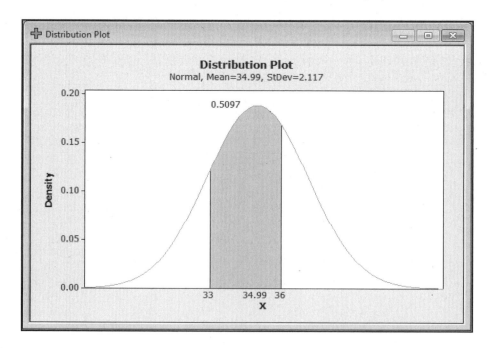

Figure 6.9 gives an example to show how well the normal curve approximates the soldier data.

Figure 6.9
Using a normal curve to approximate the soldier distribution

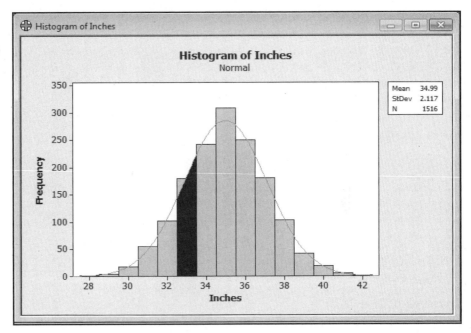

Since the rectangular bars are one unit wide, the proportion of soldiers with a chest girth of 33 is just the area of the bar centered at 33. By Table 6.1 p. 183, this area is the observed proportion $180/1516 = 0.1187$. To approximate it, we might use the area of the region under the Normal curve that approximates the bar at 33. This is the region between 32.5 and 33.5 that is darkly shaded.[†] To calculate the area of the shaded region we could use the cdf command twice and subtract. Alternatively, we could proceed as we did to construct Figure 6.8. The result is the same, namely, 0.1210 — quite close to the observed proportion of 0.1187. Both round to 12%.

As another example, Figure 6.10 approximates the proportion of soldiers with chest girths from 31 to 33 inches. The observed proportion is the sum of the areas of the bars at 31, 32, and 33. This area is $(55+102+180)/1516 = 0.2223$. The normal approximation to this is the darkly shaded area under the curve or 0.2238, which is very close to the observed proportion. Again, we could have obtained this value as we did in constructing Figure 6.8 or by using the cdf command.

[†] Note that we cannot just use the value of the pdf at 33. We need to use areas.

Figure 6.10
Using a
normal curve
to approximate
the soldier
data

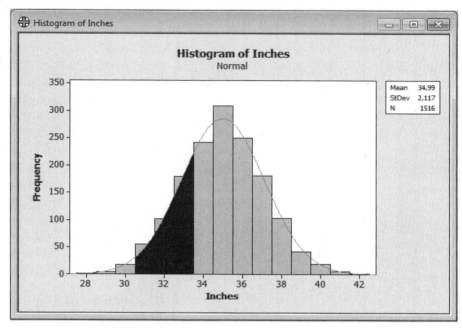

Inverse of the CDF

The cdf option calculates the area associated with a value. The inverse cdf
option does the opposite: it calculates the value associated with a given area or
probability. Figure 6.12 gives an example, using a normal curve with $\mu = 34.99$
and $\sigma = 2.117$. We want to find the value x that has an area of 0.25 below it.

To use the inverse cdf to calculate the value that has an area of 0.25 below it,

1. Reopen the **Normal Distribution** dialog box.
2. If necessary, reenter the parameter values: $\mu = 34.99$ and $\sigma = 2.117$.
3. Select the **Inverse cumulative probability** option button.
4. Select the **Input** constant option button and enter **0.25** in the corresponding
   text box.
5. Click **OK**. Minitab displays the answer as 33.5621, shown in Figure 6.11.

Figure 6.11 Inverse CDF Session Window output

```
Inverse Cumulative Distribution Function

Normal with mean = 34.99 and standard deviation = 2.117

P( X <= x )          x
       0.25   33.5621
```

Thus one-quarter of the area under this normal curve lies below 33.5621 or about 33.6. Using Figures 6.5, 6.6, and 6.7 but selecting **Probability**, **Left Tail**, and **0.25** in Figure 6.7, we can display the results graphically as in Figure 6.12.

Figure 6.12
Normal curve with
$\mu = 34.99$ and
$\sigma = 2.117$

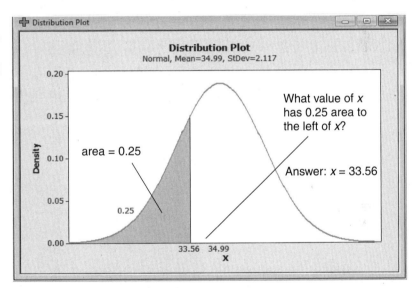

6.2   The Binomial Distribution

Suppose you bought four light bulbs. The manufacturer claims that 85% of their bulbs will last at least 700 hours. If the manufacturer is right, what are the chances that all four of your bulbs will last at least 700 hours? That three will last 700 hours, but one will fail before that?

Consider another situation. You've just entered a class in ancient Chinese literature. You haven't even learned the language yet, but they've given you a pop quiz. You'll have to guess on every question. It's a multiple-choice test; each of the 20 questions has 3 possible answers. To pass, you must get at least 12 correct. What are the chances that you'll pass?

Questions such as these can sometimes be answered with the help of the binomial distribution. However, for the binomial distribution to hold, the trials must be independent. That is, success on one trial must not change the probability that there will be a success on some other trial. Thus there must be no special reason why successes or failures should tend to occur in streaks. (Is this condition likely to be met in the light bulb example? In the Chinese literature example?)

Calculating Binomial Probabilities with PDF

A binomial distribution is specified by two parameters: n, the number of trials, and p, the probability of success on each trial. The number of successes can be 0, 1, 2,…, n. The PDF command will calculate the probability for any possible number of successes.

To find out the probability that three of your four light bulbs will be successes (last more than 700 hours) and one will fail,

1. Choose **Calc ▸ Probability Distributions ▸ Binomial**.
2. Select the **Probability** option button.
3. Enter **4** as the **Number of trials** and **0.85** as the **Probability of success**.
4. Click the **Input constant** option button and enter **3** in the corresponding box. See Figure 6.13.
5. Click **OK**.

Figure 6.13
Binomial
distribution
dialog box

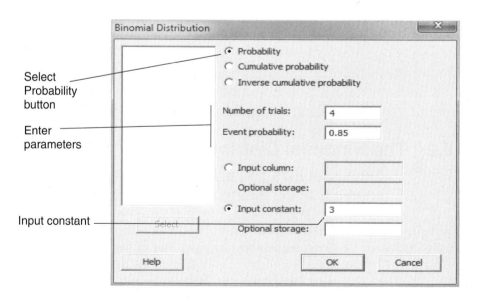

Minitab gives the answer as 0.368475 or about 37%.

If you want to compute probabilities for several x values all at once you can enter those x values into a column prior to step 1 above and use the **Input column** choice instead of the **Input constant** in step 4. It might be easier to use session commands to print a table of the complete set of probabilities. With the PDF and CDF session commands, if you don't indicate a column, Minitab prints such a table in the Session Window. For this reason, in some of the next examples we'll use session commands instead of menu commands. To perform the computation using session commands,

1. Activate the Session Window and make sure command language is enabled (**Editor ▸ Enable Commands**).
2. At the MTB > prompt, type **PDF;** (Don't forget the semicolon at the end!) and press **Enter**.
3. Type **BINOMIAL 4 0.85.** (Don't forget the period at the end!) Press **Enter**. Minitab produces the table of probabilities shown in Figure 6.14.

Figure 6.14 Using Session commands to table binomial probabilities

Probability Density Function

Binomial with n = 4 and p = 0.85

x	P(X = x)
0	0.000506
1	0.011475
2	0.097538
3	0.368475
4	0.522006

Notice that in this set of session commands you did not specify a value of 3 light bulbs, so Minitab printed a probability table for all the possibilities from 0 through 4. The value for 3 appears, as it did before, as 0.368475.

Graphical Display of the Binomial

To get a feel for a distribution it often helps to see a graphical display of the distribution. Figure 6.15 shows a histogram-type display for a binomial distribution with $n = 20$ and $p = 0.3333$.

Figure 6.15
Binomial
distribution
histogram

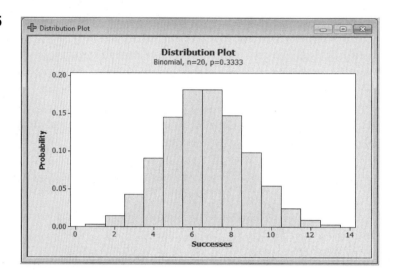

In principle, the distribution has nonzero values from 0 to 20 successes. However, the probabilities are so small that they disappear from the graph at zero and above 13 successes.

This graph is produced with **Graph ▸ Probability Distribution Plot**, selecting **Binomial** for the distribution, clicking on **View Single** and entering the parameters $n = 20$ and $p = 0.3333$. Review Figures 6.5, 6.6, and 6.7 for a similar procedure.

Cumulative Distribution Function

The CDF command calculates cumulative probabilities from 0 to any specified value. A cumulative probability is the probability that your result will be less than or equal to a given value. As an example suppose we calculate the probability you will fail the test in ancient Chinese literature. Here $n = 20$ and $p = 0.3333$. You will fail the test if you get fewer than or equal to 11 answers correct. (You will pass if you get 12 or more right.) To calculate the cumulative probability,

1. Reopen the **Binomial Distribution** dialog box.
2. Click the **Cumulative probability** option button.
3. Enter **20** as the **Number of trials**, **0.3333** as the **Probability of success**, and **11** as the **Input constant**.
4. Click **OK**. Minitab reports the answer as 0.987038.

The probability of failure is 0.987038. Thus the probability you will pass is $1 - 0.987038 = 0.012962$ or about 1.3%. As you might expect, you have a very small chance of passing the quiz.

Inverse of the CDF

As with the all distributions, the command CDF calculates the cumulative probability associated with a value. The command INVCDF does the opposite: it calculates the value associated with a given cumulative probability. To calculate the inverse CDF for a binomial with $n = 25$ and $p = 0.4$ where the cumulative probability is 0.95,

1. Reopen the **Binomial Distributions** dialog box.
2. Click the **Inverse cumulative probability** option button.
3. Enter **25** as the **Number of trials** and **0.4** as the **Probability of success**.
4. Enter **0.95** as the Input constant.
5. Click **OK**.

Figure 6.16 displays Minitab's resulting output.

Figure 6.16 Binomial inverse CDF

```
Inverse Cumulative Distribution Function

Binomial with n = 25 and p = 0.4

 x  P( X <= x )        x  P( X <= x )
13     0.922199       14     0.965608
```

Since the binomial distribution is discrete, there is no x value that produces a cumulative probability of exactly 0.95. Minitab displays the two x values which come closest to giving 0.95. There is 0.965608 probability for 14 or fewer "successes" and 0.922199 probability for 13 or fewer successes.

Session commands can be used to compute a table of cumulative probabilities for each possible number of successes. To compute cumulative probabilities for the binomial with $n = 5$ and $p = 0.4$,

1. Enter the following session commands in the Session Window (Don't forget the semicolon and period):
   ```
   CDF;
   ```

```
BINOMIAL 5 0.4.
```
  2. Press Enter after each command.

  Minitab displays the complete set of five cumulative probabilities as shown in Figure 6.17.

Figure 6.17 Binomial CDF with $n = 5$ and $p = 0.4$

```
Cumulative Distribution Function

Binomial with n = 5 and p = 0.4

x   P( X <= x )
0       0.07776
1       0.33696
2       0.68256
3       0.91296
4       0.98976
5       1.00000
```

  These are the only probabilities for which INVCDF can give an exact answer for $n = 5$ and $p = 0.4$. When there is no exact answer, Minitab always displays the values below and the values above the requested probabilities.

6.3   Normal Approximation to the Binomial

For many combinations of n and p, the binomial distribution can be well approximated by a normal distribution that has the same mean and standard deviation—that is, by a normal distribution that has $\mu = np$ and $\sigma = \sqrt{np(1-p)}$. This is a consequence of the Central Limit Theorem. Of course, with Minitab available we can find the binomial probabilities easily and exactly and do not need the approximation. For historical reasons[†] let's see how the approximations work. Perhaps this situation where we can calculate both the exact and approximate answers will help convince you that the Central Limit Theorem might give useful approximations in the numerous other situations where we cannot obtain exact answers.

Let's see how the normal approximates a binomial with $p = 1/2$ and $n = 20$. The approximating normal has $\mu = 20(1/2) = 10$ and $\sigma = \sqrt{20(1/2)(1/2)} = \sqrt{5} = 2.236$.

Since the binomial is discrete while the normal is continuous we need to be careful and compare "apples to apples." One good way is to do this graphically. We like to plot overlaid "density histograms" for the binomial and the corre-

---
[†] and for fun!

sponding normal—that is, a normal with the same mean and standard deviation as the binomial. We plot the binomial histogram with unit width bars as we did in Figure 6.15, p. 194. In this display the area of the bar represents the binomial probability for the number of successes at the midpoint of the base of the bar. We are now comparing areas to areas. To facilitate forming this display we have written a Minitab macro named **BinoNorm**. It is available for download at the Data Library at www.CengageBrain.com associated with this book. With the macro you can specify any values for n and p and see how the distributions compare.

Figure 6.18 gives an example for $n = 20$ and $p = 0.5$.

Figure 6.18
Binomial with
$n = 20$ and
$p = 0.5$ and
normal
approximation

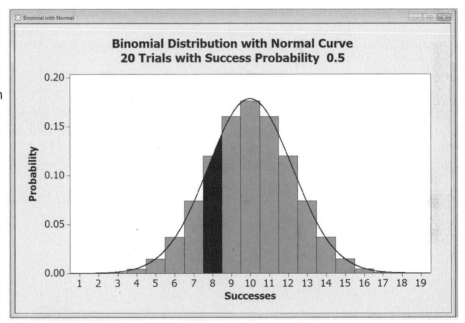

Suppose that we want to use the normal curve to approximate the probability of exactly 8 successes. The exact probability is the area of the bar at 8 successes. We approximate that rectangular area with the area under the corresponding normal curve from 7.5 to 8.5 as shown in Figure 6.19. (This is usually called the *continuity correction*.) From the graph, it appears that the *included* area shown will nearly balance out with the *excluded* area and the approximation should be quite good.

Let's use Minitab to calculate both the exact value from the binomial distribution and the approximate value from the normal distribution.

Using the binomial pdf command (as we did in Section 6.2, p. 191) we find the exact probability of 8 successes is 0.120134. Using the normal CDF with mean 10 and standard deviation 2.236, we find CDF values for 8.5 and 7.5 and subtract. We get $0.251161 - 0.131769 = 0.119392$ which agrees with the exact answer extraordinarily well.

One more example. Suppose we want to find the probability of 7, 8, or 9 successes. Figure 6.20 shows the areas we need for the exact and approximate answers. We find the exact answer by subtracting the binomial CDF for 6 successes from the binomial CDF for 9 successes. Doing so we get $0.411901 - 0.0576591 = 0.3542419$. The normal approximation uses a normal curve with mean 10 and standard deviation 2.236 and the area between 6.5 and 9.5. We obtain $0.411529 - 0.0587569 = 0.3527721$—another excellent approximation.

Of course, the approximation is not this good in all situations. Experiment with the **BinoNorm** macro to investigate for yourself the situations where the approximation should be satisfactory and those where it fails. The usual guidelines for reasonable approximations require that np and $n(1-p)$ both be 10 or larger. We pursue this further in the exercises.

Figure 6.19
Normal approximation of the probability of 8 successes

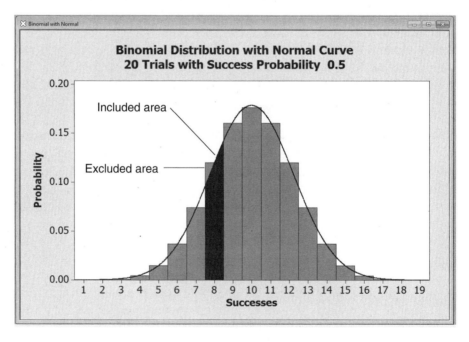

Figure 6.20
Normal approximation of 7, 8, or 9 successes

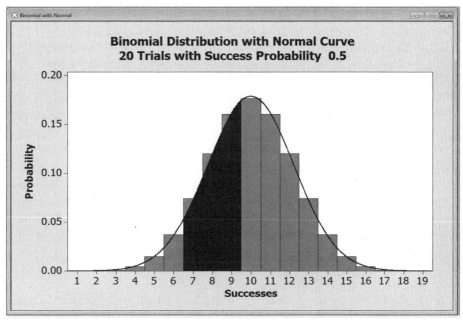

6.4   The Poisson Distribution

A famous statistician once called the Poisson distribution "the distribution to read newspapers by." He said this because so many times we read facts such as "Crime Increases 25% in Year" or "543 Automobile Fatalities Expected This Weekend." Such figures frequently can be modeled with the Poisson distribution.

The Poisson distribution arises when we count the number of occurrences of an event that happens relatively infrequently, given the number of times it could happen. For example the potential number of automobile accidents that could occur in a given county next (nonholiday) weekend might be very large. Any motorist in the area could have an accident. But the chance that any given motorist will have an accident is very small, so the actual number of accidents probably will not be too large.

Calculating Poisson Probabilities with PDF

A Poisson distribution is completely specified by just one parameter, the mean, μ. For example if we know that there are, on the average, 6 accidents per (nonholiday) weekend, then we can calculate the probability there will be, say, 10 accidents next weekend, or 20 accidents, or no accidents. The Poisson PDF command will calculate probabilities for any values you specify. To calculate probabilities for a Poisson distribution,

1. Enter **10**, **20**, **0**, and **5** into C1 and name the column **Accidents**.
2. Choose **Calc ▶ Probability Distributions ▶ Poisson**.
3. Enter **6** as the mean.
4. Enter **C1** as the **Input column** and store the results in a new column named **Probability**.
5. Click **OK**. The worksheet shows the resulting probabilities as in Figure 6.21.

Figure 6.21 Individual Poisson probabilities

	C1	C2	C3
	Accidents	Probability	
1	10	0.041303	
2	20	0.000004	
3	0	0.002479	
4	5	0.160623	
5			

As with the binomial distribution, it often helps to visualize the distribution with an appropriate graph. The sequence **Graph ▶ Probability Distribution Plot** allows you to display the Poisson distribution in various ways. Figure 6.22 shows the result of using this to graph the Poisson distribution with mean 6. In theory the probabilities go on 'forever' but they are too small to see on the graph after a count of 16.

Figure 6.22
Poisson
distribution
with mean 6

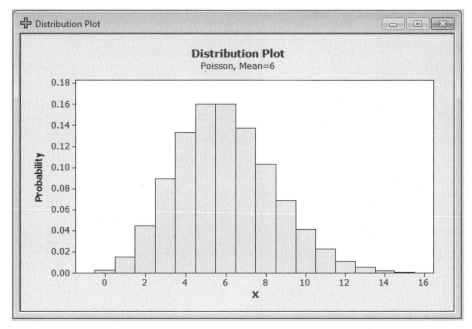

Cumulative Distribution Function

Most often we are not interested in individual probabilities but in cumulative probabilities, such as the probability of ten or fewer accidents. We then use the CDF command.[†] To use the CDF command,

1. Type the following session commands in the Session Window:
   ```
   CDF;
   POISSON 6.
   ```
2. Press **Enter** after each command and note the semicolon (;) and period.

Figure 6.23 shows the resulting table of probabilities. From this we see that if the average number of accidents per (nonholiday) weekend is six, then the probability of no accidents next weekend is 0.00248, the probability of six or fewer is 0.60630, and the probability of ten or more accidents is $1 - 0.91608 = 0.08392$.

Suppose the police decide to crack down on speeders, and that the following weekend there are only three accidents. We can imagine a headline: 'Police Crackdown Leads to 50% Reduction in Accidents.' What do you think? From Figure 6.23 the probability of three or fewer accidents for $\mu = 6$ is 0.15120. That is, there is a 15% chance of a 50% or better reduction in accidents even if the police crackdown has no real effect whatsoever.

[†] First make sure that commands have been enabled. See "Enabling Commands" p. 43.

Figure 6.23 Poisson cumulative distribution with mean 6

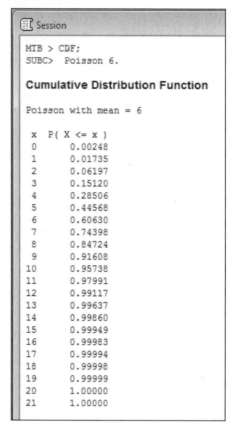

```
MTB > CDF;
SUBC>  Poisson 6.

Cumulative Distribution Function

Poisson with mean = 6

 x    P( X <= x )
 0       0.00248
 1       0.01735
 2       0.06197
 3       0.15120
 4       0.28506
 5       0.44568
 6       0.60630
 7       0.74398
 8       0.84724
 9       0.91608
10       0.95738
11       0.97991
12       0.99117
13       0.99637
14       0.99860
15       0.99949
16       0.99983
17       0.99994
18       0.99998
19       0.99999
20       1.00000
21       1.00000
```

Inverse of the CDF

Suppose, in another community, we know that the average number of accidents is 50 per weekend. We want to get some idea of how many accidents we can expect to occur in a given weekend. To calculate the probability,

1. Enter **0.25**, **0.75**, **0.005**, and **0.995** into **C1**.

2. Type the following session commands in the Session Window:

   ```
   INVCDF C1;
   POISSON 50.
   ```

3. Press **Enter** after each command. Figure 6.24 shows the results.

Figure 6.24 Inverse CDF for Poisson with mean 50

```
Inverse Cumulative Distribution Function

Poisson with mean = 50

  x   P( X <= x )        x   P( X <= x )
 44      0.221040       45      0.266866
 54      0.742306       55      0.784470
 32      0.004393       33      0.006979
 68      0.993756       69      0.995665
```

As with the binomial distribution, the Poisson does not have an exact answer for every probability we might specify. One way we could summarize this output is as follows: Since P(54 or fewer accidents) = 0.7423 and P (44 or fewer accidents) = 0.2210, then P(from 45 through 54 accidents) = 0.7423 − 0.2210 = 0.5213. Thus over 50% of the time there will be from 45 to 54 accidents, inclusive. Similarly P(from 33 through 69 accidents) = P(69 or fewer) − P (32 or fewer) = 0.9957 − 0.0044 = 0.9917. Thus over 99% of the time there will be from 33 to 69 accidents. Rarely (about 0.5% of the time) will there be fewer than 33 accidents and rarely (about 0.5% of the time) will there be more than 69.

Normal Approximation to the Poisson

The Central Limit Theorem implies that the Poisson distribution may be approximated by the corresponding normal distribution when the parameter μ is large. There is some hint of this even in Figure 6.22, p. 201, with $\mu = 6$. To see the approximation graphically, we provide the Minitab macro named **PoissonNorm**. Figure 6.25 displays the output of this macro when $\mu = 30$. (Here $\sigma = \sqrt{30} = 5.477$.)

(If you are working with Minitab right now, use the PoissonNorm macro to produce this display on your screen. Maximize the graph window so you can see as much detail as possible.) Looking at the graph it appears that the approximation would be reasonable for some values but not others. We can use the PDF and CDF commands for the Poisson to get exact answers and the CDF command for the Normal to obtain the approximate results. Table 6.2 shows a few of the values we tried.

Figure 6.25
Poisson
distribution
with mean 30
and normal
curve

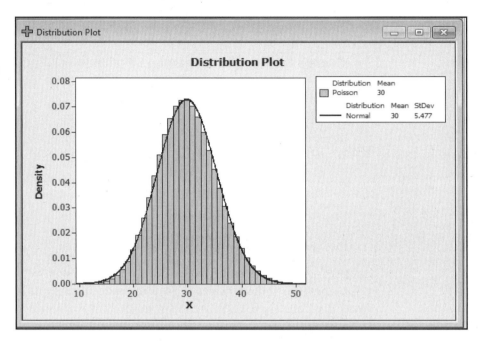

Table 6.2
Poisson
probabilities:
exact and
approximate
from normal
curve

Probability	Exact from Poisson: μ = 30	Approximate from Normal Curve: μ = 30, σ = 5.477
P(X = 23)	0.0340769	0.117657 − 0.0854429 = 0.0322141
P(X = 25)	0.0511123	0.205647 − 0.157641 = 0.048006
P(X ≤ 23)	0.114646	0.117657

As with the normal approximation to the binomial, with Minitab software available there is no compelling reason to use the Normal approximation. Minitab will produce exact answers in nearly any Poisson situation encountered.

6.5   A Summary of the Theoretical Distributions in Minitab

The distributions on the Calc ▸ Probability Distribution menu all use the same PDF, CDF, and INVCDF options, as does the Random Data option discussed in Chapter 7. The Graph ▸ Probability Distribution Plot will graph these distributions in various ways. This section displays the formula for the pdf of each of the distributions available.

Continuous Distributions

Normal: $\mu = K$, $\sigma = K$

Normal distribution with mean μ and standard deviation σ. The pdf is

$$f(x) = \frac{1}{\sqrt{2\pi}\sigma} e^{-[(x-\mu)]^2/2} \qquad -\infty < x < \infty, \, 0 < \sigma$$

Uniform: $a = K$, $b = K$

Continuous uniform distribution on a to b. The pdf is

$$f(x) = \frac{1}{b-a} \qquad a < x < b$$

Cauchy: $a = K$, $b = K$

Cauchy distribution. The pdf is

$$f(x) = \frac{1}{(\pi b)\{1 + [(x-a)/b]^2\}} \qquad -\infty < x < \infty, \, 0 < b$$

Laplace: $a = K$, $b = K$

Laplace or double exponential distribution. The pdf is

$$f(x) = \frac{1}{2b} e^{(-|x-a|)/b} \qquad -\infty < x < \infty, \, 0 < b$$

Logistic: $a = K$, $b = K$

Logistic distribution. The pdf is

$$f(x) = \frac{e^{-(x-a)/b}}{b[1 + e^{-(x-a)/b}]^2} \qquad -\infty < x < \infty, \, 0 < b$$

Lognormal: $\mu = K$, $\sigma = K$

Lognormal distribution. a variable X has a lognormal distribution if $\log(X)$ has a normal distribution with mean μ and standard deviation σ. The pdf of the lognormal is

$$f(x) = \frac{1}{x\sigma\sqrt{2\pi}} e^{-\{(\ln x)-\mu\}^2/(2\sigma^2)} \qquad 0 < x, \, 0 < \sigma$$

T: $v = K$

Student's t distribution with v degrees of freedom. The pdf is

$$f(x) = \frac{\Gamma[(v+1)/2]}{\Gamma(v/2)} \frac{1}{(1 + x^2/v)^{(v+1)/2}} \qquad -\infty < x < \infty, \, 0 < v$$

F: $u = K$, $v = K$

F distribution with u degrees of freedom for the numerator and v degrees of freedom for the denominator.

$$f(x) = \frac{\Gamma[(u+v)/2]}{\Gamma(v/2)\sqrt{v\pi}}\left(\frac{u}{v}\right)^{\frac{u}{2}}\frac{x^{(u-2)/2}}{[1+(u/v)x]^{(u+v)/2}} \qquad 0 < x, 0 < u, 0 < v$$

Chi-Square: $v = \mathbf{K}$

χ^2 distribution with v degrees of freedom. The pdf is

$$f(x) = \frac{x^{(v-2)/2}e^{-x/2}}{2^{v/2}\Gamma(v/2)} \qquad 0 < x, 0 < v$$

Exponential: $b = \mathbf{K}$

Exponential distribution. (Note: Some books use $1/b$ where we have used b.) The pdf is

$$f(x) = \frac{1}{b}e^{-x/b} \qquad 0 < x, 0 < b$$

Gamma: $a = \mathbf{K}$, $b = \mathbf{K}$

Gamma distribution. (Note: Some books use $1/b$ where we have used b.) The pdf is

$$f(x) = \frac{x^{a-1}e^{-x/b}}{\Gamma(a)b^a} \qquad 0 < x, 0 < a, 0 < b$$

Weibull: $a = \mathbf{K}$, $b = \mathbf{K}$

Weibull distribution. The pdf is

$$f(x) = \frac{ax^{a-1}e^{-(x/b)^a}}{b^a} \qquad 0 < x, 0 < a, 0 < b$$

Beta: $a = \mathbf{K}$, $b = \mathbf{K}$

Beta distribution. The pdf is

$$f(x) = \frac{\Gamma(a+b)x^{a-1}(1-x)^{b-1}}{\Gamma(a)\Gamma(b)} \qquad 0 < x < 1, 0 < a, 0 < b$$

Largest: $a = $ location parameter $= \mathbf{K}$, $b = $ scale parameter $= \mathbf{K}$

Largest extreme value distribution

$$f(x) = \frac{1}{b}e^{-(x-a)/b}\exp[-e^{-(x-a)/b}] \qquad a < x$$

Smallest: a = location parameter = **K**, b = scale parameter = **K**
Smallest extreme value distribution

$$f(x) = \frac{1}{b}e^{-(-x+a)/b}\exp[e^{-(-x+a)/b}] \qquad x < a$$

Loglogistic:
Loglogistic distribution

$$f(x) = \frac{1}{b(x-c)}\frac{e^{((\log(x-c)-a))/b)}}{[1+e^{-((\log((x-c)-a))/b)}]^2} \qquad c < x$$

Triangular: a = **K**, b = **K**, c = **K**
Triangular distribution starting at a, mode at b, ending at c

$$f(x) = \begin{cases} 2\dfrac{(x-a)}{(b-a)(c-a)} & \text{for} \quad a < x < b \\[3mm] 2\dfrac{(c-x)}{(b-a)(c-a)} & \text{for} \quad b < x < c \end{cases}$$

Discrete Distributions

Bernoulli n = **K**, p = **K**
This is only available on the Random Data submenu. It simulates n Bernoulli trials with probability p of success on each trial.

$$f(x) = p^x(1-p)^{1-x} \qquad x = 0, 1$$

Binomial: n = **K**, p = **K**

$$f(x) = \binom{n}{x}p^x(1-p)^{n-x} \qquad x = 0, 1, 2, ..., n$$

Poisson: μ = **K**
Poisson distribution. The probability of x is

$$f(x) = \frac{e^{-\mu}\mu^x}{x!} \qquad x = 0, 1, 2, ...$$

Hypergeometric: N = **K**, M = **K**, n = **K**
Hypergeometric distribution with a sample of size n from a population of size N containing M "successes."

$$f(x) = \frac{\binom{M}{x}\binom{N-M}{n-x}}{\binom{N}{n}} \qquad max(0, n-N-M) < x < min(n, M)$$

Integer: $a = \mathbf{K}$, $b = \mathbf{K}$

Discrete uniform distribution on the integers from a to b. Each integer has the same probability. With $a = 0$ and $b = 9$ you can select random digits.

$$f(x) = \frac{1}{b - a + 1} \qquad x = a, a + 1, \ldots, b$$

Negative Binomial: There are two closely related versions both based on Bernoulli trials with success probability p.

Distribution of number of trials x to produce r successes.

$$f(x) = \binom{x - 1}{r - 1} p^r (1 - p)^{x - r} \qquad x = r, r + 1, r + 2, \ldots$$

Distribution of number of failures y that occur before observing r successes.

$$f(y) = \binom{y + r - 1}{r - 1} p^r (1 - p)^y \qquad y = 0, 1, 2, \ldots$$

Note $x = y + r$

Discrete values in C and probabilities in C

Arbitrary discrete distribution. You store the values and the corresponding probabilities into two columns. Then pdf, cdf, and inverse cdf work as always with this distribution.

Exercises

6-1   Using the data in Table 6.1, p. 183:
(a) How many soldiers had chest girths of 37 inches?
(b) What proportion of soldiers had chest girths of 37 inches?
(c) What proportion had chest girths of 37 inches or less?
(d) What proportion had chest girths of 37 inches or more?
(e) Which chest size was the most common?

6-2   Table 6.3 displays the chest sizes of 5738 Scottish soldiers. (The data are available from the Data Library at www.CengageBrain.com associated with this book. The data are given in both summarized form as in Table 6.3 and also in one long column of 5738 values. filename: ChestSize)

Table 6.3 Chest size (in inches) of 5738 Scottish soldiers

Inches	33	34	35	36	37	38	39	40	41	42	43	44	45	46	47	48
Frequency	3	18	81	185	420	749	1073	1079	934	658	370	92	50	21	4	1

(a) Use Minitab's histogram with the fitted normal option to produce a histogram of these data similar to that shown in Figure 6.2, p. 184.

(b) Comment on the general adequacy of the normal approximation.

(c) What proportion of the soldiers had chest sizes between 36 and 38 inches inclusive?

(d) Verify that the mean and standard deviation of these data are 39.89 and 2.050, respectively.

(e) Using the mean and standard deviation in part (d) and a normal distribution, find the approximate proportion of soldiers that had chest sizes between 36 and 38 inches inclusive? Compare to your answer in part (c).

6-3   For a normal distribution with $\mu = 35$ and $\sigma = 2$, use the CDF command to find:

(a) The area above 33.

(b) The area below 36.

(c) The area between 33 and 35.

(d) The area between 35 and 36.

6-4   Suppose a normal distribution has $\mu = 100$ and $\sigma = 10$. In each case below make a sketch like those in Figure 6.8, p. 188. Find each of the following (Use Graph ▸ Probability Distribution Plot ▸ View Probability to make it easy!):

(a) The area below 90.

(b) The area above 110.

(c) The area below 80.

(d) The area above 120.

(e) The area between 90 and 110.

(f) The area between 80 and 120.

(g) The area between 80 and 110.

6-5   In this exercise you will compute and plot the pdf and the cdf for a normal distribution with $\mu = 100$ and $\sigma = 10$. The normal pdf is essentially zero for values smaller than $\mu - 3\sigma$ and for values larger than $\mu + 3\sigma$. Therefore we will use the values 70, 71, 72,…,130 on the data axis.

(a) Enter the pattern 70:130 into C1.

(b) Compute the Normal pdf for values in C1 with parameters 100 and 10. Place the results in C2.

(c) Compute the cdf of C1 with the same parameters and place the results in C3.

(d) Plot C2 versus C1 and C3 versus C1, each with connecting lines. Comment on the shape of the two curves.

6-6   (a) Make a histogram for the SAT math scores for the first set of data in the Grades data set (described in the Appendix, p. 527). Do you think these scores might have been a random sample from a normal distribution? Why or why not?

(b) Make a histogram for the SAT math scores for the second set of data in the Grades data set. Compare both histograms. How much do they vary in shape? Do they both look bell-shaped? Do they both look as if they are random samples from a normal population? Do they look as if they both came from the same population?

6-7   A total of 16 mice are sent down a maze, one by one. From previous experience it is believed that at any given turn the probability a mouse turns right is 0.38. Suppose their turning pattern follows a binomial distribution. Use the PDF and CDF commands to answer each of the following:

(a) What is the probability that exactly 7 of these 16 mice turn right at a given turn?

(b) That 8 or fewer turn right at a given turn at a given turn?

(c) That more than 8 turn right at a given turn?

(d) That 8 or more turn right at a given turn?

(e) That from 5 to 7 turn right at a given turn?

6-8   In this exercise you'll compute and plot some binomial probabilities in order to get a better idea what the distributions look like. Use Graph ► Probability Distribution Plot to produce the plots.

(a) First, let's fix n at 8 and vary p. Use $p = 0.01, 0.1, 0.2, 0.5, 0.8$, and 0.9. For each value of p compute the binomial probabilities and get a plot. (Use Graph ► Probability Distribution Plot ► Vary Parameters ► Binomial)

(b) How do the shapes of the plot change as p increases? Compare the two plots where $p = 0.1$ and $p = 0.9$. Do you see any relationship? What about the two plots where $p = 0.2$ and $p = 0.8$? (Use Graph ► Probability Distribution Plot ► Two Distributions ► Binomial)

(c) Next, let's fix p at 0.2 and vary n. Use $n = 2, 5, 10, 20$, and 40. For each value of n, display the plots as in part (a). How do the plots change as n increases? How about the spread? The shape? Does the 'middle' move when n increases? (Use Graph ► Probability Distribution Plot ► Vary Parameters ► Binomial)

6-9   Suppose X is a binomial random variable with $n = 16$ and $p = 0.75$.

(a) Use Minitab to calculate the mean of X using the formula $\mu = \Sigma x \Pr(X = x)$. Does the answer agree with the answer you get when you use the formula $\mu = np$?

(b) Use Minitab to calculate the variance of X, using the formula $\sigma^2 = \Sigma[x-\mu]^2 \Pr(X = x)$. Use the value of μ from part (a). Does the answer agree with the formula $\sigma^2 = npq$?

(c) Repeat parts (a) and (b) with $n = 10$ and $p = 0.5$

6-10   In the Pulse experiment (described in the Appendix, p. 529) students were asked to toss a coin. If the coin came up heads, they were asked to run in place. Tails meant they would not run in place. Thirty-five students ran in place and the rest did not. Compare the data by using CDF with the Binomial distribution with parameters 92 and 0.5.

Does it seem very likely that only 35 of 92 students would get heads if they all flipped coins? Can you make a conjecture as to what might have happened?

6-11   Acceptance Sampling. When a company buys a large lot of materials, it usually does not check every single item to see that they all are satisfactory. Instead, some companies pick a sample of items, then check these; if they do not find many defective items, they go ahead and accept the whole lot. In this problem we'll look at the kind of risk they run when they do this. Suppose the inspection plan consists of looking at 40 items chosen at random from a large shipment, then accepting the entire shipment if there are 0 or 1 defective items, and rejecting the shipment if there are 2 or more defective items.

(a) If in the entire shipment 25% of the items are defective, what is the probability the shipment will be accepted?

(b) Compute the probability of acceptance if the shipment has 0.1% defective, 0.5% defective, 1%, 2%, 3%, 5%, and 8% defective. Then sketch a plot of the probability of acceptance versus the percent defective.

(c) About what percent defective leads to a 50–50 chance of acceptance?

(d) Repeat part (b) but use a different plan, where 100 items are checked and the lot is accepted if 0, 1, or 2 are found defective, and rejected if 3 or more are found defective.

(e) Sketch both graphs from (b) and (d) on the same plot. Discuss the advantages and disadvantages of the two different plans.

6-12   (a) Make plots as in Figure 6.19, p. 198, but use $p = 0.4$ instead of $p = 0.5$. Try $n = 20$, 30 and 10. (The macro, BinoNorm, will be useful here.)

(b) Repeat part (a), using $p = 0.2$.

(c) What can you say about the normal approximation to the binomial? For what values of n and p does it seem to work best?

6-13   Suppose X has a binomial distribution with $p = 0.8$ and $n = 25$. Use Minitab to calculate each of the probabilities below exactly. Also compute the normal approximation to these probabilities. Remember to go 0.5 below and above, as appropriate, when computing the normal approximation. (This is often called using the continuity correction.) Compare the binomial results with the normal approximations. (Note that $n(1-p) \geq 10$ is *not* satisfied here.)

(a) $P(X = 21)$.

(b) $P(X \leq 21)$.

(c) $P(X \geq 24)$.

(d) $P(21 \leq X \leq 24)$.

6-14   Use Graph ▸ Probability Distribution Plot to graph the Poisson distribution for the four values of mean μ. Comment on the different shapes displayed.

(a) $\mu = 5$.

(b) $\mu = 0.5$.

(c) $\mu = 15$.

(d) $\mu = 25$.

6-15   A typist makes an average of only one error every two pages, or 0.5 error per page. Suppose these mistakes follow a Poisson distribution. Use the PDF and CDF commands to answer each of the following:

(a) What is the probability the typist will make no errors on the next page?

(b) One error?

(c) Fewer than two errors?

(d) One or more errors?

(e) Sketch a histogram of this distribution (the pdf).

(f) Comment on reasons why the Poisson distribution might or might not be a good approximation for the number of errors made by a typist.

6-16   In high-energy physics the rate at which certain particles are emitted has a Poisson distribution. Suppose, on the average, 15 particles are emitted per second.

(a) What is the probability that exactly 15 will be emitted in the next second?

(b) That 15 or fewer will be emitted?

(c) That 15 or more will be emitted?

(d) Find a number x such that approximately 95% of the time fewer than x particles will be emitted.

(e) Sketch a histogram of this distribution. (Use the Poisson macro.)

6-17   If a Poisson distribution has a large mean μ, its probabilities can be closely approximated by a normal distribution with the same mean μ and with a standard deviation $\sigma = \sqrt{\mu}$. Use the macro PoissonNorm to assess the Normal approximation to the Poisson distribution under each of the following conditions. Comment on the adequacy of the approximation in each case.

(a) $\mu = 5$.
(b) $\mu = 15$.
(c) $\mu = 25$.
(d) $\mu = 50$.

6-18   As stated in Exercise 6-17, if a Poisson distribution has a large mean μ, its probabilities can be closely approximated by a normal distribution with the same mean μ and with a standard deviation $\sigma = \sqrt{\mu}$. For each of the following conditions, compute exact values with the Poisson pdf and approximate values with the normal cdf. Remember to go 0.5 below and above, as appropriate, when computing the normal approximation. Comment on the quality of the approximation.
(a) $P(X \leq 3)$ with $\mu = 5$.
(b) $P(10 \leq X \leq 12)$ with $\mu = 15$.
(c) $P(X \leq 21)$ with $\mu = 25$.
(d) $P(38 \leq X \leq 46)$ with $\mu = 50$.

6-19   In the fall of 1971 testimony was presented before the Atomic Energy Commission that the nuclear reactor used in teaching at Penn State University was causing increased infant mortality in the surrounding town of State College. This reactor had been installed in 1965. The data in Table 6.4 for State College were presented as part of the testimony. (The file name is Reactor). In addition data for Lebanon, Pennsylvania, a city of similar size and rural character, were presented.

Table 6.4 Infant deaths

	State College			Lebanon		
Year	Live births	Infant deaths	Infant deaths per 1000 births	Live births	Infant deaths	Infant deaths per 1000 births
1962	369	4	10.8	666	16	24.0
1963	403	4	9.9	464	15	23.2
1964	365	5	13.7	668	9	13.5
1965	365	6	16.4	582	10	17.2
1966	327	4	12.2	538	8	14.9
1967	385	6	15.6	501	8	16.0
1968	405	10	24.7	439	5	11.4
1969	441	6	13.6	434	3	6.9
1970	452	8	17.7	500	8	16.0

One of the authors of this text (BJ) was asked to examine the evidence in detail and find out whether there was cause for concern. A wide variety of statistical procedures were used, but one important discussion centered on whether there had been an abnormal peak in infant mortality in State College in 1968.

(a) Suppose we assume that infant deaths follow a Poisson distribution. Over the nine-year period presented in the testimony, there were 53 deaths in State College. On the average this is 53/9 = 5.9 per year. Simulate nine observations from a Poisson distribution with $\mu = 5.9$. Repeat this simulation 10 times. Are peaks such as the one in State College in 1968 unusual?

(b) One of the strongest critics of Penn State's reactor drew the following conclusion from the data: "Following the end of atmospheric testing by the US, USSR, and Britain in 1962, infant mortality declined steadily for Lebanon, while it rose sharply for State College. Using 1962 as a reference equal to 100, State College rose to (24.7/10.8)100 = 229 by 1968 while Lebanon declined to (11.4/24.0)100 = 48. Furthermore, not only is there an anomalous rise above the 1962 levels in State College after 1963, but there are two clear peaks of infant mortality rates in 1965 and 1968. Especially the high peak in 1968 has no parallel in Lebanon."

Do you think the data support the writer's conclusions that Penn State's reactor has led to a significant rise in infant mortality? What criticisms of his argument can you make?

6-20   If you have a binomial distribution with a large value of n and a small value of p, its probabilities can be closely approximated by a Poisson distribution with mean equal to np. Note that if p is small then $1-p \approx 1$ and $\sigma = \sqrt{np(1-p)} \approx \sqrt{np} = \sqrt{\mu}$ as in a Poisson distribution. (Graph ▶ Probability Distribution Plot will be helpful for some of these.)

(a) Use $n = 30$ and $p = 0.01$ and compute the corresponding binomial and Poisson probabilities (pdfs). Do they seem to be fairly close?

(b) Plot both pdfs on the same plot. Do the pdfs agree fairly well everywhere?

(c) Compare the binomial and Poisson cdfs. How well do they agree?

(d) Repeat (a) but use $n = 30$ and $p = 0.5$. How good is the approximation now?

(e) Plot both pdfs from part (d) on the same plot. How well do they agree?

(f) Compare the binomial and Poisson cdfs. How well do they agree?

7

Simulation

7.1 Learning about Randomness

Most statistical methods described in this book use data to try to answer questions about populations or processes. The data may come from a specially conducted survey, from a carefully designed experiment, from a large database, or just be happenstance data. In all cases there is variability in data. This could be due to measurement error. Measuring devices (thermometers, IQ tests, scales, and so on) are never perfect. Variability could be due to sampling error (we take a sample from a population, study the sample, then generalize to the population). Variability could come from environmental conditions (a laboratory gradually warms up during the day, ambient noise varies from testing room to testing room, and so on). Variability could be inherent in the process. You toss a coin; sometimes it comes up heads and sometimes it comes up tails.

If you have the time, money, and awareness, you can often reduce variability. You can use very high-quality measuring devices, take a very large carefully selected sample, control environmental conditions, and so on—but not always, and never perfectly. This means you must detect the major features of your data in the presence of variability. An engineer would say you must separate the signal from the noise.

To use statistical methods wisely, you need an appreciation of random variability. Simulation can help. Simulation allows you to study the properties of statistical methods in an environment where variability is introduced in a controlled manner. Simulation is also inexpensive and fast, because data are created by a computer rather than through a real experiment or survey. Simulation does have limits, however. All simulated experiments are artificial and idealized. The better the simulation models the real world, the better will be our understanding of randomness in the real world.

Simulation is a very good learning tool. It is also used by researchers to study real problems. Economists have developed sophisticated computer models to simulate ways in which the economy might behave under different conditions. Meteorologists also have sophisticated computer models, which they use to predict weather patterns. Statisticians often use simulation to estimate theoretical quantities that are too difficult to study directly.

In this chapter we will use simulation to study properties of the distributions discussed in Chapter 6. Minitab can simulate data from all the distributions listed in Chapter 6, p. 204. The Calc ▸ Random Data submenu lists each distribution; select the one you want, supply the parameters, and Minitab will generate the data according to that distribution. First, however, we will look at a very simple random process: Bernoulli trials.

7.2 Bernoulli Trials

There are many cases in which an outcome can be classified into one of two categories: a coin falls either heads or tails; the next child born in a hospital is either a girl or a boy; the next person to walk into a store either buys something or does not; a tomato seed either germinates or does not. These two categories are often labeled, somewhat arbitrarily, as 'success' (coin falls heads, baby is a girl, and so on) and 'failure' (coin falls tails, baby is a boy, and so on).

Now suppose we imagine a sequence of, say, 20 trials: We toss a coin 20 times, or we observe the next 20 people who walk into a store. If the chance of a success remains the same from trial to trial, and the outcomes of the trials are independent of one another, we have **Bernoulli trials**. We use the letter p to represent the probability of a success on a single trial. The probability of a failure is then $1 - p$.

One way to learn about Bernoulli trials is to toss a coin 20 times, or to observe the next 20 people who walk into a store. Unfortunately even studies as simple as these can be time-consuming. So instead of actually doing a study, we can use the computer to simulate the results. Minitab's Random Data command will simulate data from many different distributions.

Let's look at the gambling game of roulette. Suppose we spin a roulette wheel that has 38 different stopping points. If it stops on any of 18 specific points, we win. Otherwise we lose. Thus the probability of a success is 18/38 = 0.4737. Suppose that on Monday evening we play this game for 15 spins. What is likely to happen? Suppose we play for another 15 spins on Tuesday evening, and for another 15 spins on Wednesday. Try simulating 15 spins of a roulette wheel, where 1 represents a win and 0 represents a loss.

Calc ▸ Random Data ▸ Bernoulli simulates Bernoulli trials, using the value you specify as the probability of a success on each trial. Each success is assigned the value 1, and each failure is assigned the value 0. To simulate 15 spins of roulette over a period of three days,

1. Restart Minitab to set all dialog boxes to their default values.
2. Choose **Calc ▸ Random Data ▸ Bernoulli**.
3. Type **15** in the Generate box.
4. Type **Monday Tuesday Wednesday** in the Store in column(s) box.

5. Enter **0.4737** in the Probability of success box. See Figure 7.1.

6. Click **OK**. Minitab generates three columns of data

Figure 7.1
Simulating three days of roulette

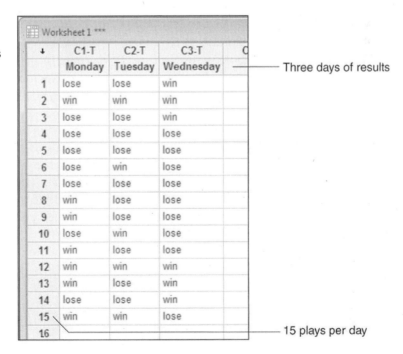

Figure 7.2 shows the results after we coded the 0's as **lose** and 1's as **win**.

Figure 7.2
Simulated roulette wins and losses

↓	C1-T	C2-T	C3-T	
	Monday	Tuesday	Wednesday	— Three days of results
1	lose	lose	win	
2	win	win	win	
3	lose	lose	win	
4	lose	lose	lose	
5	lose	lose	lose	
6	lose	win	lose	
7	lose	lose	lose	
8	win	lose	lose	
9	win	lose	lose	
10	lose	win	lose	
11	win	lose	lose	
12	win	win	win	
13	win	lose	win	
14	lose	lose	win	
15	win	win	lose	
16				— 15 plays per day

Worksheet 1 ***

Now choose **Stat ▸ Tables ▸ Tally**, enter **Monday-Wednesday** in the Variables box, and then click **OK**. Minitab tallies your results; yours will be different than the ones shown here because they are randomly generated.

Figure 7.3
Tally of simulated roulette results

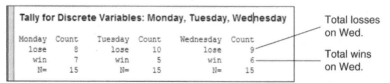

Figure 7.3 summarizes the results for simulating three evenings of gambling using Bernoulli trials with $p = 0.4737$. How did we do? On Monday: 8 wins and 7 losses. We did lousy on Tuesday: only 5 wins. Wednesday was the best: 9 wins and 6 losses. On all three days, the same underlying process generated the results, but we did not get the same results each day. That is because there is inherent variability in this process. Overall we won 18 out of 45 plays or 40%.

Random Sequences

It is important to understand that randomness has to do with the process that generates observations, not with the observations themselves. However, even if the data come from a random process, this still does not guarantee that a particular set of outcomes will look random. It is possible to spin the roulette wheel 15 times and win each time, or lose each time, or get a sequence that alternates win, lose, win, lose, win, lose, and so on. Such clearly patterned sequences are possible. They are not very likely, however, if the process that generated the observations is truly random. Using the methods of Chapter 6 and the Binomial Distribution we find the chance of 15 wins out of 15 spins is 0.0000136. As we said—not very likely!

Now suppose we collect some real data. How can we know whether we have a random sequence? The best situation is where we (or the investigator) created the randomness using randomization methods to select the survey respondents or the treatments applied. Otherwise, we can never know for certain. The best we can do is to understand how the data arose and how they were collected, and then try to determine whether a random process is a good model for the data generation process.

7.3 Simulating Data from a Normal Distribution

Let's see how data can vary when you take a sample from a normal population. We will use a simple example. For the population of males between 18 and 74 years old inclusive, in the United States, systolic blood pressure is approximately normally distributed with mean 129 millimeters of mercury and standard deviation 19.8 mmHg.[†]

Suppose we were to take a random sample of 25 men and measure their systolic blood pressure. Would the sample look normal? Would its mean be close to 129? Would you expect any observation to be over 190? (That's more than three standard deviations above the population mean.) We do not have the time or money to take a random sample of 25 men, so we will use simulation.

The command Calc ▸ Random Data ▸ Normal puts a random sample of the number of observations you specify into each column. You can specify μ and σ, but if you don't, Minitab assumes a standard normal distribution with μ = 0 and σ = 1. To simulate a random sample of 25 men,

1. Start a new worksheet.
2. Choose **Calc ▸ Random Data ▸ Normal**.
3. Generate **25** rows of data.
4. Store the results in columns named **Sample1 Sample2 Sample3**.
5. Enter a mean of **129** and standard deviation of **19.8**. See Figure 7.4.
6. Click **OK**. Minitab generates the data you requested.

Figure 7.4
Simulating data from a normal distribution

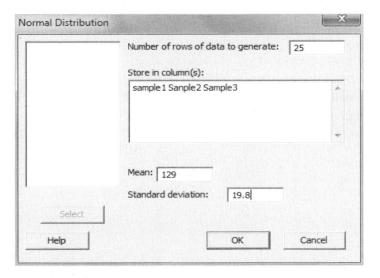

[†] M. Pagano and K. Gruvreau, *Principles of Biostatistics*. North Scituate, Mass.: Duxbury Press, 1993, p. 163.

The three samples could be displayed and compared in many ways. We'll display three dotplots on the same graph so you can compare the three graphs more easily.

1. Choose **Graph ► Dotplot ► Multiple Y's, Simple**, click **OK**, and enter **Sample1-Sample3** in the Variables box.
2. Click **Multiple Graphs** and then select **in separate panels of the same graph**.
3. On that same dialog box also select **Same X, including same bins**.
4. Click **OK**. Figure 7.5 displays the three dotplots.

Figure 7.5
Distributions of three samples

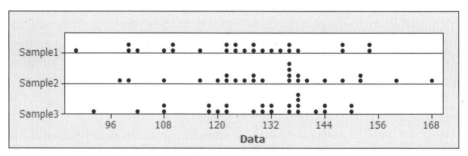

N

otice that the results for the three samples differ. Each time we take a sample, we get different "men" with different blood pressures. As a result, the distributions of the samples vary in shape; the sample means and any other statistics vary in value. This variation is called **sampling variation**.

Now let's look at some sample statistics in the three samples. Choose **Stat ► Basic Statistics ► Display Descriptive Statistics**. Enter **Sample1-Sample3** in the Variables box, and click **OK**. Figure 7.6 shows the descriptive statistics in the Session Window.

Figure 7.6   Descriptive statistics of our three samples

Descriptive Statistics: Sample1, Sample2, Sample3

Variable	N	N*	Mean	SE Mean	StDev	Minimum	Q1	Median	Q3
Sample1	25	0	124.55	3.49	17.44	87.15	110.10	125.00	135.33
Sample2	25	0	132.13	3.41	17.05	98.03	121.62	135.01	141.24
Sample3	25	0	128.61	2.95	14.75	92.45	118.96	131.26	138.13

Variable	Maximum
Sample1	154.25
Sample2	168.35
Sample3	150.41

Look at the sample means. The first is 124.55, a little more than 5 below the population mean of 129, the second is 132.13, a little more than 3 units above the population mean of 129. The third is 128.61—just slightly below the population mean. The difference between a sample mean and the population mean is called **sampling error**. Notice that here error does not mean we made a mistake—sampling error is inherent when we collect data in the real world and it is important for us to understand, measure, and account for it.

Suppose we draw samples from the population of blood pressures again, but now use a sample size of $n = 100$. Histograms for the three samples which we obtained are shown in Figure 7.7.

Figure 7.7
Distributions of three samples each with $n = 100$

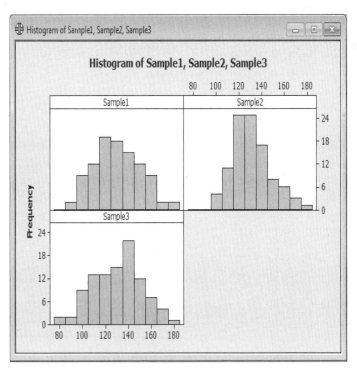

When n is larger, a random sample will, in general, look more like the population from which it was drawn; these three certainly do. The histograms are all approximately bell-shaped. There is still variability from sample to sample, but less than in Figure 7.5. If we were to use a still larger sample size, say $n = 500$ or $n = 1000$, then, in general, the histograms would look even more bell-shaped, and the sample means would be even closer to the population mean.

A Useful Technique for Simulation Studies

Suppose we wanted 1000 random samples, each containing 25 observations. There are two ways to do this in Minitab: generate 1000 columns of 25 rows each or generate 25 columns of 1000 rows each and consider each *row* as a sample. When you generate 1000 columns, you consider each column to be a separate sample. When you generate 1000 rows, you consider each row to be a separate sample. For large simulation studies it is more convenient to have samples in rows since we will want to compute various statistics from the samples and store them in columns to investigate how the statistics vary from sample to sample, that is, to investigate the sampling variation.

As an example suppose we simulate samples, each of size 25, from a normal distribution with $\mu = 129$ and $\sigma = 19.8$, and consider the largest value in each sample of 25. The distribution of the largest value in a random sample from a normal population is a difficult theoretical problem. With simulation we can get a very good idea what this distribution is. In principle, we would like to generate infinitely many samples each of size 25. We will make do with 1000 samples.

To simulate 1000 normal samples,

1. Choose **Calc ▸ Random Data ▸ Normal** and generate **1000** rows of data, storing the results in **C1-C25**, using the mean of **129** and standard deviation of **19.8**. Click **OK**.

2. Choose **Calc ▸ Row Statistics**, click the **Maximum** button, enter **C1-C25** as the input variables, and store the results in a column named **Max**. Click **OK**. See Figure 7.8. (Minitab will name the first unused column Max and put the maximum value of the 25 values in each row of columns C1-C25 into the corresponding row of the column named Max.)

Figure 7.8
Row statistics
dialog box

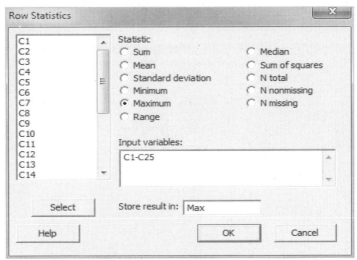

We can now analyze the data in the column Max anyway we wish. For example: Choose **Graph ▸ Histogram ▸ Simple** and enter **Max** as the variable. Click **OK**. Minitab produces the graph shown in Figure 7.9.

If you repeat this simulation your histogram will look similar but not identical since each simulation produces a new set of data. If you generate a large number of samples of size 25 as we did, your histogram should have the same general shape, center, and spread as shown in Figure 7.9.

Notice that the distribution is skewed toward higher values and that most of the values are 144 or higher. In random samples of size 25 from a normal population with mean 129 and standard deviation 19.8 it is almost a sure thing to observe one or more values of 144 or higher. Values around 160 would be quite common.[†]

[†] The *theory* concerning the distribution of the maximum of a random sample selected from a normal population is quite complicated. We will not pursue it here. Fortunately, simulations can give us some insight into difficult theoretical issues.

Figure 7.9
Distribution of
Maximum in
random
samples of
size 25 from a
normal
population
with $\mu = 129$
and $\sigma = 19.8$

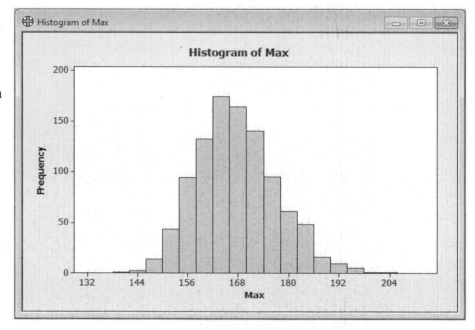

7.4   Simulating Data from the Binomial Distribution

In Chapter 6 we discussed an example concerning light bulbs. Each package contains four bulbs; the manufacturer claims that 85% of its bulbs will last at least 700 hours. Suppose we were to keep track of the next 50 packages sold. For each package we record the number of bulbs that last at least 700 hours. What might happen? Use Graph ► Time Series Plot to display the results over time. To simulate binomial data,

1. Choose **Calc ► Random Data ► Binomial**.

2. Generate **50** rows of data in **Successes**, using **4** trials with a **0.85** probability of success. See Figure 7.10.

3. Click **OK**. Minitab generates the data and places it into C1. Now create a time series plot of the data.

4. Choose **Graph ► Time Series Plot** and select C1 as the Y variable. Click **OK**. Minitab displays the graph shown in Figure 7.11.

Figure 7.10
Binomial
random data
dialog box

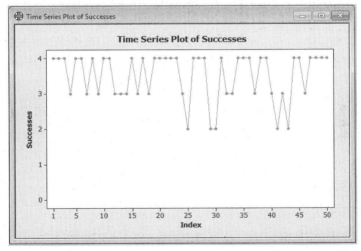

Figure 7.11
Time series
plot of
binomial data

In the first three packages all four bulbs lasted over 700 hours. In the third package only 3 bulbs did. In the twenty fifth package, only two bulbs lasted over 700 hours. In the last four batches all four lasted over 700 hours. This simulation gives you some idea of how a sequence of 50 binomial variables with $n = 4$ and $p = 0.85$ might look. Someone might try to argue that the second half of the packages seem to show a trend to shorter lifetimes. However, we know that the simulation process is the same for every package and has no memory. The number of successes in the 25 packages were generated independently. To develop your appreciation of randomness, you should repeat this same simulation several times and notice the similarities and differences among the results.

Let's use the intuition about randomness that you are developing to look at an example with data. Suppose you check the next 50 packages of 4 light bulbs from this manufacturer and get the data shown in Figure 7.12.

Figure 7.12
Time series plot of successes in 50 packages of 4 light bulbs

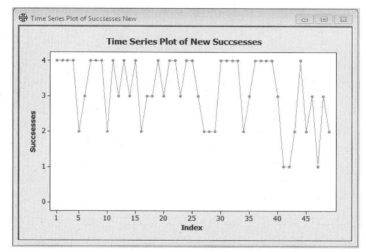

What do you see? Perhaps something has changed around package 40. The numbers seem to be smaller, on the average, than they were for the first 39 packages. The theoretical mean of a binomial variable with $n = 4$ and $p = 0.85$ is $\mu = np = 4(0.85) = 3.4$. Notice that 9 of the last 10 observations are all below the mean. Do you think these observations came from a binomial with $n = 4$ and $p = 0.85$? It looks as if something may have changed in the manufacturing process, something that decreased the lifetimes of the bulbs.

7.5   Sampling from Other Distributions

Minitab has the capability to select random samples from many theoretical distributions—all of the distributions listed in Section 6.5, p. 204. All of these work quite similarly to the normal and binomial cases that we illustrated in the last two sections. You should have no trouble sampling from any of these distributions.

7.6   Sampling Actual Populations

So far in this chapter we have talked about random samples from theoretical distributions. In doing so we assume, artificially, that the populations we sample from are infinite. However, sample surveys, for example, are taken from an actual (finite) population. Random samples from actual populations are usually taken without replacement. (Unless we are trying to assess the magnitude of measurement error, we get no new information if we elicit data from the same person or thing twice!) If the population size is large compared to the sample size we usually use the simpler model that assumes outcomes are independent. The limited dependence does not affect most answers very much. However, if the sample size is more than about 10% of the population size, much of our statistical theory changes. In particular, the standard error of the mean is *not* the usual formula, $\sigma/\sqrt{n}$. Rather, if we take a simple random sample of size n without replacement from a population of size N, the standard error of the mean is given by

$$SE_{\bar{x}} = \sigma\sqrt{\frac{N-n}{N-1}}$$

The factor $\sqrt{(N-n)/(N-1)}$ is often called the **finite population correction** factor or *fpc*. It is easy to see that the *fpc* will be close to 1 if N is much larger than n. In that case the usual formula, $\sigma/\sqrt{n}$, works quite well as an approximation to the correct standard error. Let's consider an example.

A new residential subdivision contains 100 individual properties which range in size from one-quarter acre up to four and a half acres. Table 7.1 gives a frequency table of the acreages of the 100 properties where the basic unit of area is one-quarter acre. That is, there are four properties of one-quarter acre, 11 of two-quarters acre, and so on. To illustrate random sampling from a column without replacement we will assume that a property assessor does not have Table 7.1 available but wishes to estimate the total acreage of the 100 properties. Her resources (time and money) are limited so she must base her estimate on measurements from a random sample of just 5 properties.

Table 7.1 Acreage of 100 properties

Area	1	2	3	4	5	6	7	8	9	10	12	13	14	15	16	17	18
Frequency	4	11	17	19	13	10	7	5	4	2	2	1	1	1	1	1	1

The data are arranged in two columns of a Minitab Worksheet as shown, in part, in Figure 7.13. (These data are available from the Data Library at www.CengageBrain.com associated with this book. The dataset name is Areas.mtw.)

To select our random sample without replacement:

1. Choose **Calc ▸ Random Data ▸ Sample From Columns**.

Figure 7.13
Areas
Worksheet
(Areas are in
quarter-acre
units)

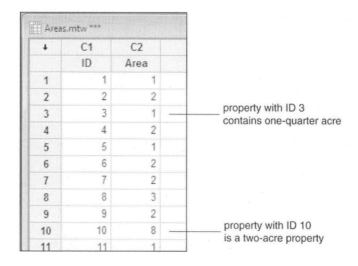

property with ID 3
contains one-quarter acre

property with ID 10
is a two-acre property

2. Sample **5** rows from **C1-C2** (or **ID** and **Area**) and store samples in new columns named '**ID Selected**' and '**Sample Area**'.[†] See Figure 7.14.

3. Click **OK**.

Figure 7.14
Sample from
columns
dialog box

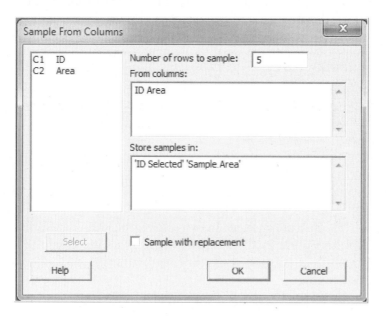

_____

[†] Remember that if column names contain spaces they must be enclosed in single or double quotation marks when they are entered into dialog boxes. If the column names are already in use, Minitab will supply the quotation marks when you select the columns.

Minitab randomly samples 5 pairs from columns C1 and C2 and places them into C3 and C4 named ID Selected and Sample Area. Our results are shown in Figure 7.15. Of course, yours will be different. We selected properties with IDs 8, 10, 36, 39, and 65 and areas 3, 8, 5, 6, and 8 for an average area of 6.0 quarter-acres. Based on this sample we would estimate the total acreage as 100×6 = 600 quarter-acres or, finally, 150 acres total.

Figure 7.15
The results of our random sample of size 5

Areas.mtw ***					
↓	C1	C2	C3	C4	C5
	ID	Area	ID Selected	Sample Area	
1	1	1	8	3	
2	2	2	10	8	
3	3	1	36	5	
4	4	2	39	6	
5	5	1	65	8	
6	6	2			
7	7	2			
8	8	3			

In this artificial example we can see how well we did with the estimation. There are, in fact, 540 quarter-acre units in the population or 135 total acres. Again the difference between the population value of 135 and the sample estimate of 150 is called sampling error.

To illustrate the possibilities, we repeated this sampling process 10,000 times. (We used a macro, AreaSampler, available at the Data Library at www.CengageBrain.com associated with this book.) Figure 7.16 shows the sampling distribution of the totals that we obtained. The mean of this distribution is 134.78 and the standard deviation is 37.60. The closeness of our result to the population mean of 135 illustrates the fact that random sampling produces an unbiased estimate of the actual total.

Figure 7.16
Histogram of totals from simulated samples of size 5

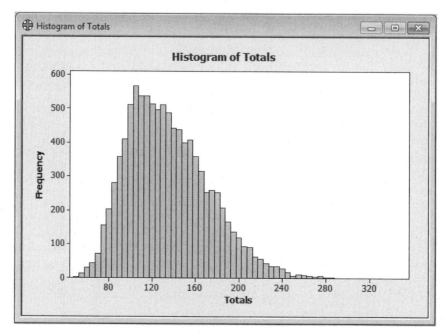

We repeated the sampling process one more time. This time we selected samples of size 20. Again we did 10,000 samples. This time we got a mean of 135.13 and a standard deviation of 17.48. Notice how the mean remained essentially the same while the standard deviation was reduced considerably. This is, of course, due to our increased sample size. Figure 7.17 displays the two histograms for $n = 5$ and $n = 20$ side by side. Both histograms use the same x and y scales to make a fair comparison. Note that for the histogram on the right side, the scale for the data axis is at the top. With the larger sample size the distribution is much less spread out and is starting to look more like a normal distribution. We pursue this further in Section 7.8, p. 231

In the real world we will not have the areas for all the properties. (If we did, there would be no reason to sample!) Instead we would have just the ID numbers, would select our sample of IDs and then go measure the areas for those (and only those) properties to get our sample of acreages.

Figure 7.17
Histogram of
totals from
simulated
samples of
size 5 and 20

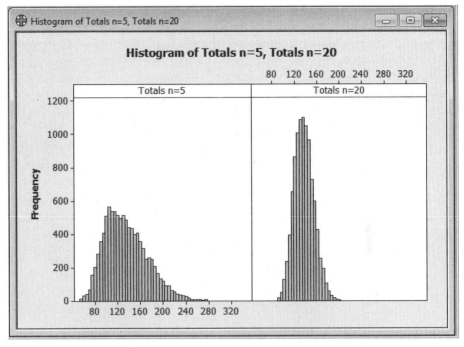

7.7   The Base for the Random Number Generator

To simulate random data, Minitab uses an algorithm that can create a very long list of numbers, which appear to be random. When you use a Random Data command, Minitab haphazardly chooses a place to start 'reading' from this list. Each time you use a Random Data command, Minitab chooses a different place to start, and therefore you get a different set of simulated data.

Occasionally you may want to generate the same set of 'random' data more than once. You can use the Calc ▸ Set Base command to generate the same set of random data at different times. The value you enter tells Minitab where to start 'reading' in its list of random numbers. (Note: The value you enter determines where in the list to start. It is not the first random number Minitab gives you.) If you use two Random Data commands in the same session (without using Set Base), the second Random Data command will continue 'reading' the list of random numbers where the first one stopped.

7.8   The Central Limit Theorem

Simulation is an excellent technique for investigating the approximation provided by the Central Limit Theorem. The Central Limit Theorem states that if $X_1, X_2, \ldots, X_n$ represents a random sample of size n from a population with mean μ and standard deviation σ, then, as n goes to infinity, the distribution of

$\sqrt{n}((\bar{X}-\mu)/\sigma)$ approaches that of a standard normal distribution. We then say that, for large n, we can approximate the distribution of $\bar{X}$ using a normal distribution with mean μ and standard deviation $\sigma/\sqrt{n}$.[†] Alternatively, we can approximate the distribution of the sample sum ΣX_i using a normal distribution with mean $n\mu$ and standard deviation $\sqrt{n}\sigma$. For example, in Section 6.3, p. 196, we saw how the normal distribution could approximate the exact binomial distribution. With Minitab's simulation and theoretical capabilities we can investigate the normal approximation in other situations.

Here is an example. Theory says that the sum of independent chi-square variables has a chi-square distribution and the degrees of freedom add. Suppose $X_1, X_2,..., X_{10}$ represents a random sample of size 10 from a chi-square distributed population with 8 degrees of freedom. Then the exact distribution of the sample sum, $Y = \Sigma X_i$, is a chi-square distribution with 10×8 = 80 degrees of freedom. To simulate the distribution of Y,

1. Choose **Calc ▸ Random Data ▸ Chi-square**.
2. Enter **10000** rows of data. (You might want to 'test' this with 1000 depending on the speed of your computer.)
3. Enter **C1-C10** for Store in column(s).
4. Enter **8** for Degrees of Freedom and click **OK**.

Ten thousand samples each of size 10 are generated and placed in rows 1 through 1000 of C1, C2,..., C10. To 'see' the population (that is, chi-square with 8 degrees of freedom),

5. Choose **Data ▸ Stack ▸ Columns**.
6. Enter **C1-C10** for **Stack the following columns:**
7. Click the radio button for **Column of current worksheet:** and enter the column name **Data** in the text box provided. Now all 100,000 individual data points are in the column named Data.
8. Choose **Graph ▸ Histogram ▸ With Fit** and enter **Data** as the graph variable.
9. Click the **Data View** button and select the **Distribution** tab.
10. Select **Gamma** from the pull-down menu and enter **4** for the **Shape parameter** and **2** for the **Scale parameter**. (Chi-square is a special gamma distribution)

[†] If the sampling is done *without replacement*, the finite population correction, *fpc*, is needed to "correct" the standard error of the sample mean. See Section 7.6, p. 226.

11. Click **OK** twice and the histogram is displayed with the theoretical Chi-squared distribution overlaid as seen in Figure 7.18.

Figure 7.18

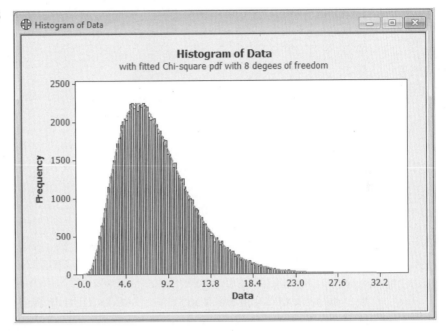

This histogram is so close to a chi-square distribution with 8 degrees of freedom that the theoretical Chi-square curve can barely be seen. If you maximize the plot on your computer screen, you might be able to see the difference more clearly.

Now to illustrate the Central Limit Theorem we need to calculate the sum of the 10 values in each of our 10,000 samples and see how the sum is distributed. To get the sum,

1. Choose **Calc ▸ Row Statistics** to open the dialog box.
2. Press the button for **Sum** as the Statistic desired.
3. Choose **C1-C10** as the input variables.
4. Enter **Sum** for the name of the column to **Store the result in**.
5. Click **OK** to get the 10,000 sums calculated and stored.

Figure 7.19 displays the histogram of the Sum variable. Yours should be similar but not identical.

Figure 7.19
Simulated
distribution of
the sum of 10
independent
chi-square
variables,
each with 8
degrees of
freedom

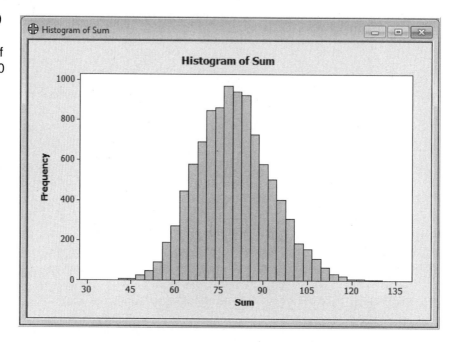

6. Now right-click the graph and choose **Add ▸ Distribution Fit**.
7. Select **Normal** distribution, enter **80** for the mean and **12.65** $(=\sqrt{10 \times 8})$ for the standard deviation and click **OK**.

Figure 7.20
Simulated
distribution of
the sum of 10
independent
chi-square
variables with
approximating
normal curve

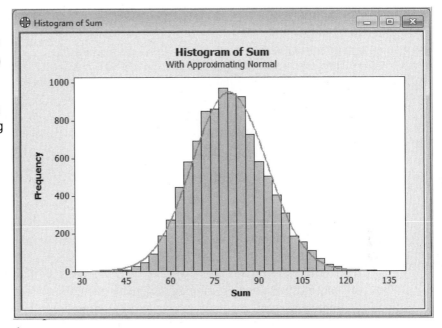

h

e appropriate normal curve is displayed on top of your histogram. Our results
are shown in Figure 7.20. The fit is not perfect but could be quite usable.
Remember that our "infinite" sample size is really only $n = 10$ and the distribution that we sampled from is skewed. We pursue this further in the exercises.
Once more you could maximize the plot on your computer screen to see the
difference more clearly.

Exercises

7-1   Suppose a basketball team wins its games, on the average, 55% of the time.
Suppose we make the simplifying assumptions that there is a 55% chance of
winning each game and that these odds are not influenced by what the team has
done so far, that is, the outcomes for the games are independent of one another.
Then we can simulate a 'season' for the team, using Minitab. Suppose a season
consists of 30 games. Simulate one season.

(a) Did the team have a winning season (win more games than it lost)?

(b) How long was the team's longest winning streak? How long was its longest
losing streak?

(c) Simulate four more seasons and answer the questions in (a) and (b) for
each season. Roughly how much variation should you expect from season
to season in this team's performance, even when its basic winning
capability is unchanged?

7-2   Suppose someone gives you a coin that is loaded so that heads comes up more often than tails. You want to know exactly how loaded the coin is. To estimate the probability of a head, p, you could toss the coin, say, 100 times.

(a) How could you simulate this, assuming that p is really 0.6? How could you estimate p from the data? Will your estimate be exactly equal to p?

(b) Suppose you asked five friends to estimate the probability of a head by having each toss the coin 100 times. Will their estimates all be equal to p? Will they all get the same estimate? How can you simulate the results for all five friends?

(c) Perform the simulation for yourself and your five friends. How much do the six estimates vary?

7-3   The World Series in baseball pits the winner of the American League against the winner of the National League. These two teams play until one team wins a total of four games.

(a) Pretend that the two teams are perfectly matched so that there is a 50–50 chance of each team's winning any given game. Further suppose the games are independent; that is, the chance of winning a particular game does not depend on whether other games have been won or lost. Simulate 20 World Series by using Calc ▸ Random ▸ Bernoulli. Place the data in C1-C7 and use 0.5 as the probability of success. Now pretend a 1 means the National League won that game and 0 means the American League won. For each series figure out which team won the series (that is, won four games first). In how many series did the American League win? The National League? How many series lasted only four games? How many lasted five games? Six games? Seven games?

(b) Repeat (a), but now pretend the National League had a probability of $p = 0.6$ of winning each game. How do these results compare with those in (a)?

7-4   Consider a simple gambling game. You toss a fair coin repeatedly. For each head you win $1; for each tail you lose $1. Use Minitab to simulate this game for 50 tosses, using 1 for a win and 0 for a loss. Convert the data to winnings, using +1 for win $1 and −1 for lose $1, with the Calculator expression $C2 = 2*C1-1$. Compute the cumulative winnings in C3 with the Calculator function Partial sums (use online Help if you want more information on this function). For example the number in row 10 is the amount of money you have after 10 tosses. (Note: This process is an example of what is called a simple random walk.) Name C3 'Winnings' and then display a time series plot of C3.

(a) Run this simulation. How did you do?

(b) How often were you ahead? Were you ahead at the end of 50 tosses?

(c) What was your longest winning streak? Losing streak?

(d) Run the simulation four more times. How do the results compare?

7-5   In samples from the normal distribution about 68% of the observations should fall between $\bar{x} - s$ and $\bar{x} + s$. About 95% should fall between $\bar{x} - 2s$ and $\bar{x} + 2s$. Simulate a random sample of size 100 from a normal distribution. What percentages do you find in these two regions? (You choose μ and σ.)

7-6   Repeat Exercise 7-5, using real data. For example, use the SAT math scores from sample A of the Grades data (described in the Appendix, p. 527).

7-7   In this exercise we look at the distribution of a linear combination of normal variables. Simulate 200 observations from a normal distribution with $\mu = 20$ and $\sigma = 5$ into C1. Simulate another 200 with $\mu = 7$ and $\sigma = 1$ into C2. Then compute the following (you can do this by using either the Calculator or Session commands; Session commands might be quicker):

```
LET C3=3*C1
LET C4=2*C2
LET C5=3*C1+10
LET C6=2*C2+5
LET C7=C1+C2
LET C8=3*C1+2*C2+3
```

Now display descriptive statistics for C1-C8 and histograms of C1-C8. Try to give formulas indicating how the means and standard deviations of the various columns are related. Use the descriptive statistics you generated to support your formulas. What sort of distributions do the various columns seem to have?

7-8   In this exercise we look at what is called a **mixture** of two normal populations.
(a) The heights of women are approximately normally distributed with $\mu = 64$ inches and $\sigma = 3$ inches. The heights of men are approximately normal distributed with $\mu = 69$ and $\sigma = 3$. Suppose you took a sample of 200 people—100 men and 100 women—and drew a histogram of the 200 heights. Do you think it would look normal? Should it look normal? Use Minitab to simulate this process. Simulate 100 men's heights and put them into C1. Simulate 100 women's heights and put them into C2. Then use the Data ▶ Stack command to put the combined sample into C3. Produce a histogram of the data in C3. Does it look normal?
(b) Part (a) is an example of a mixture of two populations. Let's try a slightly more extreme example. Take a sample of 200 observations: 100 of them from a normal population with $\mu = 5$ and $\sigma = 1$, and the other 100 from a normal population with $\mu = 10$ and $\sigma = 1$. Get a histogram of the 200 observations. Does this histogram seem to indicate that there might be two populations mixed together in the sample?

(c) In general, how is the setup in part (a) 'more extreme' than that in part (b)?

(d) Repeat part (a) but using samples of size 500 for both men and women. Is it easier to 'see' the mixture now?

7-9   Exercise 7-4 contained an example of a random walk in which each "step" was +1 or −1. Now we will look at an example in which the size of each step is selected from a normal distribution. A process such as this is often used to model the way the stock prices seem to behave. Generate 50 rows of normal data in C1 with mean 0 and standard deviation 2. Compute partial sums of the data in C1 and store it in C3. Name C3 'Price' and generate a time series plot of C3. Run this simulation five times. Compare the various patterns.

7-10   (a) Imagine flipping six fair coins and recording the number of heads obtained. Simulate this process 100 times. Use the method described in the material on p. 222. Get a histogram of your results.

(b) How often did all six coins come up heads? How often would you expect this event to occur in the 100 simulations? (PDF can help you answer this.)

(c) How often did you get more heads than tails? How often would you expect this event to occur in the 100 simulations? (PDF can help you answer this.)

(d) Does your histogram look symmetric (more or less)? Would you expect it to?

(e) Convert your observed frequencies of 0, 1, 2,..., 6 heads to relative frequencies. Compare these relative frequencies with the theoretical probabilities from the PDF command.

(f) Repeat (a) through (e), but assume the coin is biased so that the probability of a head on an individual toss is 0.9.

7-11   (a) Use **Calc ▸ Random Data ▸ Poisson** to simulate 52 weekends of accidents where the mean number of accidents per weekend is 6.

(b) In how many weekends did you get 12 or more accidents? What is the probability of getting 12 or more accidents? How many times would you expect to get 12 or more accidents in one year?

(c) What was the mode—that is, the number of accidents that occurred most often? The Tally command can help you.

(d) Was there ever a weekend that was accident-free? Would you expect that this might happen within a year?

(e) What was the total number of accidents for the year?

7-12   The Pulse experiment was run in one class. There were 92 students in the class.

(a) Open the Pulse worksheet and use the Sample From Columns to take a random sample of 20 students from this class without replacement.

(b) What proportion of the students in your sample are women? How does this compare with the full class?

(c) What is the average height of the students in your sample? How does this compare with the full class?

(d) Take another sample of 20 students and answer (b) and (c) for this sample.

7-13 The table below lists a (small) population in alphabetical order.

Name	Jamal	Jeffrey	Jennifer	Jill	Jon	Jose	Joseph	Judy

Enter these names into a column of a Minitab Worksheet.

(a) Use Minitab to select a simple random sample of size 3. Who got selected?

(b) Use the proportion of females in your sample to estimate the proportion of females in the population. How large is your sampling error?

(c) Now use Minitab to select a simple random sample of size 7. Who got selected this time?

(d) Use the proportion of females in your sample in part (c) to estimate the proportion of females in the population. How large is your sampling error now?

(e) What would be your sampling error if you were to select a simple random sample of size 8? (Think about it!)

7-14 A small population consists of 6 elements. The values associated with these elements are: 5, 8, 8, 11, 14, and 17. Consider taking random samples of size 3 from this population (without replacement). (The dataset file AllSamples might be helpful.)

(a) Calculate the mean and standard deviation for the population.

(b) List the 20 different samples that you might get.

(c) Calculate the means for each of these samples.

(d) Display a dotplot of the 20 means. (This is the sampling distribution of the mean.)

(e) Find the mean and standard deviation of the 20 means, that is, the mean and standard deviation of the sampling distribution of the mean.

(f) How do the mean and standard deviation found in part (e) compare with those found in part (a)? Do they agree with theory?

7-15 Exercise 7-14 continued.

(a) Calculate the standard deviations for the 20 samples.

(b) Display a dotplot of the 20 standard deviations found in part (a). This is the sampling distribution of the standard deviation.

(c) Calculate the mean of the 20 standard deviations found in part (a) and compare it to the population standard deviation. Is the (sample) standard deviation an unbiased estimate of the population standard deviation?

(d) Calculate the variances of the 20 samples.

(e) Calculate the mean of the 20 variances found in part (a) and compare it to the population variance. Is the (sample) variance an unbiased estimate of the population variance?

7-16   Load the Areas dataset into Minitab. You will use Minitab to select your own samples of the IDs and areas as we did in Section 7.6, p. 226.

(a) Use Calc ➤ Random Data ➤ Sample From Columns to select a sample of size 5 from the IDs and Areas columns. You will get different ones than we did!

(b) Calculate the mean of the 5 areas that you got. Compare them to our answers.

(c) Compare the mean of the 5 areas that you got to the actual value in the population of 5.4.

(d) Now select a random sample of size 20. Compare the mean of the 20 areas that you got to our values and to the actual value in the population.

7-17   Load the Areas dataset into Minitab. You will use the Minitab macro AreaSampler to select many samples of the IDs and areas.

(a) Use the AreaSampler macro to select 1000 samples of size 5 from the IDs and Areas columns. (You might want to test first by selecting 100 samples to see how long 1000 might take.)

(b) Display a histogram of the 1000 mean areas from the samples selected.

(c) Calculate the mean of the 1000 mean areas that you got. Compare them to our answers.

(d) Compare the mean that you got to the actual value in the population of 5.4.

(e) Now select 1000 random samples of size 20 and repeat parts (b), (c) and (d).

(f) Finally, select 1000 random samples of size 50 and repeat parts (b), (c) and (d).

7-18   *Discrete Uniform Distribution.* Suppose there are 30 people at a party. Do you think it is very likely that at least two of the 30 people have the same birthday? Try estimating this probability by a simulation. To simplify things, ignore leap years (assume all years have 365 days) and assume all days of the year are equally likely to be birthdays. Simulate 10 sets of 30 birthdays. Use the Calc ➤ Random Data ➤ Integer command to generate 30 observations into C1-C10 of integers 1-365. (This will sample randomly *with* replacement.) Then Tally C1-C10. How many of your 10 sets of 30 people had no matching birthdays?

7-19   *Continuous Uniform Distribution.* The uniform distribution and the *t*-distribution with 2 degrees of freedom are examples of very nonnormal distributions. Their histograms should look quite different from those for normal data sets.
(a) Simulate 1000 observations, using the Calc ▶ Random Data ▶ Uniform command and make a histogram of the resulting data. Comment on the general shape of the histogram.
(b) Do (a), but this time simulate data from a *t*-distribution, using Calc ▶ Random Data ▶ *t* with 2 degrees of freedom.

7-20   In Section 7.8, p. 231, we simulated the distribution of the sum 10 chi-squared variables, each with 8 degrees of freedom, and compared it visually to a corresponding normal distribution. Theory says that the distribution of the sum, *Y*, is a chi-square distribution with 10×8 = 80 degrees of freedom. Here you will compare the distributions numerically. You may want to refer to the material in Chapter 6, p. 194, on calculating with the cumulative and inverse cumulative normal distribution. Calculations with the chi-square distribution follow similarly. In all cases we compare answers obtained from a chi-square distribution with 80 degrees of freedom with those from a normal distribution with mean 80 and standard deviation $\sqrt{2 \times 80} \approx 12.649$. Find the exact and approximating values for
(a) $P(Y < 80)$.
(b) $P(Y < 65)$
(c) $P(Y < 50)$
(d) $P(Y > 110)$
(e) $P(Y > 120)$
(f) $P(65 < Y < 110)$
(g) $P(50 < Y < 120)$
(h) Find the 5th and 95th percentiles of both the exact and approximate distributions. How do they compare?

7-21   Imagine rolling 6 pairs of dice simultaneously. This is like choosing 12 digits independently each equally likely to be 1 through 6. What is the distribution of the total spots on the 6 pairs of dice? Evaluating this distribution exactly is virtually impossible but you can do it with simulation.
(a) Use Calc ▶ Random Data ▶ Integer to simulate 1000 samples each of 12 dice rolls. Place the samples in 12 columns and 1000 rows.
(b) Stack all the 12,000 values in one column and tally the results selecting percents to be displayed. How do your results compare to theory?
(c) Use Calc ▶ Row Statistics to obtain your simulated dice totals in a column named Total.

(d) Display the distribution of Total in a bar chart. Describe the shape of the distribution. Is the Central Limit Theorem starting to be visible even though n is only 12 and the population is highly non-normal?

7-22   *Chi-Square Distribution.* The mean, standard deviation, and shape of the chi-square distribution change as the number of degrees of freedom changes. For a chi-square distribution with n degrees of freedom, the mean is n and the standard deviation is $\sqrt{2n}$.

(a) Simulate 1000 observations from a chi-square distribution with 1 degree of freedom into C1. Use Stat ▶ Basic Statistics ▶ Display Descriptive Statistics to compute the sample mean and the standard deviation. Are they approximately 1 (= df) and 1.414 (= $\sqrt{2}$ = $\sqrt{2df}$), respectively? Now make a histogram and sketch the shape of the distribution. The data are all constrained to be positive. Review the Histogram material in Chapter 4, p. 120, about how to handle this situation correctly.

(b) Repeat (a), but use 2 degrees of freedom.

(c) Repeat (a), but use 5 degrees of freedom.

(d) Repeat (a), but use 10 degrees of freedom.

7-23   *Chi-Square Distribution.* The chi-square distribution arises in several ways in statistics. Here we will use simulation to illustrate three of those ways. For all histograms in this exercise be careful about intervals that go below zero. The data are all constrained to be positive. Review the Histogram material in Chapter 4, p. 120, about how to handle this situation correctly.

(a) If X is from a standard normal distribution (mean 0 and standard deviation, X^2 has a chi-square distribution with 1 degree of freedom. Simulate 1000 standard normals into a column, then square each of them. Then simulate 1000 chi-square observations with 1 degree of freedom into another column. Now make a histogram of the X^2s and a histogram of the chi-squares. The two distributions should be very similar.

(b) Another way the chi-square distribution arises is from the variance of samples from a normal distribution. We will simulate samples of size 4. Simulate 1000 rows of standard normal observations into C1-C4. Use Calc ▶ Row Statistics to compute the sample standard deviations of the 200 rows. Square the results to get variances. Make a histogram of the variances. Compare this to a histogram of 1000 chi-square observations with 3 degrees of freedom.

(c) Still another source of the chi-square distribution is as follows: Simulate 1000 Poisson observations with μ = 5 into C1-C3. Now compute the chi-square test, which we will discuss in Chapter 12, using the following commands:

```
LET C4 = (C1+C2+C3)/3
LET C6 = (C1-C4)**2+(C2-C4)**2+(C3-C4)**2
LET C7 = C6/C4
```

Now make a histogram of C7 and compare it to a histogram of a chi-square with 2 degrees of freedom.

7-24   The *t* Distribution. The *t* distribution arises most naturally as the distribution of $(\bar{x} - \mu)/(s/\sqrt{n})$ when random sampling from a normal population. To illustrate the development of a *t* distribution, simulate 200 rows of normal data into C1-C3. You choose μ and σ. Use Calc ▶ Row Statistics to calculate row means and standard deviations ($\bar{x}$ and *s*) for C1-C3. Use Calc ▶ Calculator to compute $t = (\bar{x} - \mu)/(s/\sqrt{3})$. Make a histogram of *t*. Compare this histogram to one you obtain by using the Calc ▶ Random Data ▶ *t* command to simulate directly 200 observations from a *t* distribution with 2 degrees of freedom. Describe the shapes of the two distributions.

8

One-Sample Confidence Intervals and Tests for Means

We often want to know something about a population or process. The population could be all the 10-year-old girls in Philadelphia, and we might want to know the mean blood pressure. The population could be all the light bulbs produced by a plant last month, and we might want to know what proportion are defective. In most cases we cannot afford to study the entire population. So we take a sample—perhaps 200 girls or 80 light bulbs—and we use information learned from the sample to estimate what we want to know about the population. If our sample is taken appropriately, then our estimate may be quite good.

In the example with 10-year-old girls we want to know the mean blood pressure of the population. This mean is denoted by the letter μ. We can use the mean of the sample, $\bar{x}$, as a guess or estimate of μ. In the light bulb example we want to know the proportion of defective bulbs in the population. This population proportion is typically denoted by p. We can estimate p by the proportion of defectives in the sample of 80 bulbs. This sample proportion is often denoted by $\hat{p}$.

But how do we get a sample? One good way is to take what is called a simple random sample. A simple random sample of size n has an important property: Every sample of size n has the same chance of being selected.

Here is one straightforward way to get a simple random sample of 200 girls: Start by writing the name of each 10-year-old girl in Philadelphia on a slip of paper. We might get a good list from school records. Then put all the slips in a large box. Mix them up and draw one slip. Mix up the remaining slips and draw a second one, and so on until we have 200 slips. We could then contact these girls and measure their blood pressure. In this example everything seems fairly simple. Unfortunately in most studies it is very difficult to get a truly random sample. In Section 8.5, p. 256, we will briefly discuss some ways to spot non-randomness.

In this chapter we study different methods of learning about the mean of a population. These methods require two things: We must have a random sample and we must have a normal population. The first requirement is very important. The second, however, is not as important for our methods, and can be relaxed in many cases. In Section 8.5 we'll briefly discuss the ways a nonnormal population might affect your conclusions and how you can spot nonnormality using a sample.

8.1 How Sample Means Vary

Imagine drawing a random sample from a population where the variable of interest has a normal distribution. Calculate the mean of that sample. Imagine doing this 1000 times so that we have 1000 values of $\bar{x}$. Figure 8.1 shows the results, using 1000 samples, each with 9 observations.[†] (There is nothing special about choosing 9 observations. Any sample size would illustrate sampling variation.) To generate 1000 samples of size 9 from a normal distribution with mean 80 and standard deviation 5,

1. Restart Minitab to set all dialog boxes to their defaults.
2. Choose **Calc ▸ Random Data ▸ Normal** and generate **1000** rows of data, stored in **C1-C9**, using mean **80** and standard deviation **5**. Click **OK**.
3. Choose **Calc ▸ Row Statistics** and calculate the row **Means** of **C1-C9**, storing them in a new column named **Xbars**. Click **OK**.
4. Choose **Data ▸ Stack ▸ Columns** and stack all of the data in columns **C1-C9** into one new column named **Xvalues**. Click **OK**.
5. Choose **Graph ▸ Histogram** and create histograms of **Xvalues** and **Xbars** on the same graph. Be sure to click the **Multiple Graphs** option and select **Overlaid on the same graph**. Click **OK**. Also click the **Scale** option button, the **Y-scale** tab, and select either **Percent** or **Density** since these columns have very different lengths. Click **OK** twice. Our results using the density option are shown in Figure 8.1. Yours should be similar but not identical.

[†] We would like to do this *infinitely* many times but 1000 will illustrate the theory nicely.

Figure 8.1   Simulated distribution of the sample mean and of individual values

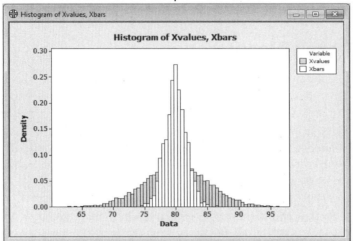

The two histograms, one of the 9000 individual data values and one of the 1000 sample means, are on the same scale so we can compare them more easily. There is one striking difference: The distribution of the sample means is much less spread out than that of the individual observations. Let's calculate descriptive statistics for both variables and do a numerical comparison. Figure 8.2 displays the results.

Figure 8.2   Descriptive statistics for Xvalues and Xbars

Descriptive Statistics: Xvalues, Xbars

Variable	N	N*	Mean	SE Mean	StDev	Minimum	Q1	Median	Q3
Xvalues	9000	0	79.987	0.0529	5.018	60.869	76.588	80.037	83.372
Xbars	1000	0	79.987	0.0534	1.688	74.688	78.891	79.998	81.082

Variable	Maximum
Xvalues	97.182
Xbars	86.625

Most of the sample means, the Xbars, are fairly close to the population mean, $\mu = 80$. All of them are between 74.688 and 86.625. In real-life sampling we usually have just one sample, not 1000. That one sample would yield just one sample mean, not 1000 of them. We could be "lucky" and get a sample mean that's near 80—most are reasonably close—or we could be really "unlucky" and get a sample mean that's down around 74.

Notice that the mean of the 1000 $\bar{x}$'s is 79.987, which is very close to the population mean of 80. Even if we take samples from nonnormal distributions, the mean of the population of $\bar{x}$'s will always be equal to the mean of the population of individual observations. We say $\bar{x}$ is an unbiased estimate of μ. Some of the exercises allow you to explore this fact. In symbols we write

$$\mu_{\bar{x}} = \mu$$

As we noticed before, the distribution of $\bar{x}$'s differs from that of the individual observations in one very important aspect—it is less spread out. In fact for random samples of size n from any population,[†]

$$\sigma_{\bar{x}} = \frac{\sigma}{\sqrt{n}}$$

In our simulation $\sigma = 5$ and $n = 9$ so that $\sigma_{\bar{x}} = 5/\sqrt{9} = 1.667$. This is the theoretical standard deviation for the distribution of $\bar{x}$'s. From Figure 8.2, the standard deviation of our particular simulated collection of 1000 x's is 1.688, which is quite close to what theory tells us will happen.

Notice that in Figure 8.1 rarely did a sample mean deviate from μ by more than $2\sigma_{\bar{x}} = 3.33$. This important fact forms the basis for the confidence intervals and tests described in this chapter.

8.2   Confidence Interval for μ

Suppose we want to estimate the mean μ of a population. We might take a random sample, calculate the sample mean, $\bar{x}$, and use this as an estimate of μ.

Figure 8.2 shows how close the population mean and sample mean are likely to be. A confidence interval uses this idea in a more formal way.

A **95% confidence interval** is an interval, calculated from the sample, that is very likely to cover the unknown mean μ. To be more precise, if we were to use the formula for a 95% confidence interval on many sets of data, then 95% of our intervals would surround the unknown mean we are trying to estimate but 5% would not cover it. Unfortunately in practice we can never know which intervals are the successful ones and which are the failures. The value 95% is called the **confidence level** of the interval. The confidence levels most commonly used in practice are 95% and 99%. For illustration, let's use 95%. A 95% confidence interval for μ is an interval (L, U) which has two properties:

- $\Pr(L < \mu < U) = 0.95$ regardless of the values of the unknown parameters μ and σ and

---

[†] This result assumes the population is infinite or that we are sampling *with* replacement. It will be approximately true if the population size is very large relative to the sample size. Notice that the sample size does *not* have to be large for this to hold.

- Both L and U are statistics. That is, they can be calculated from the observed data alone. Their computation cannot involve any unknown parameters.

To develop a confidence interval for μ, we start with the fact that with random samples from a normal distribution, $(\bar{x} - \mu)/(\sigma/\sqrt{n})$ has a standard normal distribution. So

$$Pr\left(-1.96 < \frac{\bar{x} - \mu}{\sigma/\sqrt{n}} < 1.96\right) = 0.95$$

Now take the inequality $-1.96 < (\bar{x} - \mu)/(\sigma/\sqrt{n}) < 1.96$ and rearrange it to the form $\bar{x} - 1.96(\sigma/\sqrt{n}) < \mu < \bar{x} + 1.96(\sigma/\sqrt{n})$. At first glance it appears that we may take the endpoints of the confidence interval as:

$$L = \bar{x} - 1.96(\sigma/\sqrt{n})$$
$$U = \bar{x} + 1.96(\sigma/\sqrt{n})$$

However, since σ is not a known quantity, this choice fails the requirement that L and U both be statistics, that is, computable from the data. To get around this difficulty we consider the distribution of

$$t = \frac{\bar{x} - \mu}{s/\sqrt{n}}$$

where s is the sample standard deviation. If the variable t has a known distribution we can replace the -1.96 and $+1.96$ in the derivation above by the appropriate numbers and proceed as before. Fortunately, the distribution of this quantity was discovered long ago and is called Student's t distribution. It's properties are well-known to statisticians (and Minitab programmers!). Thus the correct results for the endpoints of the confidence interval are:

$$L = \bar{x} - t_c(s/\sqrt{n})$$
$$U = \bar{x} + t_c(s/\sqrt{n})$$

where t_c is chosen from Student's t distribution with $n-1$ degrees of freedom so that there is 0.95 probability between $-t_c$ and t_c. The choice of confidence level and sample size will both affect the value of t_c. It is also known that, for large sample sizes, the t distribution is effectively a standard normal distribution. In that case $t_c \approx 1.96$ for a 95% confidence level and $t_c \approx 2.56$ for 99% confidence. Of course, Minitab will do all of the proper calculations for us for any sample size.

We use the Student's t confidence interval procedure and the 1-Sample t command. The **Stat ▸ Basic Statistics ▸ 1-Sample t** sequence of commands calculates and prints a confidence interval for the population mean. If you do not specify a confidence level, Minitab uses 95%. Here is an example.

The Cartoon data set contains Otis scores (similar to IQ scores) for 179 people including a sample of 53 professional hospital workers. The professionals are identified by ED = 1. Suppose we want to find a 95% confidence interval for the mean Otis score of all professional hospital workers. First we need to subset the data. You may want to review the methods for subsetting discussed on p. 76. To subset the data,

1. Open the **Cartoon** worksheet.
2. Choose **Data ▸ Copy ▸ Columns to Columns**.
3. Enter **Otis** in the Copy from columns box. Minitab will copy data just from that column.
4. Click the **Store Copied Data in Columns** list arrow and select **In current Worksheet, in Columns**.
5. Enter **ProOtis** into the columns box.
6. **Deselect** the **Name the columns containing the copied data**.
7. Click the **Subset the Data** button.
8. Under **Specify Which Rows To Include**, select **Rows that match** and press the **Condition...** button.
9. Fill in the **Condition:** text box with **Ed=1**.
10. Click **OK** three times. Minitab copies the Otis data into a column named ProOtis, but only for professionals.

To compute the confidence interval for mean Otis score for professionals,

1. Choose **Stat ▸ Basic Statistics ▸ 1-Sample t**.
2. Enter **ProOtis** as the **Sample in columns:** and click **OK**. See Figure 8.3. Minitab displays the results in the Session Window, shown in Figure 8.4.

Figure 8.3   1 Sample t dialog box

Enter column(s)
for data to be
used

Use these options
if data is already
summarized

Click Options to
change the
confidence level

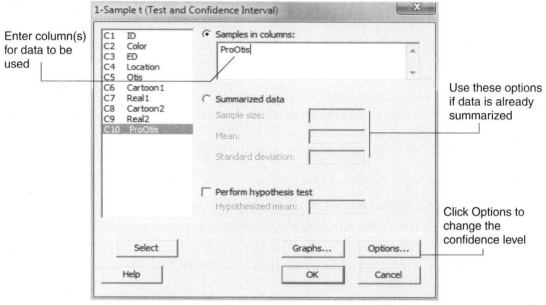

Figure 8.4 shows the results. Based on this output we might say, "We estimate the mean to be about 104.9, and we are 95% confident that it is somewhere
between 101.1 and 108.7."

Here SE Mean is the standard error of the mean. It is calculated by using
$s/\sqrt{n}$, where s is the sample standard deviation.

Figure 8.4   Confidence interval for mean Otis score for professionals

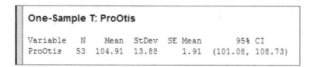

One-Sample T: ProOtis

Variable	N	Mean	StDev	SE Mean	95% CI
ProOtis	53	104.91	13.88	1.91	(101.08, 108.73)

Simulated Confidence Intervals for the Mean

To understand the meaning of confidence intervals it is helpful to use simulation
and graphics. In a simulation we get to play "mother nature" and know the true
values of all parameters. Suppose we are selecting a sample of size 9 from a
normal distribution with mean 80 and standard deviation 5. Once we have the
sample in hand we calculate a 95% confidence interval for the (unknown)

population mean using $\bar{x} \pm t_c (s/\sqrt{n})$ as always. Note we use the *sample* standard deviation in this calculation—not the population standard deviation which we would not know in real sampling. Now we can see if our particular interval "covers" the true mean of 80. The 95% confidence level says that if we repeat this process many times about 95% of our intervals will cover the true mean and about 5% will miss. With simulation we can sample then calculate a confidence interval many times and display our overall results graphically.

Figure 8.5 gives the results of 100 confidence intervals under these same conditions. For this particular simulation, 6 of the 100 intervals missed and 94 or 94% "hit", that is, covered or surrounded the true population mean. The misses in the plot are shown as dashed lines. All of the solid lines are hits.

All of the intervals are centered on the respective sample means which, as we know, vary according to a certain normal distribution. Note that the lengths of the intervals also vary—some are quite short while others are longer. This is because the lengths are proportional to the sample standard deviations and the sample standard deviations also vary from sample to sample.

This simulation and plot was created with a Minitab macro named GMeanCI which is available from the Data Library at www.CengageBrain.com associated with this book. The macro plots the misses in solid red so that they are easy to spot. You might also maximize the graph to full screen to see the details. In the Exercises we ask you to use this macro to investigate confidence intervals of your own.

Figure 8.5 100 simulated confidence intervals for the mean

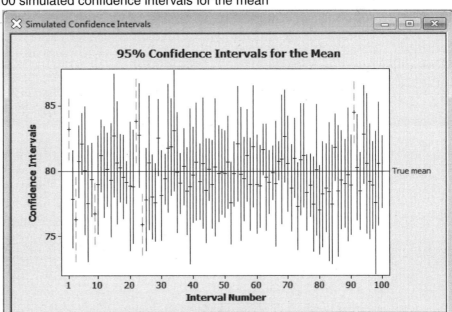

8.3   Test of Hypothesis for μ

Suppose a machine is designed to roll aluminum into sheets that are 40.0 thousandths of an inch thick. To check that the machine stays in adjustment, the operator periodically takes five measurements of sheet thickness. On the latest occasion, the results were

40.1 39.2 39.4 39.8 39.0

This sample has a mean of 39.5, which is below the desired or target mean of 40.0. Is this discrepancy just due to random fluctuation or does it indicate that the machine is not adjusted to the correct level? A partial answer to this question may be obtained with an significance test of the mean. We take a null hypothesis of H_0: $\mu = 40.0$ and an alternative hypothesis of H_1: $\mu \neq 40.0$.

We can again use Minitab's 1-Sample t command. Suppose we test whether or not the aluminum sheet machine mentioned in the previous section is correctly adjusted; that is, whether $\mu = 40$ or not. The test used is called Student's t test and is performed by the Minitab's Stat ▶ Basic Statistics ▶ 1-Sample t command, selecting the Perform hypotheses test option. For each column Minitab tests the null hypothesis, H_0: $\mu = K$, where K is the hypothesized population mean. The default alternative hypothesis is H_1: $\mu \neq K$. If you don't specify K, Minitab uses $K = 0$. The command calculates the test statistic:

$$t = \frac{\bar{x} - K}{s/\sqrt{n}}$$

Here $\bar{x}$ is the sample mean, n is the sample size, s is the sample standard deviation, and K is the hypothesized or target value of the population mean, μ. You can also click the Alternative list arrow to do a one-sided test.

To perform the t test with the Aluminum sheet data,

1. Enter the numbers **40.1 39.2 39.4 39.8 39.0** into a column and name the column **Thickness**.
2. Choose **Stat ▶ Basic Statistics ▶ 1-Sample t**. Enter **Thickness** as the variable.
3. Select **Perform hypothesis test** and enter **40** as the **Hypothesized mean**. See Figure 8.6.
4. Click **OK** and Minitab displays the results in the Session Window, shown in Figure 8.7.

Figure 8.6 1-Sample *t* test dialog box

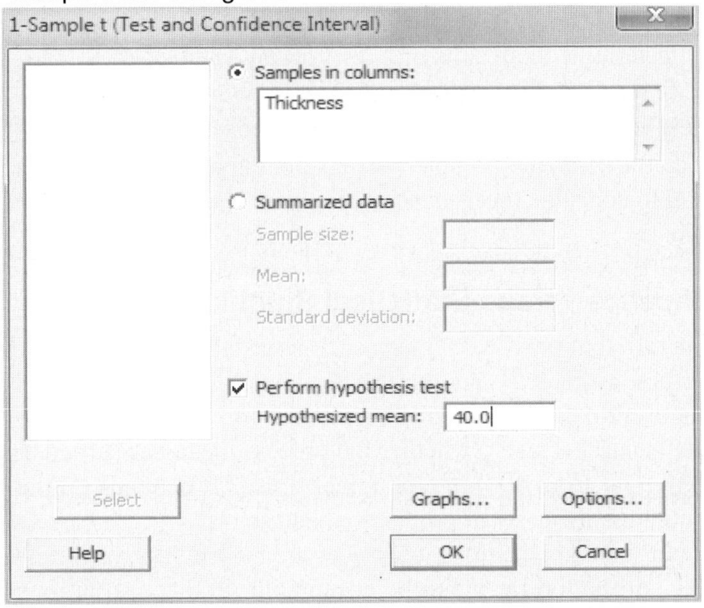

Figure 8.7 shows the results that are displayed in the Session Window. the ***p*-value** (or **observed significance level**) shows how unusual the observed test statistic is assuming the null hypothesis is true. Here the *p*-value of 0.067 is small but not below the traditional cutoff values of 0.05 or 0.01. The evidence that the rolling machine is out of adjustment is not statistically significant.

Figure 8.7 1-Sample *t* test results

```
One-Sample T: Thickness

Test of mu = 40 vs not = 40


Variable    N    Mean   StDev  SE Mean        95% CI          T      P
Thickness   5  39.500   0.447   0.200  (38.945, 40.055)  -2.50  0.067
```

The number −2.50, labeled T in the output, is the *t* statistic used to do the test, namely

$$t = \frac{\bar{x} - 40}{s/\sqrt{n}} = \frac{39.5 - 40}{0.447/\sqrt{5}} = -2.50$$

If the null hypothesis is true, that is, if the population mean is 40, then this statistic follows a *t* distribution with $n-1 = 4$ degrees of freedom and Minitab uses this distribution to compute the *p*-value.

In many textbooks, statistical tests are done by specifying what is called an α (alpha) level or significance level. Once we know the p-value, however, we can determine the results of a test for any value of α. If the p-value is less than α, we reject H_0; if it is greater than α, we do not reject H_0. For example if we did a test here using $\alpha = 0.05$, we would not reject H_0, because 0.067 is not less than 0.05. Based on this sample, there is not enough evidence at the 5% significance level to conclude that the mean thickness is different from 40 thousandths of an inch.

Practical Significance versus Statistical Significance

It is rare that practical significance and statistical significance coincide. Here our best estimate is that the process mean is about 39.5. The t test showed that this result is not statistically significant for $\alpha = 0.05$. This in itself gives us no clue whatsoever as to whether the process mean has shifted seriously off target. Is 39.5 far from 40.0 in a practical sense? It depends on the use of the rolled aluminum, the costs of production, and so on.

In some applications tolerances are very tight. Even results as close as 39.9 may still be too far off to be acceptable. In other situations it may not make much practical difference as long as the sheets are at least as thick as 39.3. In other words a statistical test of significance tells us only whether or not the observed data are unusual under the hypothesized situation. It tells us nothing about what practical course of action we should take. This is the heart of the reason why we prefer confidence intervals over tests of significance. From Figure 8.7 we see that a 95% confidence interval for these data goes from 38.94 and 40.06. Thus we are fairly confident that the true mean thickness is between 38.94 and 40.06. If all we really need is a mean thickness of 39.3 or more, then for practical purposes, the machine may be all right. Note also that we are discussing *means* here. It may be that each *individual* sheet must have a thickness of at least 39.3 thousandths of an inch to be acceptable. This is a very different requirement.

8.4   Confidence Intervals and Tests for Proportions

Proportions are special means—means of zero-one variables. So much of our discussion of means carries over to proportions. One important difference is that, with random sampling, the mean and standard deviation of a sample proportion are both determined by the population proportion. Let p and $\hat{p}$ denote the population and sample proportions, respectively, and let n be the sample size. Then statistical theory says that,

$$\mu_{\hat{p}} = p$$

$$\sigma_{\hat{p}} = \sqrt{\frac{p(1-p)}{n}}$$

These facts together with an extension of the normal approximation to the binomial distribution, see Section 6.3, p. 196, lead to the simplest version of confidence intervals for a proportion p, namely,

$$\hat{p} \pm z_c \sqrt{\frac{\hat{p}(1-\hat{p})}{n}}$$

where z_c is chosen from a standard normal distribution so that the percentage of area between $-z_c$ and z_c is the required confidence level. More detailed theory using the binomial distribution allows Minitab to calculate an "exact" confidence level. The differences in the resulting intervals are slight except for small samples.

Here is an example. In August 2003, the Gallup Poll asked 507 randomly chosen adults "Do you think the possession of small amounts of marijuana should be treated as a criminal offense?" Two hundred and thirty-eight people responded no. Let p be the proportion of adults who would respond no in the general population. Our guess is that p would be around $\hat{p} = 238/507 \approx 0.46.9$ or about 47%. To obtain a confidence interval for p,

1. Choose **Stat ▸ Basic Statistics ▸ 1 Proportion**. See Figure 8.8.
2. Click the **Summarized data** option and fill in the Number of events as **238** and Number of trials as **507**.
3. Select **Perform hypothesis test** and enter **0.5** for **Hypothesized proportion**.
4. Click **OK** to obtain the results in the Session Window as shown in Figure 8.9.

Figure 8.8   1-proportion test and confidence interval dialog box

With raw binary
data, zeros and
ones or Yes/No
enter the column
data here

Use these boxes
for summarized
data

Click the Options
button to change
the confidence
level, null
hypothesis
value, or to use
the normal
approximation

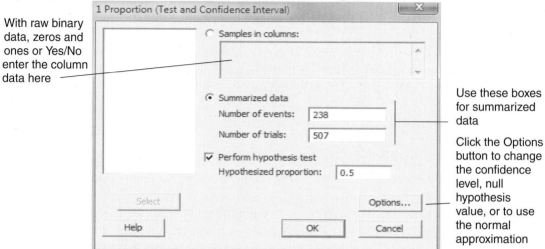

From Figure 8.9 we see that we can be 95% confident that the proportion of all adults who would have answered no in the poll would be somewhere between 43% and 51%. Notice that Minitab also carried out a significance test of the default null hypothesis H_0: $p = 0.5$. Click the Options button to change the confidence level or the null hypothesis value. The Options dialog box also gives you the choice to force Minitab to produce the approximate confidence interval and p-value based on $\hat{p} \pm z_c\sqrt{\hat{p}(1-\hat{p})/n}$. The 1-Proportion command will also work with raw two-valued text data such Yes/No or Male/Female.

Figure 8.9   Confidence interval for a proportion

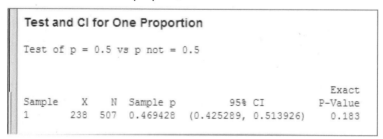

8.5   Departures from Assumptions

Many procedures in statistical inference, including those described in this chapter, are based on the assumption that the data are a random sample from a normal distribution. Here we take a quick look at how you can spot departures from these assumptions. We treat the most serious problem first—lack of randomness.

Nonrandomness

Most data are not produced from a random sample from any population or process. In particular, observations taken close together in time or in space are often more alike than those taken further apart. For example:

- People from the same part of the country tend to be more alike than people from different regions.
- Light bulbs manufactured on the same day tend to be more alike than those manufactured on different days.
- Adjacent pieces from the same roll of tape tend to have more nearly the same degree of stickiness than do pieces of tape made on different days or by different machines.
- The measurements made by one inspector are often more alike than measurements made by a different inspector.

Observations taken in close proximity are often correlated. Observations that are correlated do not form a simple random sample. One of the best ways to look for correlations is to make plots. For example if measurements are made one after the other, plot them in that order. If they are made in bunches, plot them so the bunches are readily identifiable. If you see a clear pattern, you probably do not have a random sample.

Figure 8.10 gives six examples. In each case the observations are plotted on the vertical axis. The horizontal axis is "order." Here order could be observation number, time order, spatial order, order in groups, or some other order that helps us see possible nonrandom nature in the data. The first plot shows no evidence of nonrandomness, but all the other plots have particular nonrandom patterns as stated on the graph.

Figure 8.10 Random scatter and various departures from randomness

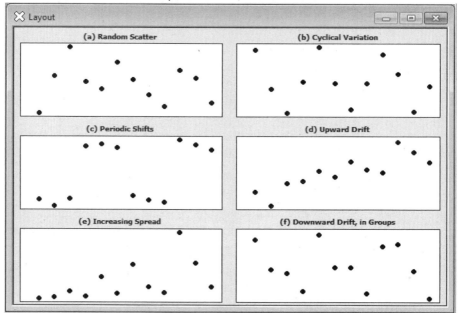

Nonnormality

The techniques in this chapter, and in many of the chapters that follow, are based on statistical theory that, to be exactly valid, requires the population from which our sample is drawn to have a normal distribution. This would appear to rule out practical applications of these techniques, since no real population is ever exactly normal. But, in fact, it does not. In many cases techniques based on the normal distribution give very good answers even if the population distribution is not exactly normal.

For example suppose we construct a 95% t confidence interval for μ, using 20 observations. If the distribution sampled from is not normal, then the true confidence level may not be 95%. It might be 94% or 90% or 96%, depending on what the actual shape of the population is, but it is not likely to be very far from 95%. Similarly, suppose we do a t test using these data and get a p-value of, say, 0.031. If the population is not normal, the correct p-value may not be 0.031. It might be 0.038 or 0.047 or 0.026 — not be very far from 0.031.

Table 8.1 shows the results of some confidence interval simulations with nonnormal populations. In each case we assumed the population was normal and used the 1-sample t procedure to calculate 95% confidence intervals from the sample. We repeated this 10,000 times and kept track of the "hits" and "misses" to see how close or far off our results would be relative to the supposed confidence level of 95%. The results in the first row are based on sampling from a t distribution with 15 degrees of freedom. This distribution is quite close to a normal distribution. It is symmetric around zero and "bell-shaped." Never-the-less, the 1-sample t procedure produced confidence intervals that covered the "mean" of zero 95.21% of the time with a sample of size 10. In the second row the population was a t distribution with 5 degrees of freedom. This distribution is also symmetric around zero and "bell-shaped," but it is much more spread out than a normal distribution. The actual confidence levels were still quite close to 95% for both sample sizes considered.

The last two rows of the table deal with asymmetric populations based on the chi-square distribution. Chi-square with 5 degrees of freedom is especially asymmetric. Even in that case the actual confidence levels were not too far from 95% for the two sample sizes considered. In the Exercises we ask you to reproduce these results and consider some other cases.

Table 8.1 Actual confidence levels for nominal 95% t confidence intervals

"Population"	Sample Size	
	10	25
t with 15 df	95.21	95.37
t with 5 df	94.49	95.29
Chi-square with 15 df	94.27	94.62
Chi-square with 5 df	93.28	93.72

We should mention that even though these procedures usually work well for nonnormal populations, in some cases the nonparametric methods in Chapter 13 are better (or more powerful, in statistical terminology). The confidence interval constructed by a nonparametric procedure may be shorter than the t confidence interval. A shorter interval gives us a more precise estimate of μ. A nonparametric test may be more likely to reject the null hypothesis when the null hypothesis is indeed false.

There are many methods that can help us decide whether a population is normal. Some of the sophisticated methods use formal tests. Here we will look at two simple graphical techniques.

Suppose we take a sample from the population and display a histogram. If the population is normal, the histogram should have approximately the shape of a normal curve. If the sample size is large, this approximation should be very good. Of course if we have just a few observations, say 10 or 15, then it will be very difficult to see a clear shape in the sample histogram. In fact, with 10 or 15 observations, histograms would be meaningless. A dotplot might give us a little information but a special graph, called a normal probability plot, is a better tool.

Normal Probability plots

A **normal probability plot**, plots the sample data versus the values we would expect to get if the data came from an ideal normal sample. If the data is from a normal population, this plot is approximately a straight line. If the plot exhibits curvature, the population is not normal.

Try this for the Otis scores. To create a normal probability plot and perform a formal hypothesis test,

1. Activate the **Cartoon** worksheet.
2. Choose **Stat ▸ Basic Statistics ▸ Normality Test**. Enter **Otis** in the variable text box. Leave the other options as is. See Figure 8.11.
3. Click **OK**. Minitab produces the output shown in Figure 8.12.

Figure 8.11 Normality test dialog box

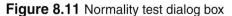

Enter the variable to be tested for normality

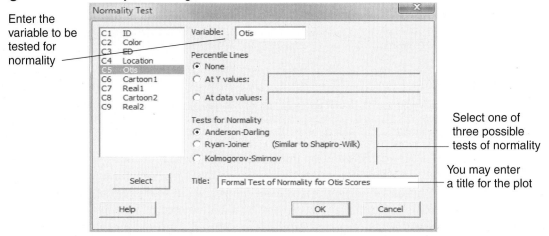

Select one of three possible tests of normality

You may enter a title for the plot

Minitab's normality test command includes a formal significance test where the null hypothesis is that the data come from a normal distribution. The alternative hypothesis is general: that the distribution is not normal. The default test is called the Anderson-Darling test. The p-value for this test is shown in the upper right hand of Figure 8.12 to be 0.139—not small enough for us to reject normality. Notice in Figure 8.11 that you may alternatively choose the Ryan-Joiner or Kolmogorov-Smirnov tests. Just looking at the Normal Scores plot, we are faced with the problem of judging a picture. Does this plot look straight? There certainly is some curvature. But is it curved enough to doubt the normality of the Otis scores? Good judgment of graphs comes from experience. Some people find it easier to learn how to judge probability plots than histograms.

Some of the exercises at the end of this section show how you can use simulation to improve your judgment of the nonnormality of histograms and normal probability plots.

Figure 8.12 Normal probability plot and Anderson-Darling normality test of Otis scores

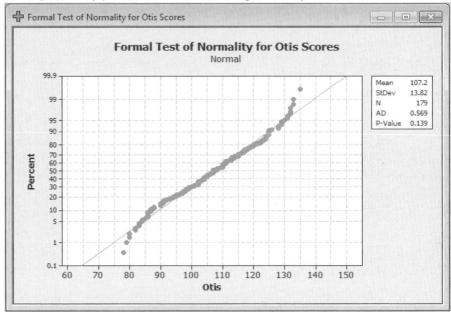

8.6   Hypothesis Tests with the Minitab Assistant

Much of what we have discussed so far in this chapter has been "automated" in the new **Minitab Assistant** available in Release 16. Six choices are available under the Assistant pull-down menu as shown at the right.

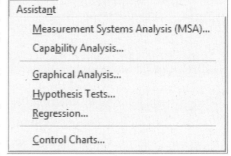

Click on Hypothesis Tests and the graphic shown in Figure 8.13 appears. Here you see an interactive decision tree to help you choose the correct tool. Most of the choices will be self-explanatory once you have covered the material in chapters 8, 9 and 10.[†]

[†] The exceptions are the ones that refer to %Defective. This terminology comes from the quality control literature and simply means that you are dealing with proportions and binary Yes/No types of variables. With the Assistant, all of the choices, options, and results are expressed as percentages rather than fractions.

Figure 8.13 The initial Assistant screen for Hypothesis Tests

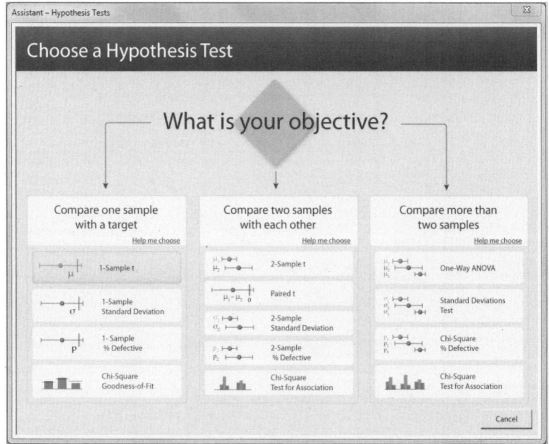

To use the Assistant to carry out the hypothesis test that we performed in Section 8.3, p. 252,

1. **Reload** the Thickness data (p. 252) either from a worksheet that you saved or by entering it once more by hand. (It's really small!)
2. Choose **Assistant ▸ Hypothesis Tests**.
3. **Click** on the rectangle of the **Choose a Hypothesis Test** screen that says **1-Sample t.** The dialog box shown in Figure 8.14 appears.
4. Enter **Thickness** for the data column.
5. Enter **40** for the Target (null hypothesis) value.
6. Choose **Is the mean thickness different from 40** and Click **OK**.

Figure 8.14 Assistant dialog box for 1-Sample t Test

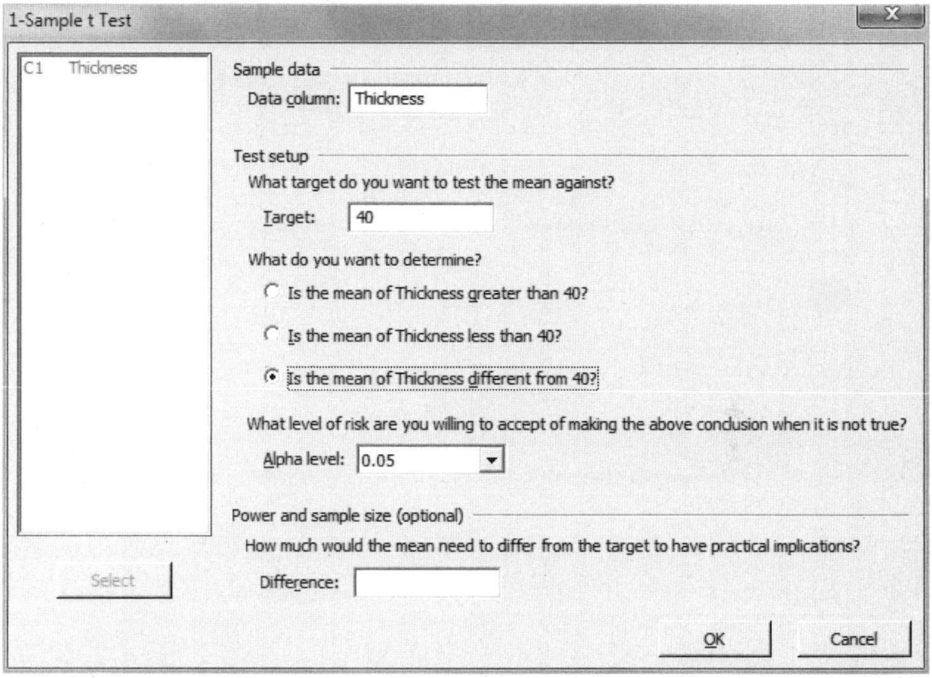

Three reports are produced when you click OK: A **Summary Report** shown in Figure 8.15, a **Report Card** shown in Figure 8.16, and a **Diagnostic Report** shown in Figure 8.17. Each report contains several different pieces of information relative to the 1-sample hypothesis test and the corresponding confidence interval. Some of the information is numerical, some graphical, and some textual. Color (not shown here) is used to enhance the displays. Some of the information provides guidelines to help ensure your analysis is successful. Other information provides interpretation of the output. Additional information will be displayed if you enter a value for the **Difference** in the Power and sample size portion of the dialog box shown at the bottom of Figure 8.14.

Figure 8.15 Assistant Summary Report

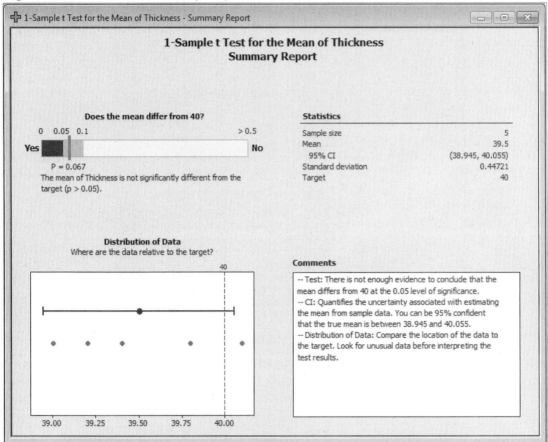

The confidence interval and p-value reported here agree with those given in Figure 8.7, p. 253 and the verbal description in the Comments section agrees with our earlier conclusions.

Figure 8.16 Assistant Report Card

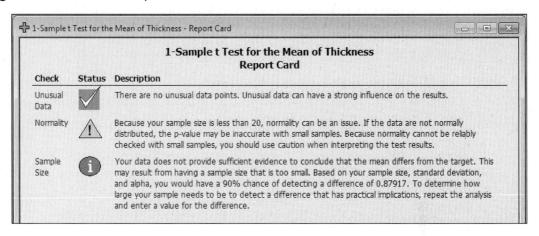

The Report Card gives additional information about the test and confidence interval. For example, it notes that there is no way to reliably check on normality with a sample of size 5.

Figure 8.17 Assistant Diagnostic Report

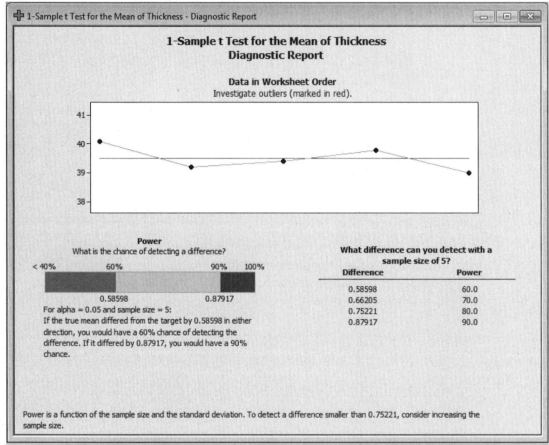

No outliers are detected. If there were, they would be plotted with red symbols on the top graphic of the Diagnostic Report. We will return to the Assistant in subsequent chapters.

Exercises

Exercises for Section 8.1

8-1
    (a) Run the steps that produced Figure 8.1, p. 246, again, but now use a sample of size $n = 2$ instead of 9. Discuss the output.

    (b) Run the same steps, using $n = 9$ and $n = 25$. Compare the output for $n = 2, 9$, and 25.

(c) Compute means and standard distributions of the Xbars for the three sample sizes and for the individual values. Do they agree approximately with the theory which says $\sigma_{\bar{x}} = \sigma/\sqrt{n}$?

8-2   Figure 8.1, p. 246, helped illustrate the fact that the means of random samples from normal distributions are themselves normal. An equally important fact is that the means of samples of at least moderate size from most nonnormal distributions are also themselves approximately normal. This approximation gets better as the sample size n gets larger. Simulate 1000 observations from a continuous uniform distribution and put the results in columns C1-C6 (use Calc ▶ Random Data ▶ Uniform). Then compute the means for samples of sizes $n = 1, 2, 3, 4, 5$, and 6 (use Calc ▶ Row Statistics, and first place the mean of C1 into C11, then of C1-C2 into C12, then of C1-C3 into C13, C1-C4 into C14, and so on). (Using Edit Last Dialog button, 🖼, will save you a lot of rework!) Name C11-C16, Mean1, Mean2, and so forth.
(a) Create dotplots of C11-C16 on the same graph and interpret.
(b) Compute the means and standard deviations of Mean1 through Mean6.
(c) Compare the results obtained in part (b) with theory that says $\sigma_{\bar{x}} = \sigma/\sqrt{n}$. It may be shown here that $\sigma = 0.288675$.

8-3   In this exercise we will take a look at how the distribution of the sample mean relates to the distribution of the individual observations in samples from some nonnormal distributions.
(a) Use Calc ▶ Random Data ▶ Uniform, using 0 and 1 as your endpoints, to simulate 1000 observations from a distribution that is uniform on the interval 0 to 1. Put the observations into C10. Construct a dotplot and histogram of C10.
(b) How do these displays convey the feature that leads to the descriptive name uniform distribution? Discuss why we need to be careful when choosing cell boundaries in histograms, especially when the data are constrained to lie in a particular interval. (See "Histograms for Positive Data" p. 120.)
(c) Again use the uniform distribution on 0 to 1, but now simulate 1000 samples, each of size 5, into C1-C5. Then use Calc ▶ Row Statistics to place row means of these 1000 samples into C6. How does the distribution of the 1000 means in C6 relate to the distribution of the 1000 individual observations in C10? Use the Descriptive Statistics and Dotplot commands to look at the differences.
(d) Compare the shape of the distribution of sample means in C6 and individual observations in C10. Does the shape of the distribution of the sample means look familiar?
(e) Theory tells us that the mean of C6 should be approximately equal to the mean of C10. Is it? Theory also tells us that the standard deviation of C6

should be approximately equal to the standard deviation of C10 divided by $\sqrt{n}$. Is it? Why did we divide by $\sqrt{n}$?

8-4   In Exercise 8-3 we looked at data from a uniform distribution. Now we will look at another nonnormal distribution, the chi-square distribution with 3 degrees of freedom.

(a) Simulate 1000 observations from the chi-square distribution and construct displays as follows. Generate 1000 rows of data, using the chi-square distribution and 3 degrees of freedom. Place the data in C10. Then create dotplot and histogram displays.

(b) How does the shape of this chi-square distribution compare to the normal curve?

(c) Again use the chi-square distribution with 3 degrees of freedom, but now simulate 1000 samples, each of size 5, into C1-C5. Then use Calc ▸ Row Statistics to put the means of these 1000 samples into C6. Now answer the questions in parts (c)-(e) of Exercise 8-3.

Exercises for Section 8.2

8-5   The output in Figure 8.4, p. 250, gives a 95% confidence interval for μ. Use this output to calculate, by hand, a 90% confidence interval. You may use Minitab to obtain the proper multiplier from the t distribution.

8-6   (a) Display a histogram for the SAT verbal scores in sample A of the Grades data (described in the Appendix, p. 527). Calculate $\bar{x}$ and obtain a 95% confidence interval for the mean score, μ.

(b) Also display a 90% and then a 99% confidence interval for mean SAT verbal scores. How do they compare with the 95% confidence interval in part (a)? Do they have the same centers? Do they have the same widths?

8-7   Suppose 9 men are selected at random from a large population. Assume the heights of the men in this population are normal with $\mu = 69$ inches and $\sigma = 3$ inches. Simulate the results of this selection 20 times and in each case find a 90% confidence interval for μ. Use Random Data ▸ Normal to generate 9 rows of data in C1-C20 with mean 69 and standard deviation 3. Using C1-C20 as the variables, compute 1-Sample t confidence intervals each with 90% confidence.

(a) How many of your 20 intervals contain μ?

(b) Would you expect all 20 of the intervals to contain μ? Explain.

(c) Do all the intervals have the same width? Why or why not?

(d) Suppose you took 95% intervals instead of 90%. Would they be narrower or wider?

(e) How many of your intervals contain the value 72? The value 70? The value 69?

(f) Suppose you took samples of size $n = 100$ instead of $n = 9$. Would you expect more or fewer intervals to cover 72? 70? 69? What about the width of the intervals for $n = 100$? Would they be narrower or wider than for $n = 9$?

(g) Suppose you calculated 90% confidence intervals for 20 sets of real data. About how many of these intervals would you expect to contain μ? Could you tell which intervals were successful and which were not? Why or why not?

8-8   In this exercise you will use the macro GMeanCI (available from the Data Library at www.CengageBrain.com associated with this book) to simulate normal samples and the resulting confidence intervals.

(a) Use the macro GMeanCI to simulate and display 100 confidence intervals for the mean when sampling from a normal distribution with mean 80 and standard deviation 5 with samples of size 9. Compare your results to those shown in Figure 8.5, p. 251.

(b) Repeat part (a) for another 100 samples of size 9. Compare the new results to those obtained in part (a).

(c) If you combine the results of parts (a) and (b) what percentage of your intervals covered the true mean?

(d) Repeat part (a) but now take 100 samples of size 25. What changes do you note?

8-9   Minitab has a command Stat ▶ Basic Statistics ▶ 1 Sample Z that performs inferences about μ assuming you *know* σ. This exercise takes a brief look at what happens if you specify an incorrect value for σ. For this Exercise, use the Minitab macro, ZMeanCI, that is available from the Data Library at www.CengageBrain.com associated with this book. The macro generates random samples from a normal distribution with mean 30 and standard deviation 2 then computes Z confidence intervals based on $\bar{x} \pm 1.96(\sigma/\sqrt{n})$ with $\sigma = 2$. It asks you for your *assumed* sigma and reports the percentage "hits" by the intervals. In each case repeat the simulation about 1000 (or more) times to get good estimates of the hit percentage.

(a) Use a sample size of 5 and an assumed sigma of 2—the correct value! Did you have about 95% of your intervals covering μ?

(b) Repeat part (a), but now use 1 for σ. Now what percentage of your intervals cover μ? Is it close to the "advertised" or nominal 95%?

(c) Repeat part (a), but now use 3 for σ. Now what percentage of your intervals cover μ?

(d) Repeat part (a), but now use 1.5 for σ. Now what percentage of your intervals cover μ?

(e) Repeat part (a), but now use 2.5 for σ. Now what percentage of your intervals cover μ?

(f) Based on your results from Parts a-e of this question, would you say that knowing σ precisely is important or unimportant to the reliable use of a Z confidence interval?

8-10   Repeat Exercise 8-9 but now use a sample size of 25. Does the performance of the Z confidence interval improve? Be specific and justify your answer.

Exercises for Section 8.3

8-11   Do a test to see whether there is evidence that the preprofessionals who partici-pated in the Cartoon experiment (described in the Appendix, p. 524) have Otis scores that differ significantly from the national norm of 100. First, open the Cartoon worksheet and copy the column containing the Otis scores into a new column, using only rows where ED = 0. Then produce a histogram of just those Otis scores. What do you think? Now do the appropriate test, using the Test mean option with the 1-Sample t command, $\alpha = 0.05$, and null mean $\mu = 100$.

8-12   Imagine choosing $n = 16$ women at random from a large population and measur-ing their heights. Assume that the heights of the women in this population are normal, with $\mu = 64$ inches and $\sigma = 3$ inches. Suppose you then test the null hypothesis H_0: $\mu = 64$ versus the alternative that H_1: $\mu \neq 64$, using $\alpha = 0.10$. Simulate the results of doing this test 20 times by choosing Calc ▶ Random Data ▶ Normal and generating 16 rows of data in C1-C20 with 64 as the mean and 3 as the standard deviation. Then use Stat ▶ Basic Statistics ▶ 1-Sample t with the Test mean option and specify a null mean of 64.

(a) In how many tests did you fail to reject? That is, how many times did you make the "correct decision"?

(b) How many times did you make an "incorrect decision" (that is, reject H_0)? On the average how many times out of 20 would you expect to make the wrong decision?

(c) What is the attained significance level (p-value) of each test? Are they all the same?

(d) Suppose you used $\alpha = 0.05$ instead of $\alpha = 0.10$. Does this change any of your decisions to reject or not? Should it in some cases?

8-13   As in Exercise 8-11 simulate choosing 16 women at random, measuring their heights, and testing H_0: $\mu = 64$ versus H_1: $\mu \neq 64$, but this time assume that the population really has a mean of $\mu = 63$, instead of 64. Thus use the Normal command with $\mu = 63$ and $\sigma = 3$ to simulate the samples. Use $\alpha = 0.10$. Do this for a total of 20 tests.

(a) In how many tests did you reject? That is, how many times did you make the "correct decision"? How many times did you make an "incorrect decision"?

(b) What we are investigating is called the **power of a test**—that is, how well the test procedure does in detecting that the null hypothesis is wrong, when indeed it is wrong. Repeat the above simulation, but now assume the true population mean is $\mu = 62$. (Continue to use $\sigma = 3$ and the same null hypothesis.) How often did you make the correct decision (reject) in these 20 tests? On the average would you expect to make more "correct" decisions if the true mean were 62 or if the true mean were 63?

Exercises for Section 8.4

8-14   In August 2003, the Gallup Poll asked 507 randomly chosen adults "Do you think the possession of small amounts of marijuana should be treated as a criminal offense?" Two hundred and thirty-eight people responded no. Let p be the proportion of all adults who would answer no.

(a) Use the normal approximation method, namely $\hat{p} \pm z_c \sqrt{\hat{p}(1-\hat{p})/n}$, to find a 95% confidence interval for p "by hand."

(b) Use Minitab's command 1 Proportion and subcommand Options to obtain the answer in part (a).

(c) Compare your answer in (a) to the "exact" answer that Minitab produces. Do they differ much? How much?

8-15   Table 8.1 on p. 259 displayed simulation results of actual confidence levels associated with t confidence intervals under nonstandard conditions. These are estimates of true but unknown confidence levels based on 10,000 "trials." When the true population is a chi-square distribution with 15 degrees of freedom we estimated the true confidence level at 94.27%. This means there were 9427 "hits" in our 10,000 trials. Let p denote the true confidence level in fractional terms.

(a) Use these data to obtain a 95% confidence interval for p.

(b) Use these data to test the hypothesis that $p = 0.95$ versus $p \neq 0.95$. Be sure to click the Options button to set the null hypothesis value properly.

8-16   Exercise 5-17, p. 180, gave a Gallup Poll estimate of 24% for the percentage of American families for which drug abuse had ever been a problem. This was based on a sample of size 1,017 taken October 6-8, 2003. Assuming this comes from a simple random sample (it is actually a more complex probability sample), find a 95% confidence interval for the percentage of American families so affected.

8-17   Approximate binomial confidence intervals. The normal approximation to the binomial distribution (see Section 6.3) can be used as the basis for computing approximate confidence intervals. For binomial data the observed proportion of successes $\hat{p}$ provides an estimate of p. The standard deviation of $\hat{p}$ is estimated by $\sqrt{\hat{p}(1-\hat{p})/n}$. Thus the normal approximation indicates that we can compute a 90% confidence interval for p with the formula $\hat{p} \pm z_c \sqrt{\hat{p}(1-\hat{p})/n}$.

   (a) Use Calc ▶ Random Data ▶ Binomial to simulate 50 binomial observations with $n = 20$ and $p = 0.3$. Place the data in C1. Perform the following calculations (You may use Session commands or the menu Calc ▶ Calculator):

```
LET C2 = C1/20
LET C10 = 1.645*SQRT(C2*(1-C2)/20)
LET C3 = C2 - C10
LET C4 = C2 + C10
```

   This places the upper and lower confidence limits into C3 and C4. Enter 1:50 into C5 and then plot C3 versus C5 and C4 versus C5, overlaying the graphs on the same plot. Examine the plot. Do all the intervals have the same width? How many intervals were successful in catching the correct value of μ.

Exercises for Section 8.5

8-18   Suppose 9 men are selected at random from a large population. Assume the heights of the men in this population have mean of 69 inches but the population is not normal. Simulate the results of this selection 20 times as follows. Use Random Data ▶ t with 10 degrees of freedom to generate 9 rows of data in C1-C20. Now use either Session commands or the menu Calc ▶ Calculator to transform each column C1-C10 as C1=69+3*C1, C2=69+3*C2, and so forth. You will have to do this 20 times! Sorry. Each of the columns C1-C20 now represents a sample from a symmetric, nonnormal distribution centered at 69 inches. Now using C1-C20 as the variables, compute 1-Sample t confidence intervals each with 95% confidence.

   (a) Did your confidence intervals "work" 95% of the time? In particular, how many of your 20 intervals contain μ?

(b) Would you expect all 20 of the intervals to contain μ? Explain.

8-19   Table 8.1 on p. 259 displays estimates of actual confidence levels when you use t confidence intervals methods under nonstandard conditions—that is, when the population you sample from is not normal. Use the Minitab macro named MeanCI to answer the following questions. (The macro is available from the Data Library at www.CengageBrain.com associated with this book.)

(a) Use the MeanCI macro to estimate the actual confidence levels under the conditions shown in Table 8.1 on p. 259. If possible, use 10,000 replications. You may want to try 1000 replications first to see how quickly your computer will finish.

(b) Use the MeanCI macro to estimate the actual confidence level when selecting samples of size 5 from a t distribution with 15 degrees of freedom. How does your "hit rate" compare to the advertised 95%?

(c) Use the MeanCI macro to estimate the actual confidence level when selecting samples of size 25 from a t distribution with 1 degree of freedom. How does your "hit rate" compare to the advertised 95%? (In spite of the reasonably large sample size, this is a very difficult case as the t distribution with 1 degree of freedom is quite pathological!)

(d) Use the MeanCI macro to estimate the actual confidence level when selecting samples of size 5 from a chi-square distribution with 15 degrees of freedom. How does your "hit rate" compare to the advertised 95%?

(e) Use the MeanCI macro to estimate the actual confidence level when selecting samples of size 25 from a chi-square distribution with 2 degrees of freedom. How does your "hit rate" compare to the advertised 95%? (In spite of the reasonably large sample size, this is a difficult case as the chi-square distribution with 2 degrees of freedom is very asymmetrical.)

8-20   Use the Pulse data set. Display a normal probability plot of the Weight variable. Does the plot show evidence of nonnormality?

8-21   This exercise is designed to give you some idea of what normal probability plots can look like. It will help you learn to decide whether a particular plot looks definitely nonnormal.

(a) Use Calc ▶ Random Data ▶ Normal to simulate 20 observations from a normal distribution with mean 0 and standard deviation 1. Make a normal probability plot. Repeat this a total of five times.

(b) Repeat (a), but use $n = 50$ observations.

8-22   (a) Simulate 50 observations from a normal population with $\mu = 0$, $\sigma = 1$. Get a histogram and a probability plot.
              (b) Repeat part (a) until you have five histograms and five probability plots. How do the histograms and probability plots compare?

8-23   Simulate five normal probability plots to compare with the Otis scores. (Choose appropriate values for n, μ, and σ.) Compare these with the plot in Figure 8.12, p. 261. Is there any evidence that the Otis scores may not be a random sample from a normal population?

8-24   Simulate 100 observations from a normal distribution with mean 0 and standard deviation 1 into C1. Then make a normal probability plot and comment on the appearance of the plot under each of the following circumstances:
              (a) Use the entire set of 100 simulated observations.
              (b) Use only the observations between −1 and 1, selected by copying C1 into C2 using rows where C1 is between −1 and 1.
              (c) Repeat (b), but select the observations between −2 and 2.
              (d) Append two "outliers" by copying C1 into C3 and inserting the values −4 and 4 into the worksheet.
              (e) Repeat (d), but put the outliers at −5 and 5. Then do (d) again, with the outliers at −6 and 6.

Exercises for Section 8.6

8-25   (a) Using the Thickness data (p. 252) and the Assistant menu, carry out all of the hypothesis test steps displayed in Figures 8.13 and 8.14.
              (b) Do your results agree with Figures 8.15, 8.16, and 8.17?
              (c) How are colors used to describe the various results?

8-26   Repeat Exercise 8-16 using the Assistant menu. (Hint: The Assistant uses *percentages* rather than proportions.)

8-27   (a) Use the Assistant menu to repeat the analysis of the Gallup Poll data on page 255. (Hint: The Assistant uses *percentages* rather than proportions.)
              (b) Do your test results agree with the results in Section 8.4, p. 254?
              (c) Describe the results presented in the Summary Report.
              (d) Describe the results presented in the Report Card.
              (e) Describe the results presented in the Diagnostic Report

9

Comparing Two Means: Confidence Intervals and Tests

Many important questions involve comparisons between two populations.

- Did the region with the high advertising budget have substantially greater sales than the region with the low budget?
- How much stronger are the bonds made with the new glue than those made with the old glue?
- Is there evidence that women are paid less than men for the same jobs? If so, how much less?
- How much difference is there between the survival rate for patients on the new drug and the survival rate on the standard treatment?

In this chapter we introduce some methods by which two populations can be compared. The remarks in Chapter 8 concerning random samples and normal populations apply here also. That is, although the methods we present are "exact" only if the populations have normal distributions, they still work quite well for most populations. However, the nonparametric methods of Chapter 13 are more powerful for some nonnormal populations.

We begin by discussing the important distinction between samples of paired data and independent samples.

9.1 Paired and Independent Data

A Pennsylvania medical center collected some data on the blood cholesterol levels of heart-attack patients. A total of 28 heart-attack patients had their cholesterol levels measured 2 days after their attacks, 4 days after, and 14 days after. In addition cholesterol levels were recorded for a control group of 30 people who had not had heart attacks. Table 9.1 displays the data set.

This data set contains both paired and independent variables. The columns headed '2 days after' and '4 days after' are paired because the numbers in a given row are related—they are both for the same patient. For example the first patient had a '2 days after' score of 270 and a '4 days after' score of 218.

Table 9.1 Blood cholesterol levels after a heart attack

Experimental Group			Control Group
2 days after	4 days after	14 days after	
270	218	156	196
236	234	*	232
210	214	242	200
142	116	*	242
280	200	*	206
272	276	256	178
160	146	142	184
220	182	216	198
226	238	248	160
242	288	*	182
186	190	168	182
266	236	236	198
206	244	*	182
318	258	200	238
294	240	264	198
282	294	*	188
234	220	264	166
224	200	*	204
276	220	188	182
282	186	182	178
360	352	294	212
310	202	214	164
280	218	*	230
278	248	198	186
288	278	*	162
288	248	256	182
244	270	280	218
236	242	204	170
			200
			176

Because the observations are paired, we can determine how much each patient's cholesterol changed during that period.

Now let's look at an example of independent or unpaired data. The cholesterol levels in the last column of Table 9.1 are for a control group of 30 people. The people in this group were not linked in any known way to any of the patients in the experimental group. Thus we say that the "2 Days After" data and the control group data are independent.

The distinction between paired and independent data is relatively clear in this example, as it is in most situations. However, it is an important concept you should keep in mind because it determines which statistical procedures are appropriate for a given set of data.

9.2   Difference Between Two Means: Paired Data

A shoe company wanted to compare two materials, A and B, for use on the soles of boys' shoes. We could design experiments to compare the two materials in two ways. One way would be to recruit 10 boys (or more if our budget allowed) and randomly divide the boys into two groups of 5. Then give one group shoes made with material A and give the other 5 boys shoes made with material B. After a suitable length of time, say three months, we could measure the wear on each boy's shoes. This would be a situation of independent samples. Now you would expect a certain variability among 10 boys—some boys wear out shoes much faster than others. A problem arises if this variability is large. It might completely hide an important difference between the two materials.

The other method, a paired design, attempts to remove some of this variability from the analysis so we can see more clearly any differences between the materials we are studying. Again suppose we started with the same 10 boys, but this time had each boy test both materials. There are several ways we could do this. Each boy could wear material A for three months, then material B for three months. Or we could give each boy a special pair of shoes with the sole on one shoe made from material A and the other from material B. This latter procedure produced the data shown in Table 9.2. Higher values indicate more wear.

Table 9.2 Wear for boys' shoes, using a paired design

Boy	Material A	Material B
1	13.2	14.0
2	8.2	8.8
3	10.9	11.2
4	14.3	14.2
5	10.7	11.8
6	6.6	6.4
7	9.5	9.8
8	10.8	11.3
9	8.8	9.3
10	13.3	13.6

To plot the data to get a picture of the variability,

1. Set Minitab settings to their defaults.
2. Enter the data shown in Table 9.2 into C1-C3 of a new Minitab worksheet.
3. Choose **Graph ▸ Scatterplot ▸ Simple** and create two graphs: **Material A** versus **Boy** and **Material B** versus **Boy**.
4. Use the **Multiple Graphs** button to Overlay the plots on the same graph. Click **OK**. See Figure 9.1.

Figure 9.1 Boys' shoe wear plot

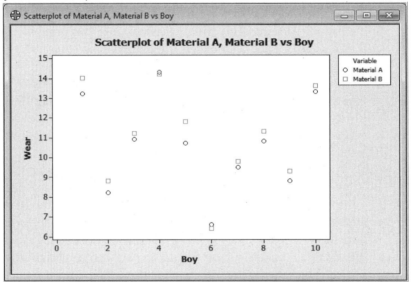

The plot shows that wear varied greatly from boy to boy. Boy 6 had the lowest wear—both shoes were well below 8. Boy 4 had the highest wear—both shoes were well above 14. For a given boy, however, both shoes had about the same amount of wear. Using statistical terminology, we would say, "The among-boy variability was high, but the within-boy variability was low."

One important feature of the plot is that in 6 of the 10 cases, Material B wear (displayed as open circles) is clearly lower than the Material B wear (open square symbols). In the remaining 4 cases the wear characteristics of the two materials are very close.

Confidence Intervals

In addition to looking at the original data, we can calculate a new quantity for each boy: the difference between the two materials. Both confidence intervals and tests for paired data analyses use this difference.

To calculate the difference between the two materials and produce the confidence interval,

1. Use the Calculator to create a new column, **C4**, that contains the difference, **C3-C2**. Name C4 **Difference**.
2. Choose **Stat ▸ Basic Statistics ▸ 1-Sample t**. Enter **Difference** as the variable and click **OK**. Figure 9.2 shows the results.

Figure 9.2   Confidence interval for differences in paired data

```
One-Sample T: Difference

Variable     N    Mean   StDev   SE Mean       95% CI
Difference   10   0.410  0.387    0.122    (0.133, 0.687)
```

Examine the worksheet to see that Difference has two negative values, −0.2 and −0.1, and eight positive values. These two negative values are from the two boys who had more wear with material A. The positive values correspond to the boys who had more wear with B.

The mean of the differences is 0.410. That is, on the average, material B did 0.41 wear units worse than A, or equivalently, A did 0.41 wear units better than B. To get some idea of the uncertainty in this estimate, we look at the confidence interval. This interval says we are 95% confident that the mean wear difference is between 0.133 and 0.687.

These numbers are a little hard to relate to. Suppose we express them as percentages. We could use the sequence **Calc ▸ Calculator** to calculate (average difference)/(average wear for A) as shown in Figure 9.3.[†] When you click OK, the result appears as a single number in column C5 of 0.0385701 or about a 4% decrease in wear for Material A versus Material B.

[†] You may enter all of this information by either typing or by clicking the column names and math and statistical functions needed.

Figure 9.3 Using the Minitab Calculator

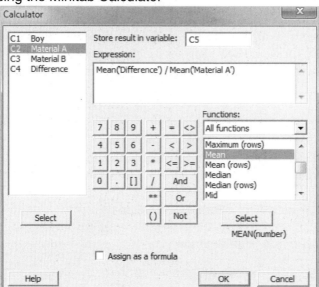

The paired-data analysis helped us see three things: There is a difference between the two materials. The difference is relatively consistent—in 8 of 10 cases material A was better than material B. The difference is quite small.

Paired *t* Test

We just saw that a confidence interval for paired data could be done by first computing the differences, then computing a (one sample) *t* interval on the differences. The *t* test for paired data can be done in a similar manner. Or you can use Minitab's Paired t command. To run a paired *t* test on the shoe data,

1. Choose **Stat ▸ Basic Statistics ▸ Paired t**
2. Enter **Material B** as the first sample and **Material A** as the second sample. See Figure 9.4.
3. Click **OK**. Figure 9.5 shows the results.

Figure 9.4 Paired *t* dialog box

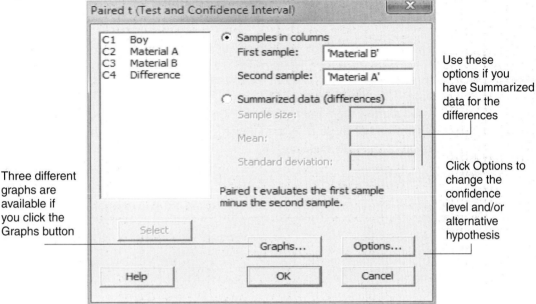

This tests the null hypothesis that the average difference between the two materials for shoes is zero.

Figure 9.5 Paired *t* test and confidence interval results

```
Paired T-Test and CI: Material B, Material A

Paired T for Material B - Material A

                N    Mean   StDev  SE Mean
Material B      10  11.040  2.518    0.796
Material A      10  10.630  2.451    0.775
Difference      10   0.410  0.387    0.122


95% CI for mean difference: (0.133, 0.687)
T-Test of mean difference = 0 (vs not = 0): T-Value = 3.35  P-Value = 0.009
```

We see, as before, that the estimate of the mean difference is 0.410. The *p*-value for the test is 0.009. With such a small *p*-value we have strong evidence that the mean change in the population is not zero. We would reject our null hypothesis if we had used $\alpha = 0.05$ or $\alpha = 0.01$ or any value of down to 0.009. Now we need to ask the subject matter specialist whether or not a difference of 0.41 wear units is practically significant.

9.3   Difference Between Two Means: Independent Samples

The data in Table 9.3 are taken from a study of Parkinson's disease, a disease that, among other things, affects a person's ability to speak. Eight of the people in this study had received one of the most common operations to treat the disease. This operation seemed to improve the patients' condition overall, but how did it affect their ability to speak? Each patient was given several tests. The results of one of these tests are shown in Table 9.3. The higher the score, the more problems with speaking.

Table 9.3 Speaking ability for patients with Parkinson's disease

Patients who had operation	Patients who did not have operation	
2.6	1.2	1.5
2.0	1.8	1.6
1.7	1.8	1.3
2.7	2.3	1.5
2.5	1.3	2.7
2.6	3.0	2.0
2.5	2.2	
3.0	1.3	

To display these data,

1. Open the **Parkinsn** worksheet.
2. Choose **Graph ▸ Dotplot ▸ Multiple Y's, Simple** and enter **Op** and **NoOp** as the variables.
3. Click **OK**. Minitab displays the dotplots shown in Figure 9.6.

Figure 9.6   Dotplots of speaking ability for those with and without the operation

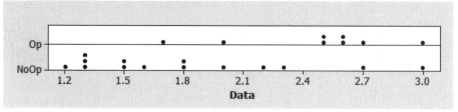

The samples overlap quite a bit. At this point it is not clear if the operation makes a difference.

Now suppose we want to do a formal test to see whether there is a statistically significant difference in the means and calculate a confidence for that difference. The 2-Sample t command does both of these things, giving the results of a *t* test and a confidence interval to compare two independent samples. If the confidence level is not specified, Minitab uses 95%. To run the two-sample test,

1. Choose **Stat ▸ Basic Statistics ▸ 2-Sample t**.
2. Select the **Samples in different columns** button.
3. Enter **Op** in the First box and **NoOp** in the Second box. See Figure 9.7.
4. Click **OK**. Figure 9.6 shows the results.

Figure 9.7 Two-sample *t* test and confidence interval dialog box

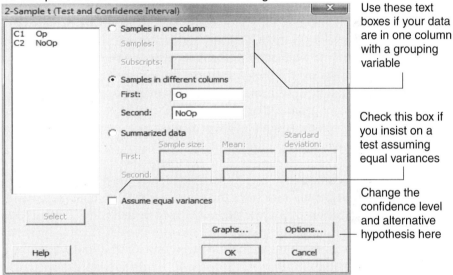

Now we can see that the means for two treatments are statistically significantly different. The *p*-value for the test is 0.0073, and we are 95% confident that the difference between the means is between 0.19 and 1.07 units.

Figure 9.8   Two-sample t test and confidence interval results

```
Two-Sample T-Test and CI: Op, NoOp

Two-sample T for Op vs NoOp

           N    Mean   StDev   SE Mean
Op         8   2.450   0.411      0.15
NoOp      14   1.821   0.556      0.15


Difference = mu (Op) - mu (NoOp)
Estimate for difference:  0.629
95% CI for difference:   (0.192, 1.065)
T-Test of difference = 0 (vs not =): T-Value = 3.02  P-Value = 0.007  DF = 18
```

Let's look for a moment at the practical significance of these results. The test says there is statistically significant evidence that the two population means differ. If we look at the sample means in Figure 9.6 and the dotplot in Figure 9.4, we see the direction of the difference. The patients who had the operation had more severe speech problems than those who did not have the operation. This, of course, does not prove that the operation causes difficulty in speaking. Perhaps the operation was performed only on patients who were already in poor condition.

The procedure used by 2-Sample t is slightly different from that described in some textbooks. Textbooks often use the pooled t, a procedure that assumes both populations have the same variance or, equivalently, the same standard deviation. The name *pooled* is used because the standard deviations from the two samples are pooled to get an estimate of the common standard deviation.

Suppose there are n_1 observations in the first sample, with mean $\bar{x}_1$ and standard deviation s_1. Suppose n_2, $\bar{x}_2$, and s_2 are the corresponding values for the second sample. Then the confidence interval is

$$(\bar{x}_1 - \bar{x}_2) \pm t_c \sqrt{\frac{s_1^2}{n_1} + \frac{s_2^2}{n_2}}$$

where t_c is the value from a t table corresponding to 95% confidence and degrees of freedom defined below.

The t statistic used in the test is

$$t = \frac{\bar{x}_1 - \bar{x}_2}{\sqrt{\frac{s_1^2}{n_1} + \frac{s_2^2}{n_2}}}$$

The degrees of freedom are obtained as:

$$df = \frac{(s_1^2/n_1) + (s_2^2/n_2)}{\dfrac{(s_1^2/n_1)^2}{n_1 - 1} + \dfrac{(s_2^2/n_2)^2}{n_2 - 1}}$$

When we use the pooled method, the pooled estimate of the common standard deviation is calculated by

$$s_p = \sqrt{\frac{(n_1 - 1)s_1^2 + (n_2 - 1)s_2^2}{n_1 + n_2 - 2}}$$

The pooled confidence interval is

$$(\bar{x}_1 - \bar{x}_2) \pm t_c s_p \sqrt{\frac{1}{n_1} + \frac{1}{n_2}}$$

where t_c is the value from a t table corresponding to $n_1 + n_2 - 2$ degrees of freedom. The pooled test statistic is

$$t = \frac{\bar{x}_1 - \bar{x}_2}{s_p \sqrt{\dfrac{1}{n_1} + \dfrac{1}{n_2}}}$$

If you use the pooled procedure when it is not appropriate—that is, when the standard deviations of the two populations are not equal—you could be seriously misled. For example you might falsely claim to have evidence that the two populations differ when they really do not. Of course you always have a chance of making such an error (called a Type I error) when you do a statistical test. In fact this is exactly what measures. When you do a test at $\alpha = 0.05$, you are supposed to have a 5% chance of claiming there is a difference when in actuality there is none. But if you use the pooled procedure when the population standard deviations are not equal, your chances of a Type I error may be very different from 5%. How different depends on how unequal the standard deviations and the sample sizes are.

The 2-Sample t command allows you to assume equal variances, which lets you do a pooled analysis. If you do not assume equal variances when you safely could have (that is, when the standard deviations of the two populations are equal), then, on the average, your analysis will be slightly conservative. That is, you will get a slightly larger confidence interval and you will be slightly less likely to reject a true null hypothesis. This conservatism essentially disappears with moderately large sample sizes (say, if both n_1 and n_2

are greater than 30). So in most cases, and especially for large sample sizes, it's better not to assume equal variance. If the standard deviations are equal, you've lost little, and if they're unequal, you may have gained a lot. We will explore this further in the Exercises.

Data in a Different Format

When you used the 2-Sample t command in the previous section, the data for the two samples were in separate columns. There is another way to organize data, a way that is more commonly used in statistics software. Table 9.4 shows the Parkinson's disease data in both forms. The first form, which we'll call unstacked, has each group in a separate column. The second puts all the data (this is often called the **response variable**) in one column and uses a second column to say which group each observation belongs to.

You can do a two-sample test and confidence interval when data are in the stacked form by clicking the Samples in one column option button in the 2-Sample t dialog box. See Figure 9.7, p. 283. The methods and output are exactly the same as for the Samples in different columns option; only the form of the data is different. In the Appendix, most data sets that contain two independent samples have the data in stacked form.

Table 9.4 Parkinson's disease data organized two different ways

Unstacked Data		Stacked Data	
Operation	No operation	Speaking ability	Operation
2.6	1.2	2.6	Op
2.0	1.8	2.0	Op
1.7	1.8	1.7	Op
2.7	2.3	2.7	Op
2.5	1.3	2.5	Op
2.6	3.0	2.6	Op
2.5	2.2	2.5	Op
3.0	1.3	3.0	Op
	1.5	1.2	NoOp
	1.6	1.8	NoOp
	1.3	1.8	NoOp
	1.5	2.3	NoOp
	2.7	1.3	NoOp
	2.0	3.0	NoOp
		2.2	NoOp
		1.3	NoOp
		1.5	NoOp
		1.6	NoOp
		1.3	NoOp

Table 9.4 Parkinson's disease data organized two different ways (Continued)

Unstacked Data		Stacked Data	
Operation	*No operation*	*Speaking ability*	*Operation*
		1.5	NoOp
		2.7	NoOp
		2.0	NoOp

Let's look at the Pulse experiment. Suppose we are interested in knowing how much difference, on the average, there is between the pulse rates of males and females. We will use the Pulse1 variable because these pulse rates were taken before anyone exercised.

The assumption of equal standard deviations seems plausible in this case, so we will assume equal variances to do the pooled *t* procedure. To compare the two groups with a boxplot and then use 2-Sample *t*,

1. Open the **Pulse** worksheet.
2. First code the numerical Sex column to more meaningful text values of Female and Male.
3. Now choose **Stat ▶ Basic Statistics ▶ 2-Sample t** and click the **Samples in one column** option button.
4. Enter **Pulse1** in the Samples box and **Sex** in the Subscripts box.
5. Click the **Graphs** option button and select **Boxplots of data**.
6. Select the **Assume equal variances** check box. See Figure 9.9.
7. Click **OK**. Figure 9.10 shows the boxplots and Figure 9.11 displays the numerical results in the Session Window.

Figure 9.9 Two-sample *t* dialog box, samples in one column

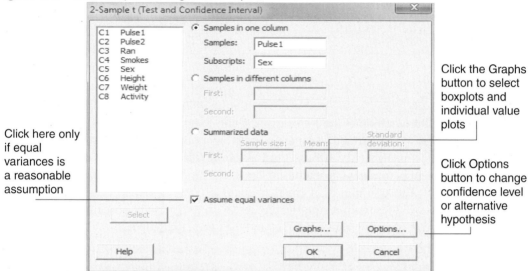

Click here only
if equal
variances is
a reasonable
assumption

Click the Graphs
button to select
boxplots and
individual value
plots

Click Options
button to change
confidence level
or alternative
hypothesis

Figure 9.10 Boxplot of pulse1 by sex

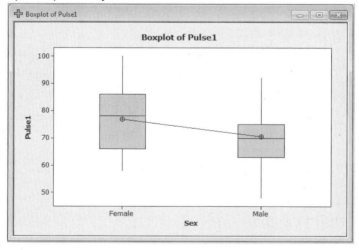

Figure 9.11 Two-sample *t* results with pooled estimate of sigma

```
Two-Sample T-Test and CI: Pulse1, Sex

Two-sample T for Pulse1

Sex       N   Mean   StDev  SE Mean
Female   35   76.9   11.6     2.0
Male     57   70.42  9.95     1.3


Difference = mu (Female) - mu (Male)
Estimate for difference:  6.44
95% CI for difference:  (1.91, 10.96)
T-Test of difference = 0 (vs not =): T-Value = 2.82  P-Value = 0.006  DF = 90
Both use Pooled StDev = 10.6093
```

From the display in Figure 9.10 the women seem to have a higher pulse rate, but there is an overlap with the men's rate. The test, however, has a *p*-value of 0.0058. This tells us there is statistically significant evidence that the mean pulse rates of the males and females are different even if we use an α as small as 0.01. We also note that the observed standard deviations of the males and females are reasonably close: 9.95 and 11.6. Thus the data do not seem to contradict our assumption of equal standard deviations.

9.4   Two-Sample Procedures with the Minitab Assistant

As in Chapter 8, the Minitab Assistant may be used to "automate" several of the two-sample procedures. Review the Assistant Hypothesis Tests screen on page 262. There are choices in the second column for comparing two means or two standard deviations from independent samples, the difference between means with paired samples, two sample proportions (%Defective) and chi-square that will be covered later. We'll use the Assistant to repeat the analysis of the paired shoe wear data from Section 9.2, p.277. When you click on the Assistant's Hypothesis Tests paired t choice the following dialog box appears:

Figure 9.12 The Assistant Hypothesis Tests Paired t dialog box

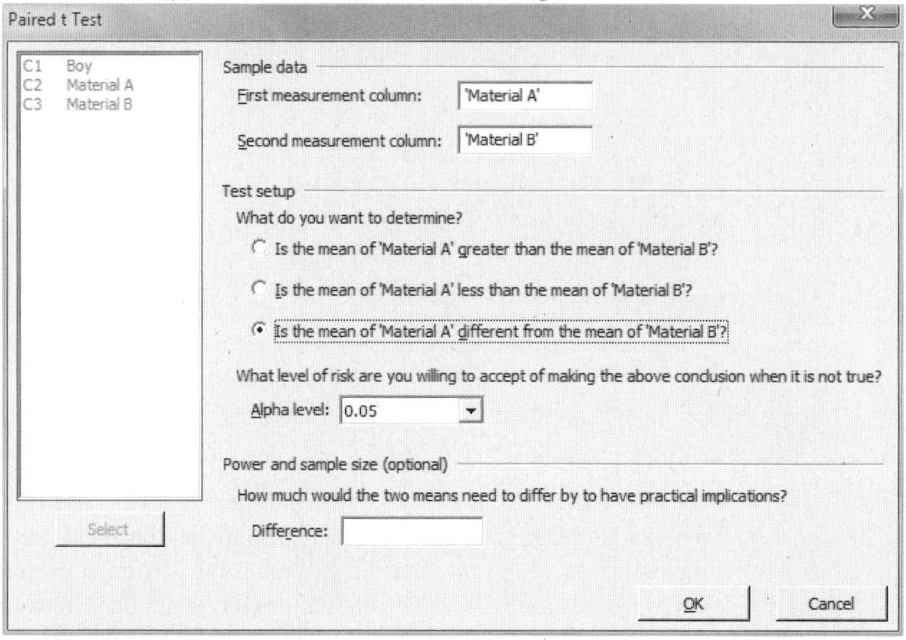

Enter the appropriate First and Second measurement columns and select the two-sided alternative hypothesis as shown in Figure 9.12. Notice how this dialog box uses the actual column names to help you specify the alternative hypothesis.

When you click OK, three screens are displayed. From the Summary Report screen, shown in Figure 9.13, we see several things. The mean wear of the two different types of shoe sole material are statistically significantly different (p-value 0.009). The estimated mean difference of Material A minus Material B is −0.41 and the 95% confidence interval for this difference is −0.69 to −0.13.

Figure 9.13 The Assistant Paired t Summary Report

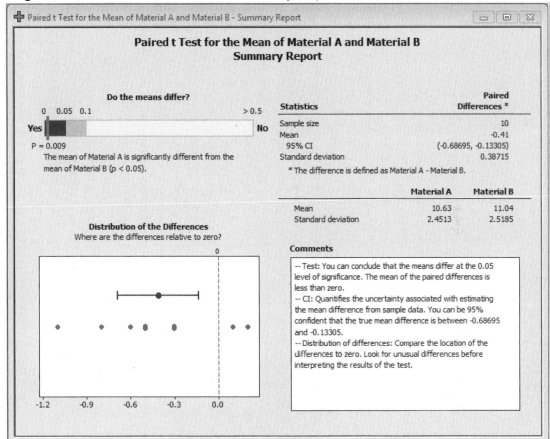

The Diagnostic report in Figure 9.14 does not indicate any problems with outliers nor trends in the data when plotted in worksheet order. The Report Card in Figure 9.15 reminds us that the sample size of 10 is too small to check for normality.

Figure 9.14 The Assistant Paired t Diagnostic Report

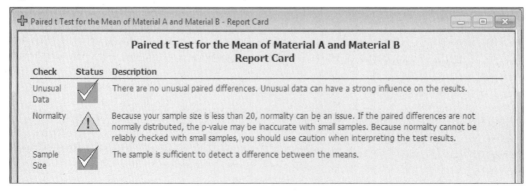

Figure 9.15 The Assistant Paired t Report Card

Paired t Test for the Mean of Material A and Material B - Report Card

Paired t Test for the Mean of Material A and Material B
Report Card

Check	Status	Description
Unusual Data	✓	There are no unusual paired differences. Unusual data can have a strong influence on the results.
Normality	⚠	Because your sample size is less than 20, normality can be an issue. If the paired differences are not normally distributed, the p-value may be inaccurate with small samples. Because normality cannot be reliably checked with small samples, you should use caution when interpreting the test results.
Sample Size	✓	The sample is sufficient to detect a difference between the means.

Exercises

Exercises for Section 9.1

9-1   For the cholesterol data in Table 9.1, p. 276, determine which of the following variables are paired and which are not:
(a) The '2 Days After' and '14 Days After.'
(b) The '4 Days After' and 'Control Group.' The '4 Days After' and '14 Days After.'
(c) The '14 Days After' and 'Control Group.'

9-2   In each of the following problems indicate whether the data are paired or independent samples. Justify your answer.
(a) A national fast-food chain used a special advertising campaign at half of its stores. The half was chosen randomly. It then compared sales at those stores with sales at its other stores.
(b) A total of 24 sets of mice were used in a study of a nutrition supplement. Each set consisted of two mice from the same litter. In each set one randomly chosen mouse was given the supplement and the other one was not. After three weeks the weight gain of each mouse was recorded for analysis.
(c) A tool manufacturer took a sample of 10 new wrenches. He randomly chose 5 of the wrenches and treated them with a special chemical. The other 5 were left untreated. The manufacturer then compared the strength of the treated and the untreated tools.
(d) A total of 50 people were used in a study to compare two treatments for the rash produced by poison ivy. A small patch of skin on the left arm of each person was exposed to poison ivy. Once a rash had developed, treatment A was used. The improvement was assessed and recorded. Two months later the same 50 people were again exposed to poison ivy. This time treatment B was used.

9-3   A number of data sets are described in the Appendix. Find three examples of paired data and three examples of impaired data. (Do not use the examples that were used in this chapter.)

9-4   In the Cartoon data (described in the Appendix, p. 524), which of the following comparisons involve paired data and which involve independent samples?
(a) Difference between the colors and black and white, as measured by the immediate cartoon score, and by the immediate realistic score.
(b) Difference between the Otis scores in hospital A and hospital B.

(c) Difference between the Otis scores of preprofessionals and professionals in hospital C.

(d) Difference between immediate cartoon and realistic scores for those participants who saw color slides.

(e) Difference between immediate and delayed cartoon scores for those participants who saw color slides.

Exercises for Section 9.2

9-5   Use the results shown in Figure 9.2, p. 279, to calculate by hand a 90% confidence interval for Difference. Also calculate a 99% confidence interval by hand. You may use a simple calculator.

8-6   (a) Use the Cholesterol data stored in the Cholest data file to estimate how much change in cholesterol level there is between the second and fourth day following a heart attack. Also do a plot like the one in Figure 9.1, p. 278.

(b) Estimate the amount of change between the second and fourteenth day following the attack. Notice that there are some missing observations in the data. Minitab commands omit missing cases from the analysis. What effect might this have on your conclusions?

9-7   The 179 participants in the Cartoon experiment (described in the Appendix, p. 524) each saw cartoon and realistic slides. Open the Cartoon worksheet.

(a) Do an appropriate test to see whether there is any difference between the two types of slides. Use the immediate scores.

(b) Estimate the difference between the two types of slides with a 90% confidence interval. Use a histogram to display the difference.

9-8   The file named Helium Football contains data from an experiment conducted to see if footballs filled with helium could be kicked farther than footballs filled with ordinary air. Two identical footballs, one air-filled and one helium-filled, were used outdoors on a windless day at The Ohio State University's athletic complex. Each football was kicked 39 times by a novice punter. The two footballs were alternated with each kick and the experimenter recorded the distance traveled by each ball.

(a) Analyze as paired data and test the hypothesis that, on average, helium-filled footballs will travel farther than air-filled footballs.

(b) Plot the difference, Helium minus Air, over time to look for trends.

(c) Plot the individual Helium and Air distances over time to look for trends.

Exercises for Section 9.3

9-9   The following questions refer to the results shown in Figure 9.8, p. 284.
    (a) Use this output to calculate, by hand, a 95% confidence interval for the mean speaking ability for people who did not have the operation. Use Minitab to verify your answer.
    (b) Using the 2-sample t formulas, find a 90% confidence interval for the mean difference between people who had the operation and those who did not. (Note: Round the degrees of freedom to the nearest value that is in your t table.)

9-10   Repeat the analysis of the Parkinson's disease data reported in Figure 9.8, p. 284, using pooled procedures. How do the results change? Be specific.

9-11   One of the authors, JC, ran the following experiment early in the semester of his large class in business statistics. The class was randomly divided into two groups. Each student was asked to guess the age of their professor—the author. However, one of the two groups was given a hint, namely the current ages of the professor's three children. The professor's secretary hypothesized that the students without the hint would guess lower that those who got the hint. Let μ_{hint} denote the mean of the guessed ages for the hypothetical population of business students who got the hint and let μ_{nohint} be the mean for the other group.
    (a) What is the null hypothesis?
    (b) What is the correct alternative hypothesis?
    (c) Table 9.5 gives the summary statistics for the experiment. Use the 2-Sample t command to carry out the required test. Be sure to use the correct alternative hypothesis.
    (d) Do the results support the secretary's conjecture?

Table 9.5 Results of the age guessing experiment

Group	n	Mean	Standard Deviation
Hint	224	60.205	3.450
No hint	259	51.556	6.254

9-12   Physicists are constantly trying to obtain more accurate values of the fundamental physical constants, such as the mean distance from the earth to the sun and the force exerted on us by gravity. These are very difficult measurements, and require the utmost ingenuity and care. Table 9.6 and Table 9.7 show results obtained in a Canadian experiment to measure the force of gravity. The first group of 32 measurements was made in August, and the second group in December of the following year. (*Note:* The force of gravity, measured in centimeters per second squared, can be obtained from these numbers by first dividing each number by 1000, then adding 980.61; thus the first measurement converts to 980.615.)

Table 9.6 Measurements made in August 1958

5	20	20	25	25	30	35
15	20	20	25	25	30	35
15	20	25	25	30	30	
15	20	25	25	30	30	
20	20	25	25	30	30	

(a) Use the data from August to estimate the force of gravity. Also find a 95% confidence interval.

(b) Repeat (a) for the December measurements.

Table 9.7 Measurements made in December 1959

20	30	35	40	40	45	55
25	30	35	40	40	45	60
25	30	35	40	45	50	
30	30	35	40	45	50	
30	35	40	40	45	50	

(c) Now compare the measurements made in August to those made in December. Do a dotplot and an appropriate test, using $\alpha = 0.05$. Is there a statistically significant difference between the two groups? What is the practical significance of this result?

(d) In between these two sets of measurements, it was necessary to change a few key components of the apparatus. Might this have made a difference in the measurements? If so, can you draw any guidelines for sound scientific experimentation? What if the experimenters had done all their measurements at one time? Might they have misled themselves about the accuracy of their results? What if still other parts of the apparatus were changed? Might the measurements change even more?

9-13   A small study was done to compare how well students with different majors do in an introductory statistics course. Seven majors were tested: biology, psychology, sociology, business, education, meteorology, and economics. At the end of the course the students were given a special test to measure their understanding of basic statistics. Then a series of t tests were performed to compare every pair of majors. Thus biology and psychology majors were compared, biology and sociology majors, psychology and sociology majors, and so on, for a total of 21 *t* tests. Simulate this study, assuming that all majors do about the same. Assume there are 20 students in each major and that scores on the test have a normal distribution with $\mu = 12$ and $\sigma = 2$. Use Calc ▶ Random Data ▶ Normal and generate 20 rows of data, stored in C1-C7, with a mean of 12 and standard deviation 2.

This puts a sample for each of the seven majors into a separate column.

(a) What is the null hypothesis?

(b) What are the 21 pairs of majors for the 21 *t* tests?

(c) Do the 21 *t* tests.

(d) In how many of the tests did you reject the null hypothesis at $\alpha = 0.10$?

(e) Because this study was simulated, the true situation is known—there are no differences. However, you probably did find a significant difference in at least one pair of majors. This illustrates the "hazards" of doing a lot of comparisons without making proper adjustments in the procedure. Try to think of some other situations where you might do a lot of statistical tests. For example suppose a pharmaceutical firm had 16 possible new drugs that they wanted to try out in the hope that at least one was better than the present best competing brand. What are some of the consequences of doing a lot of statistical tests?

9-14 A seventh-grade student named Kristy was given an experiment to run. She was to complete the same basic task a number of times. Half the time she was to do the task in a quiet room, and half the time she was to do it while chewing gum, with the radio and TV turned up loud. Table 9.8 shows the data.

Table 9.8 Quiet versus loud

	Time to Complete Task									
Quiet (Trials 1-10)	60	50	40	21	28	17	19	14	12	12
Noisy (Trials 11-20)	36	30	35	21	20	15	18	9	10	10

(a) Make a dotplot and use Descriptive Statistics to compare the two sets of times. Discuss the results. Is one environment appreciably better than the other?

(b) Do an appropriate test.

(c) The 20 trials were done in the order indicated: first, trial 1, then trial 2, and so on. Create a time series plot to display the data. What do you see?

(d) Discuss the planning of the experiment and how it could have been improved. Would you change your conclusions for part (a) after seeing the plot in (b)? Can you offer any advice to future analysts who might be comparing data like these?

9-15   Comparisons between two groups can be made for several variables in the Pulse data (in the Appendix, p. 529).

(a) Make stem-and-leaf displays of the first pulse rates, separating the smokers from the nonsmokers.

(b) Repeat part (a), using boxplots.

(c) Does there seem to be any overall difference between the smokers and nonsmokers? Explain.

(d) Do an appropriate test.

Exercises for Section 9.4

9-16   The Worksheet file named Maze Times contains the times (in seconds) for 19 males and 39 females to complete a maze. Use the Minitab Assistant to investigate whether there is any difference in mean completion times between males and females. (The data are available in the file named Maze Times in the Data Library at www.CengageBrain.com associated with this book.)

(a) First unstack the males and females into separate columns since the Assistant only works with data in separate columns.

(b) What are the main results displayed in the Summary Report?

(c) Does the Diagnostic Report show any outliers? If so, which data points are they?

(d) Based on the Report Card display, should we be concerned about normality of the either of the two populations? Why or why not?

9-17   Repeat the analysis of Exercise 9-12 using the Assistant.

9-18   Repeat the analysis of Exercise 9-8 using the Assistant.

10

Analysis of Variance

10.1 Analysis of Variance with One Factor

One-way analysis of variance is used to compare data from several populations. We will start with an example.

The flammability of children's sleepwear has received a lot of attention over the years. There are standards to ensure that manufacturers don't sell children's pajamas that burn easily. But a problem always arises in cases like this. How do you test the flammability of a particular garment? The shape of the garment might make a difference. How tightly it fits might also be important. Of course the flammability of clothing can't be tested on children. Perhaps a metal manne-quin could be used. But how should the material be lighted? Different people putting a match to identical cloth will probably get different answers. The list of difficulties is endless, but the problem is real and important.

One procedure that has been developed to test flammability is called the Vertical Semirestrained Test. The test has many details, but basically it involves holding a flame under a standard-size piece of cloth that is loosely held in a metal frame. The dryness of the fabric, its temperature, the height of the flame, how long the flame is held under the fabric, and so on are all carefully con-trolled. After the flame is removed and the fabric stops burning, the length of the charred portion of the fabric is measured and recorded.

Once we have a proposed way to test flammability, one important question is, "Will the laboratories of the different garment manufacturers all be able to get about the same results if they apply the same test to the same fabric?" A study was conducted to answer this question. A small part of the data is shown in Table 10.1.

We want to look at displays of these data and use one-way analysis of variance to study the differences among laboratories. Figure 10.1 shows a plot of the individual values from the five labs together with the means of the five labs. The Individual values have been jittered in the horizontal direction so that they are all visible. (We will see how easy it is to create this graph in the next Section.)

Table 10.1 An interlaboratory study of fabric flammability

| | Laboratory | | | | |
	1	2	3	4	5
	2.9	2.7	3.3	3.3	4.1
	3.1	3.4	3.3	3.2	4.1
	3.1	3.6	3.5	3.4	3.7
	3.7	3.2	3.5	2.7	4.2
	3.1	4.0	2.8	2.7	3.1
Length of charred portion of fabric	4.2	4.1	2.8	3.3	3.5
	3.7	3.8	3.2	2.9	2.8
	3.9	3.8	2.8	3.2	3.5
	3.1	4.3	3.8	2.9	3.7
	3.0	3.4	3.5	2.6	3.5
	2.9	3.3	3.8	2.8	3.9
Sample means	3.34	3.30	3.30	3.00	3.65

Figure 10.1   Individual value plot of Charred versus Lab

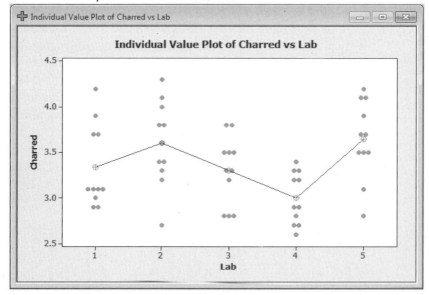

This display tells us a lot by itself. Even within a given laboratory, the char length varies from specimen to specimen. All of the specimens are supposedly identical because they were cut from the same bolt of cloth, but there were probably still some differences between them. In addition the test conditions surely differed slightly from specimen to specimen—the tautness of the material, the exact length of time the flame was held under the specimen, and so on. We must expect some variation in results even when all measurements are made in the same laboratory, under apparently identical conditions. The variation within a laboratory can be thought of as random error. There are several other names for this variation: within-group variation, unexplained variation, residual variation, or variation due to error.

There is a second source of variation in the data: variation due to differences among labs. This variation is called among-group variation, variation due to the factor, or variation explained by the factor. Here the factor under study is laboratory. (In statistics the word **treatment** is sometimes used instead of **factor**.) If we look at the dotplots again, we see some variation among the five labs. Lab 4 got mostly low values, whereas labs 2 and 5 got mostly high values. Instead of using dotplots, we could compare the sample means for the five labs, given at the bottom of Table 10.1. The five sample means do vary from lab to lab, but again, some variation is to be expected. The question is, "Do the five sample means differ any more than we would expect just from random variation"?

Put another way, "Is the variation we see among the groups significantly greater than the variation we would expect to see, given the amount of variation within groups?" Analysis of variance is a statistical procedure that gives an answer to this question.

The One-Way Analysis of Variance Procedure

In one-way analysis of variance we want to compare the means of several populations. We assume that we have a random sample from each population, that each population has a normal distribution, and that all of the populations have the same variance, σ^2. In practice the normality assumption is not too important, the equal-variances assumption is not important (provided the number of observations in each group is about the same), but the assumption of random samples is very important. If we do not have random samples from the populations, or something that is very close to it, our conclusion can be far from the truth.

The main question is, "Do all of the populations have the same mean?" Suppose that a is the number of populations we have, and that μ_1 is the mean of the first population, μ_2 is the mean of the second population, μ_3 is the mean of the third, and so on. Then the null hypothesis of no differences is $H_0: \mu_1 = \mu_2 = \mu_3 = \ldots = \mu_n$.

To test this null hypothesis, we can use the Minitab One-way command. The One-way command expects the response variable in a single column with factors in a separate column; the One-way (Unstacked) expects responses in separate columns. The Fabric worksheet stores the data in five columns, as shown in Table 10.1. The Fabric1 worksheet stores the data in one column with Lab subscripts in a second column. To perform a one-way analysis of variance on the Charred column, with Lab as a factor,

1. Open the **Fabric1** Worksheet.
2. Choose **Stat ▸ ANOVA ▸ One-way**.
3. Enter **Charred** as the response and **Lab** as the factor. See Figure 10.2.
4. Click on the **Graphs** button and check the **Individual value plot** check box.
5. Click **OK**. Minitab produces the numerical output shown in Figure 10.3 and the graph of Figure 10.1.

Figure 10.2   One-Way Analysis Of Variance dialog box

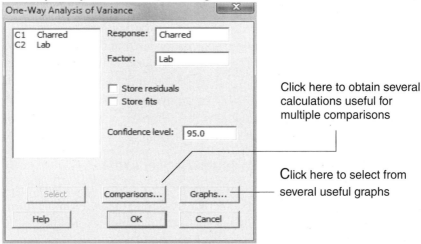

Figure 10.3 shows the results of analyzing the data in Table 10.1, p. 300. The first part of the output is called an analysis of variance table. In it the total sum of squares is broken down into two sources—the variation due to the factor (here it is differences among the five labs) and the variation due to random error (here it is variation within labs). Thus (SS TOTAL) = (SS Lab) + (SS ERROR). In this example $11.219 = 2.987 + 8.233$. Each sum of squares has a certain number of degrees of freedom associated with it. These are used to do tests. The degrees of freedom also add up:

(DF TOTAL) = (DF Lab) + (DF ERROR)

Figure 10.3 Results from one-way analysis of variance command

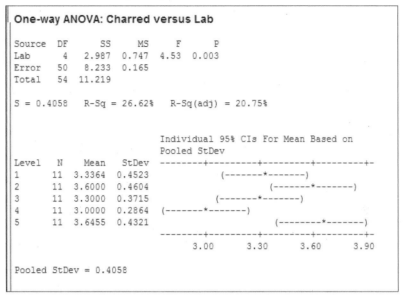

```
One-way ANOVA: Charred versus Lab

Source  DF     SS     MS     F      P
Lab      4   2.987  0.747  4.53  0.003
Error   50   8.233  0.165
Total   54  11.219

S = 0.4058   R-Sq = 26.62%   R-Sq(adj) = 20.75%


                                 Individual 95% CIs For Mean Based on
                                 Pooled StDev
Level   N    Mean    StDev   --------+---------+---------+---------+-
1      11  3.3364  0.4523              (-------*-------)
2      11  3.6000  0.4604                      (-------*-------)
3      11  3.3000  0.3715            (-------*-------)
4      11  3.0000  0.2864  (-------*-------)
5      11  3.6455  0.4321                       (--------*-------)
                          --------+---------+---------+---------+-
                            3.00      3.30      3.60      3.90

Pooled StDev = 0.4058
```

Formulas for the Three Sums of Squares. In Table 10.2, x_{ij} is the jth observation in the sample from group (lab) i, $\bar{x}_i$, and n_i are the sample mean and sample size for sample i, a is the number of groups (labs), n is the total number of observations, and $\bar{x}$ is the mean of all n observations.

Table 10.2 Sums of squares table

Source	DF	SS
Factor	$a-1$	$\sum_i n_i(\bar{x}_i - \bar{x})^2$
Error	$n-a$	$\sum_i \sum_j (x_{ij} - \bar{x}_i)^2$
Total	$n-1$	$\sum_i \sum_j (x_{ij} - \bar{x})^2$

Figure 10.3 gives mean squares (abbreviated MS). Each mean square is just the corresponding sum of squares divided by its degrees of freedom. The next column gives the quotient: F = (MS FACTOR)/(MS ERROR). This F-ratio is a useful test statistic. It is very large when MS FACTOR is much larger than MS ERROR—that is, when the variation among the labs is much greater than the variation due to random error (within the labs). In such cases we would reject the null hypothesis that the labs all have the same average char length. Is our observed F-ratio large? That's what the p-value tells us, as it did in the t tests discussed in Chapters 8 and 9. Here the p-value is 0.003. Thus for

$\alpha = 0.05$ or even 0.01, we would reject the null hypothesis and conclude we have evidence that there are some statistically significant differences among the five labs.

The table in Figure 10.3 also summarizes the results separately for each lab. The sample size, sample mean, sample standard deviation, and a 95% confidence interval are given for each lab. Each individual 95% confidence interval is calculated as

$$\bar{x}_i \pm t_c s_p / \sqrt{n_i}$$

Here s_p = Pooled StDev = $\sqrt{\text{MS Error}}$ is the pooled estimate of the common standard deviation σ, and t_c is the multiplier from the t distribution corresponding to 95% confidence based on the degrees of freedom associated with MS Error. These intervals give us some idea of how the population means differ.

10.2 Analysis of Variance with Two Factors

Table 10.3 presents data from an experiment to study the driving abilities of two types of drivers, inexperienced and experienced. Experience is our first factor. Twelve drivers of each type took part in the study. Three road types were used: first class, second class, and dirt. Road type is our second factor. Four of the inexperienced drivers were assigned at random to each road type, and four of the experienced drivers were assigned at random to each road type. This gives a total of 2×3 = 6 treatment combinations or cells in our design, with four observations per cell. Each subject drove a one-mile section of road. The number of steering corrections the driver had to make was recorded. This is the response variable.

The Drive worksheet stores these data in the following format: C1 contains driving experience; 1 = inexperienced and 2 = experienced. C2 is used for road condition; 1 = first class, 2 = second class, and 3 = dirt. The numbers of driving corrections are stored in C4. C3 stores information about whether it was day or night. For now we want to focus just on the daytime data, so you'll use only the first 24 rows.

We can use the Stat ▸ Tables ▸ Descriptive Statistics command to create a table of the means for each category. This table gives us a good idea of how the drivers did on each road type. You might want to review "Tables of Summary Statistics", p. 136, before proceeding here.

Table 10.3 Driving performance data

	Road Type		
	First Class	*Second Class*	*Dirt*
	4	23	16
Inexperienced	18	15	27
Driver	8	21	23
	10	13	14
	6	2	20
Experienced	4	6	15
Driver	13	8	8
	7	12	17

To delete the nighttime data and produce the table of means,

1. Open the **Drive** worksheet.
2. Delete rows **25-48**. (To delete rows quickly, drag the pointer down the row headers from row 25 to 48 in the Data window and press the Delete key.)
3. Choose **Stat ▸ Tables ▸ Descriptive Statistics**.
4. Enter **Expernc** for rows and **Road** for columns as the Categorical variables.
5. To display means of the correction data, click the **Associated variables** button, enter **Corrects** as the associated variable, and click the **Means** check box. Click **OK**.
6. Click the **Categorical variables** button and deselect the **Counts** check box to avoid cluttering the table too much.
7. Click **OK** twice. Minitab displays the table shown in Figure 10.4 in the Session Window.

These steps produce a working table suitable for preliminary analysis. To obtain a presentation quality table, we should first rename the response column as Corrections, the Expernc column as Experience, and recode the categorical variables to meaningful text such as Inexperienced and Experienced and First Class, Second Class, and Dirt (p. 88). Then we would need to order these text values properly (p. 127). Minitab would then produce a table that is quite readable by humans and suitable for a report.

Figure 10.4   A summary table of means

```
Tabulated statistics: Expernc, Road

Rows: Expernc    Columns: Road

              1      2      3     All

1         10.00  18.00  20.00   16.00
2          7.50   7.00  15.00    9.83
All        8.75  12.50  17.50   12.92

Cell Contents:  Corrects  :   Mean
```

On the average, inexperienced drivers made 16.00 steering corrections, whereas experienced drivers made only 9.83. Further if we look at the individual cell means, we notice that the inexperienced drivers did worse than the experienced drivers on all three road types.

Suppose we now look at road type. On the average the number of steering corrections goes up as the road quality goes down, from 8.75 to 12.50 to 17.50. The pattern does not quite hold up if we look at the individual cells. The experienced drivers did slightly better on second-class roads than on first-class roads. However, the decrease is small and could be due merely to random variation.

Minitab can easily produce a graph that displays these **main effect** means. To create the main effect graph,

1. Choose **Stat ▸ ANOVA ▸ Main Effects Plot**.
2. Enter **Corrects** as the Response and **Expernc** and **Road** as the Factors.
3. Click OK and Minitab displays the graph in Figure 10.7.

Figure 10.5   Main effects plot for driving corrections

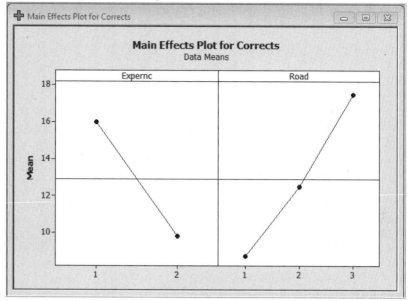

Additive Models

Now just as we did in one-way analysis of variance, we can express the total variation in the data as the sum of the variations from several sources:

(Total variation in data) = (variation due to driving experience)
+ (variation due to road type)
+ (variation due to random error)

If the variation due to driving experience is enough greater than the variation due to random error, we will have statistically significant evidence of a difference between the means of the two types of drivers. Similarly if the variation due to road type is much greater than the variation due to random error, we will have statistically significant evidence of a difference among the means for the three road types. We will give formulas for these when we discuss the model with interaction in the next section. To compute the analysis of variance,

1. Choose **Stat ▸ ANOVA ▸ Balanced ANOVA**.
2. Enter **Corrects** as the response variable and **Expernc** and **Road** in the Model text box. Click **OK**. Figure 10.6 shows the output.

The output in Figure 10.6 gives the breakdown of the total variation. Thus (SS Total) = (SS Expernc) + (SS Road) + (SS Error). As in One-way each sum of squares has a certain number of degrees of freedom associated with it, and these also add up. Each mean square is the corresponding sum of squares divided by its degrees of freedom. The value of F = (MS for the factor)/(MS Error). And the p-value gives the statistical significance of the corresponding F-statistic. Both p-values are small. Thus there is evidence to say that both factors, driving experience and road type, have an effect on steering corrections.

Figure 10.6   Two way analysis of variance with an additive model

```
ANOVA: Corrects versus Expernc, Road

Factor    Type    Levels  Values
Expernc   fixed        2  1, 2
Road      fixed        3  1, 2, 3


Analysis of Variance for Corrects

Source   DF         SS       MS      F       P
Expernc   1     228.17   228.17   8.56   0.008
Road      2     308.33   154.17   5.78   0.010
Error    20     533.33    26.67
Total    23    1069.83


S = 5.16398    R-Sq = 50.15%    R-Sq(adj) = 42.67%
```

Minitab also prints a table of factors that lists what type each factor is, the number of levels it has, and the actual values of the levels. In this case there are two types of factors, fixed and random, but in this chapter we will use only fixed factors. Random factors are discussed in Chapter 16. Note that you could also do this analysis using the Two-way ANOVA command, but if you did, Minitab would omit the table of factors and you would need to select the Fit additive model check box to get the same analysis.

Models with Interactions

There is another source of variations that we have not fully discussed: **interaction** between the two factors. The driving performance data set does not have a significant interaction, and we will have you verify that in an exercise.

Now we will look at another experiment. Table 10.4 presents data from an experiment designed to study the effects of two factors on the quality of pancakes. The two factors were the amount of whey and whether or not a supplement was used. There are four levels of whey (0%, 10%, 20%, and 30%) and two levels of supplement (used and not used), giving a total of 4×2 = 8 treatment combinations or cells. Three pancakes were baked, using each treatment combination. Each pancake was then rated by an expert; the three ratings were averaged to give one overall quality rating. The higher the quality rating, the better the pancake. This was done three times for each treatment combination, giving a total of 3×8 = 24 overall quality ratings.

Table 10.4 Quality of pancakes

	Amount of whey			
	0%	10%	20%	30%
No Supplement	4.4	4.6	4.5	4.6
	4.5	4.5	4.8	4.7
	4.3	4.8	4.8	5.1
Supplement	3.3	3.8	5.0	5.4
	3.2	3.7	5.3	5.6
	3.1	3.6	4.8	5.3

The Pancake worksheet contains the data. The first column identifies whether or not a supplement was used (1 = no, 2 = yes). The second column identifies the amount of whey (0, 10, 20, 30), and the quality ratings are in the third column. First, let's create a table of means.

1. Open the **Pancake** worksheet.
2. Choose **Stat ▸ Tables ▸ Descriptive Statistics**.
3. Enter **Supplement** for rows and **Whey** for columns as the Categorical variables.
4. To display means of the quality data, click the **Associated variables** button, enter **Quality** as the associated variable, and click the **Means** check box. Click **OK**.
5. Click the **Categorical variables** button and deselect the **Counts** check box to avoid cluttering the table too much.
6. Click **OK** twice. Minitab displays the table shown in Figure 10.7 (Once more this is a working table—not a presentation quality table.)

First, let's look at the average effect of each factor. These are often called the **main effects**.

Pancake quality increases, on the average, as the percentage of whey increases. The six pancakes with 0% whey have a mean quality of 3.8; the six with 10% whey have a mean quality of about 4.2; for 20% the mean is 4.9; and for 30% it is 5.1. The average effect of using the supplement is not as great as the effect of whey: The mean quality of the 12 pancakes with no supplement is just a little higher than the mean quality for the 12 with the supplement. The Stat ▸ ANOVA ▸ Main Effects Plot command allows you to produce a useful plot of these means shown in Figure 10.8.

Figure 10.7   Main effect means for pancake data

```
Tabulated statistics: Supplement, Whey

Rows: Supplement    Columns: Whey

              0      10      20      30     All

1          4.400   4.633   4.700   4.800   4.633
2          3.200   3.700   5.033   5.433   4.342
All        3.800   4.167   4.867   5.117   4.487

Cell Contents:   Quality  :   Mean
```

We also can look at the individual cells in Figure 10.7. The Stat ▸ ANOVA ▸ Interactions Plot command allows you to produce the plot in Figure 10.9. Choose Quality as the response and Supplement and Whey as the factors.

Figure 10.8   Main effects plot for pancake quality

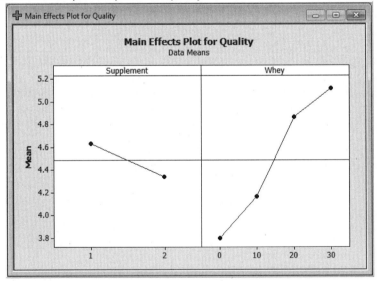

Figure 10.9   Interactions plot for pancake data

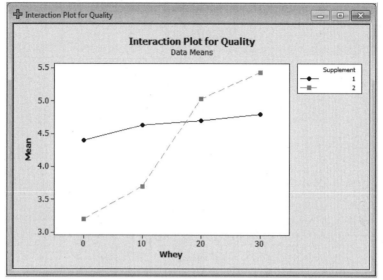

Now the effect of the supplement is not clear: Supplement increases quality when whey is 20% or 30%, but seems to lower quality when whey is 0% or 10%. Thus whether or not the supplement increases quality depends on the amount of whey. When the effect of one factor depends on the level of another, we say the two factors **interact**. Here supplement and whey appear to interact. If the two factors did not interact, the two lines in the plot of Figure 10.9 would be roughly parallel. To include an interaction term in the model, enter the name of the first factor, then an asterisk, then the name of the second factor. To perform this analysis of variance on the pancake data,

1. Choose **Stat ▸ ANOVA ▸ Balanced ANOVA**.
2. Enter **Quality** as the response and enter **Supplement   Whey   Supplement*Whey** in the Model box. See Figure 10.10.
3. Click **OK**. Figure 10.11 shows the output.

The total variation is now partitioned into four sources:
(Total variation) = (variation due to rows)
+ (variation due to column)
+ (variation due to interaction)
+ (variation due to error)

Figure 10.10 Balanced analysis of variance with interaction dialog box

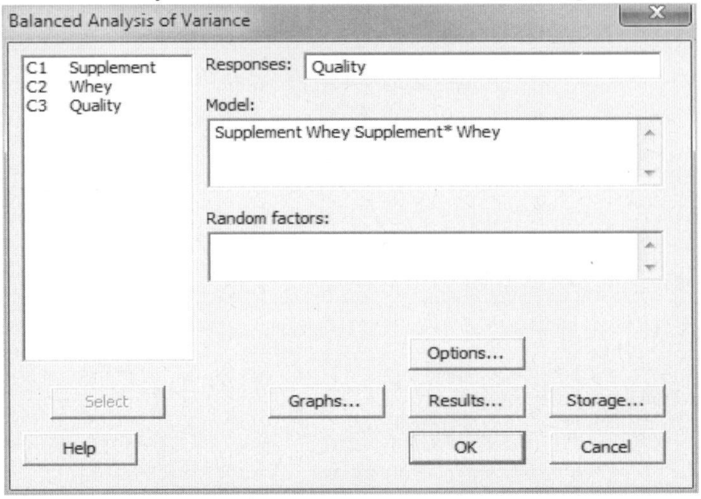

Notice that the sums of squares (and the degrees of freedom) in Figure 10.11 also add up as in this formula.

Figure 10.11 Balanced analysis of variance with interaction results

```
ANOVA: Quality versus Supplement, Whey

Factor       Type    Levels  Values
Supplement   fixed       2   1, 2
Whey         fixed       4   0, 10, 20, 30


Analysis of Variance for Quality

Source           DF      SS       MS       F      P
Supplement        1   0.5104   0.5104   17.01  0.001
Whey              3   6.6912   2.2304   74.35  0.000
Supplement*Whey   3   3.7246   1.2415   41.38  0.000
Error            16   0.4800   0.0300
Total            23  11.4062

S = 0.173205   R-Sq = 95.79%   R-Sq(adj) = 93.95%
```

Now let's look at the tests. We always test for interaction first. If interaction is present, then we must be careful how we interpret the other two tests. The *p*-value for interaction is 0.000, so there is strong evidence that the effect of the supplement depends on the amount of whey used. Or, put another way, the effect of whey depends on whether or not the supplement is used.

Let's look at the two other tests. These are often called tests for the average or main effects of the two factors; here they are Supplement and Whey. The *p*-value for Supplement is very small, 0.001, so there is statistically significant evidence of an overall effect of Supplement. The *p*-value for Whey is also very small, so there is significant difference due to Whey.

Let's look at Figure 10.7, p. 310, again. Notice that the overall mean for no supplement is 4.633, which is larger than the overall mean of 4.342 for supplement. This is consistent with our test: Supplement has an effect. But the test does not tell the whole story. In fact it tends to distort the real story. If we look at the individual cells, we see that sometimes supplement is better than no supplement, and sometimes it is worse. On the average no supplement is better than supplement. In general whenever there is a significant interaction, you should look at the individual cells to see what's really going on. Otherwise the tests on the main effects may be very misleading.

Formulas for the Sums of Squares. We will use the following notation: *a* is the number of levels of the first factor (that is, the number of rows in the table of data); *b* is the number of levels of the second factor (that is, the number of columns in the table of data); *n* is the number of observations in each cell; x_{ijk} is the *k*th observation in cell (i, j) (that is, in the cell in row *i* and column *j*); $\bar{x}_{ij.}$ is the mean of the *n* observations in cell (i, j); $\bar{x}_{i..}$ is the mean of the *bn* observations in row *i*; $\bar{x}_{.j.}$ is the mean of the observations in column *j*; and $\bar{x}$ is the mean of all *abn* observations in the table.

Table 10.5 Sums of squares

Source	DF	SS
Rows	$a-1$	$bn\sum_i(\bar{x}_{i..}-\bar{x})^2$
Columns	$b-1$	$an\sum_j(\bar{x}_{.j.}-\bar{x})^2$
Interaction	$(a-1)(b-1)$	$n\sum_i\sum_j(\bar{x}_{ij.}-\bar{x}_{i..}-\bar{x}_{.j.}+\bar{x})^2$
Error	$ab(n-1)$	$\sum_i\sum_j\sum_k(x_{ijk}-\bar{x}_{ij.})^2$
Total	$abn-1$	$\sum_i\sum_j\sum_k(x_{ijk}-\bar{x})^2$

The additive model we fit to the driving performance uses the same formulas for the rows, columns, and total that are used in this table. Since SS always add up, the SS for the error term in the additive model is (SS for Interaction) + (SS for Error) in this table. The same is true for the DF. In general if you omit any term from a model in analysis of variance, the SS and DF for that term are added into the error term.

10.3 Randomized Block Designs

In a randomized block design (abbreviated RBD) there is one factor we want to study, just as in a one-way design. But now we try to reduce some of the variability in the data by grouping the material, people, locations, time periods, or whatever into relatively homogeneous blocks. The treatments we wish to compare are then assigned at random within each block, with each treatment appearing exactly once in each block.

The data in Table 10.6 give the elasticity of billiard balls made under three different conditions. The experiment was carried out as follows: Ten batches of melted plastic were prepared. (These are the 10 blocks of material.) Each batch was divided into three equal portions. One portion was chosen at random and set aside as a control. A second portion was chosen at random from the remaining two, and was mixed with additive A. The third portion was mixed with additive B. In this way the experimenters hoped to balance out any variations in the plastic from batch to batch. Elasticity was measured on a scale from 0 to 100, with the higher numbers representing greater elasticity. (Higher elasticity is considered more desirable.)

Table 10.6 Elasticity of billiard balls

Batch	Control	Additive A	Additive B
1	51	75	39
2	45	89	43
3	49	73	51
4	66	84	34
5	53	66	54
6	41	85	43
7	58	73	42
8	56	71	41
9	60	78	37
10	63	65	44

These data are stored in the Billiard worksheet. Elasticity values are stored in C1. A second column, C2, was used for batch number. A third column, C3, was used for additive; 0 = no additive, 1 = additive A, and 2 = additive B. Let's first produce a table of means. To produce the means for the Billiard worksheet,

1. Open the **Billiard** worksheet.
2. Choose **Stat ▸ Tables ▸ Descriptive Statistics**.
3. Enter **Batch** for rows and **Additive** for columns as the Categorical variables.

4. To display means of the quality data, click the **Associated variables** button, enter **Elastic** as the associated variable, and click the **Means** check box. Click **OK**.

5. Click the **Categorical variables** button and deselect the **Counts** check box to avoid cluttering the table too much.

6. Click **OK** twice. Figure 10.12 displays the table of means.

Since there is just one number in each cell, the cell "means" are the individual data points. The column means give us the average for each treatment. Additive A seems to have improved elasticity over the control, but additive B seems to have reduced it.

Figure 10.12 Table of means for the billiard ball data

```
Tabulated statistics: Batch, Additive

Rows: Batch    Columns: Additive

              0      1      2     All

1         51.00  75.00  39.00   55.00
2         45.00  89.00  43.00   59.00
3         49.00  73.00  51.00   57.67
4         66.00  84.00  34.00   61.33
5         53.00  66.00  54.00   57.67
6         41.00  85.00  43.00   56.33
7         58.00  73.00  42.00   57.67
8         56.00  71.00  41.00   56.00
9         60.00  78.00  37.00   58.33
10        63.00  65.00  44.00   57.33
All       54.20  75.90  42.80   57.63

Cell Contents:  Elastic  :  Mean
```

Now just as we did in one-way analysis of variance, we can express the total variation in the data as the sum of the variation from several sources:

(Total variation in data) = (variation due to batch)
+ (variation due to additive)
+ (variation due to random error)

If the variation due to additive is much greater than the variation due to random error, we will have statistically significant evidence of a difference among the three levels of additive. To perform the analysis of variance,

1. Choose **Stat ▸ ANOVA ▸ Two-Way**.
2. Enter **Elastic** as the response, **Batch** as the row factor, **Additive** as the column factor, and then click **OK**. See Figure 10.13.

Figure 10.13 gives the breakdown of the total variation. Thus (SS Total) = (SS Batch) + (SS Additive) + (SS Error). As in One-way, each sum of squares has a certain number of degrees of freedom associated with it, and these also add up. Each mean square is the corresponding sum of squares divided by its degrees of freedom. The value of F = (MS for the factor)/(MS Error). And the value of p gives the statistical significance of the corresponding F-statistic. The p-value for additive is small, 0.000. Thus there is evidence that additive matters.

Figure 10.13 Analysis of variance for billiard ball elasticity data

```
Two-way ANOVA: Elastic versus Batch, Additive

Source     DF      SS       MS       F      P
Batch       9     82.30     9.14    0.12   0.999
Additive    2   5654.87  2827.43   36.62   0.000
Error      18   1389.80    77.21
Total      29   7126.97

S = 8.787   R-Sq = 80.50%   R-Sq(adj) = 68.58%
```

The p-value for batch is not small. In general we really don't care whether the blocking variable is significant. We do blocking to reduce variation so that we are more likely to detect differences in treatment. The fact that batch is not significant tells us we probably could have run this experiment as a completely randomized one-way analysis and still discovered that the effect of additive is statistically significant.

Formulas for the Sums of Squares. In Table 10.7, a is the number of treatments, b is the number of blocks, x_{ij} is the observation in block i given treatment j, $\bar{x}_{i.}$ is the mean of all a observations in block i, $\bar{x}_{.j}$ is the mean of all b observations given treatment j, and $\bar{x}$ is the mean of all ab observations.

Table 10.7 Sum of squares

Source	DF	SS
Blocks	$b-1$	$a\sum_i (\bar{x}_{i.} - \bar{x})^2$
Treatments	$a-1$	$b\sum_j (\bar{x}_{.j} - \bar{x})^2$
Error	$(b-1)(a-1)$	$\sum_i\sum_j (x_{ij} - \bar{x}_{i.} - \bar{x}_{.j} + \bar{x})^2$
Total	$ba-1$	$\sum_i\sum_j (x_{ij} - \bar{x})^2$

10.4 Residuals and Fitted Values

Analysis of variance can be viewed in terms of fitting a model to the data. This leads to a fitted value and a residual for each observation. A **fitted value** is our best estimate of the underlying population mean corresponding to that observation. The **residual** is the amount by which an observation differs from its fitted value. (The concepts of fitted values and residuals are discussed in more depth in Chapter 11, on regression models.)

One-Way Designs

In one-way analysis of variance the fitted values are just the group or cell means. A residual is then the difference between an observed value and the corresponding group mean. Consider, for example, the Fabric data. The observations, fitted values, and residuals for the first two laboratories are shown in Table 10.8.

Table 10.8 Fitted values and residuals for fabric data

	Laboratory 1			Laboratory 2	
Obs	*Fit*	*Residual*	*Obs*	*Fit*	*Residual*
2.9	3.34	−0.44	2.7	3.60	−0.90
3.1	3.34	−0.24	3.4	3.60	−0.20
3.1	3.34	−0.24	3.6	3.60	0.00
3.7	3.34	0.36	3.2	3.60	−0.40
3.1	3.34	−0.24	4.0	3.60	0.40
4.2	3.34	0.86	4.1	3.60	0.50
3.7	3.34	0.36	3.8	3.60	0.20
3.9	3.34	0.56	3.8	3.60	0.20
3.1	3.34	−0.24	4.3	3.60	0.70
3.0	3.34	−0.34	3.4	3.60	−0.20
2.9	3.34	−0.44	3.3	3.60	−0.30

You can generate these Fitted values and Residuals by either selecting the Store residuals and Store fits check boxes in the analysis of variance dialog box and then producing the plots. Alternatively, you may click the Graphs button on the analysis of variance dialog box and then select any of the plots shown there. See Figure 10.2, p. 302, for example.

It is good practice to make various plots of the residuals to check for unequal variances, nonnormality, dependence on other variables, and so on. We often make the following plots:

- Plot of the residuals versus the fitted values.
- Histogram or Dotplot of residuals.

- Normal probability plot of residuals.
- Plot of the residuals versus the order in which the data were collected.
- Plot of the residuals versus other variables (when data on other variables are available).

Two-Way Designs with an Interaction Term

In two-way analysis of variance with an interaction term, the fitted values are the cell means. A residual is then the difference between the observed value and the corresponding cell mean. We will use the Pancake data in Table 10.4, p. 309, as an example. Figure 10.7, p. 310, gave the cell means. The observations, fitted values, and residuals from the cells in which whey = 0% and 10% are shown in Table 10.9.

Table 10.9 Fitted values and residuals with an interaction term in the model

	0% Whey			10% Whey		
	Obs	Fit	Residual	Obs	Fit	Residual
No Supplement	4.4	4.40	0.00	4.6	4.63	−0.03
	4.5	4.40	0.10	4.5	4.63	−0.13
	4.3	4.40	−0.10	4.8	4.63	0.17
Supplement	3.3	3.20	0.10	3.8	3.70	0.10
	3.2	3.20	0.00	3.7	3.70	0.00
	3.1	3.20	−0.10	3.6	3.70	−0.10

Additive Models and Randomized Block Designs

Experiments with a two-factor design or a randomized block design are conducted differently, but when you analyze the data, you use the Two-way command for both. Also if you fit an additive model to the two-factor data, the fitted values and residuals are calculated in exactly the same way.

Fitted value for cell (i, j) = (mean of data in row i)
+ (mean of data in column j)
− (mean of all data)

Consider the Driving Performance data in Table 10.3. Means are given in Figure 10.4. The fitted value for the first cell, inexperienced drivers on first-class roads, is $16.000 + 8.750 − 12.917 = 11.833$. The fitted value for the next cell, inexperienced drivers on second-class roads, is 16.000 + 12.500 − 12.917 = 15.583. See Table 10.10.

Table 10.10 Fitted values and residuals for driving performance data

	First-Class Roads			*Second-Class Roads*		
	Obs	*Fits*	*Residuals*	*Obs*	*Fits*	*Residuals*
	4	11.833	−7.833	23	15.583	7.417
Inexperienced	18	11.833	6.166	15	15.583	−0.583
Drivers	8	11.833	−3.833	21	15.583	5.417
	10	11.833	−1.833	13	15.583	−2.583

Exercises

Exercises for Section 10.1

10-1   Use Dotplot and One-way to see whether the Otis scores differed by a statistically significant amount for the three education groups—preprofessional, professional, student—in the Cartoon experiment (described in the Appendix, p. 524).

10-2   Plywood is made by cutting a thin continuous layer of wood off a log as it is spun around. Several of these thin layers are then glued together to make plywood sheets. Thirty logs were used in a study. A chuck was inserted into each end of a log. The log then was turned and a sharp blade was used to cut off a thin layer of wood. Three temperature levels were used. Ten logs were tested at each temperature. The torque that could be applied to the log before the chuck spun out was measured. Open the Plywood1 worksheet and use Dotplot and One-way to see how the temperature level affects the amount of torque that can be applied. See Table 10.11.

Table 10.11 Log data

Temperature	Torque									
60°F	17.5	16.5	17.0	17.0	15.0	18.0	15.0	22.0	17.5	17.5
120°F	15.5	17.0	18.5	15.5	17.0	18.5	16.0	14.0	17.5	17.5
150°F	16.5	17.5	15.5	16.5	16.0	13.5	16.0	13.5	16.0	16.5

10-3   In a simple pendulum experiment, a weight (bob) is suspended at the end of a length of string. The top of the string is supported by some sort of stable frame. Under ideal conditions the time (T) required for a single cycle of the pendulum is related to the length (L) of the string by the equation

$$T = \frac{2\pi}{g}\sqrt{L}$$

where, as usual, $\pi = 3.14159...$ and g is the pull (acceleration) of gravity. If the time required for a cycle and the length of the pendulum are measured, this equation can be solved to give an estimate of g. In theory neither the length of the pendulum nor the type of bob should have any effect on the estimate of g. In practice, however, things that are not supposed to make any difference often do. So it is often a good policy, when doing an experiment, to vary, in a carefully balanced manner, things that are not supposed to matter. Therefore the experiment was run using four different lengths of string and two types of bobs. Table 10.12 shows estimates of g, in centimeters per second squared, as obtained by college professors during a short course on the use of statistics in physics and chemistry courses.

Table 10.12 Estimates of g from pendulum experiment

	Length of pendulum (cm)			
	60	70	80	90
Heavy bob	924	994	970	1000
	973	969	975	1017
	955	968	970	1055
Light bob	966	973	985	960
	949	997	999	1041
	955	988	994	962

(a) Open the Pendulum worksheet and use Dotplot to display the data and do a one-way analysis of variance to see whether length affects the mean estimate of g. Use only the data obtained with the heavy bob. (Weight=1 is heavy bob.)

(b) Repeat part (a), using only the data obtained with the light bob.

10-4   In the Restaurant survey (described in the Appendix, p. 530) the various restaurants were classified by the variable Typefood as fast food, supper club, and other. Open the Restaurant (recoded and value ordered) worksheet and use appropriate displays and the One-way command to determine the following:

(a) Do these three groups spend approximately the same percent of their sales on advertising? (Use the variable Ads.)

(b) Do they spend the same percent on wages? (Use the variable Wages.)

(c) Do they spend the same percent on the cost of goods? (Use the variable Costgood.)

10-5   Very small amounts of manganese are important to a good diet. Unfortunately measurement of these small amounts is quite difficult. To help researchers evaluate their ability to measure such small amounts, the National Institute of Standards and Technology (NIST; formerly the National Bureau of Standards) distributes samples of cow liver together with an accurate chemical analysis of the amount of manganese in each sample. The data given here are from one part of the evaluation. Eleven pieces were taken from one cow's liver. The experimenters wanted to know whether the amount of manganese varied from piece to piece. Of course even if the pieces were exactly the same, there would still be some differences in the recorded amounts just because of errors in making the measurements. Therefore the question posed was, "Do the 11 recorded amounts vary more than you would expect from measurement error alone?" To get some idea of how large measurement error was, the experimenters measured each piece twice. The amount of manganese (in parts per million) is given in Table 10.13. Open the Liver worksheet and use One-way to see whether there is a statistically significant difference among the 11 pieces.

Table 10.13 Manganese data

					Piece					
1	2	3	4	5	6	7	8	9	10	11
10.02	10.41	10.25	9.41	9.73	10.07	10.09	9.85	10.02	9.92	9.7
10.03	9.79	9.80	10.17	10.75	9.76	9.38	9.99	9.51	10.01	10.0

10-6   Steel makers find that small differences in the amount of oxygen in steel make an important difference in the quality of the steel they produce. Since small amounts of oxygen are difficult to measure accurately, NIST agreed to make very careful measurements on some homogeneous material, then to distribute the samples to steel manufacturers that could use them to check whether their own instruments were making accurate measurements. The researchers at NIST took a long "homogeneous" steel rod and cut it into 4-inch pieces. They randomly selected 20 of these pieces and labeled them 1 through 20. Then they made two very careful measurements of the amount of oxygen in each piece. Measurements were made over a five-day period in January. On each day the measurements were made in the order shown in Table 10.14. The table also provides the data, which are stored in the worksheet Oxygen. Bear in mind that finding and measuring quantities as small as 5 parts of oxygen in 1 million parts of steel is very difficult and is subject to a wide variety of unsuspected sources of error.

(a) Is there any evidence that the 20 pieces are not homogeneous? Use One-way.

(b) Is there any evidence of day-to-day variation?

(c) There is a major unsuspected source of error in these measurements. Can you find it?

Table 10.14 Steel data

Day	Piece number	Amount of oxygen (parts per million)	Day	Piece number	Amount of oxygen (parts per million)
1	17	5.6	3	19	6.2
1	11	5.9	3	17	5.1
1	10	6.8	3	9	3.3
1	15	7.5	4	13	5.4
1	5	4.7	4	18	6.1
1	6	4.0	4	2	5.7
1	19	4.4	4	10	6.1
2	4	6.6	4	4	5.7
2	7	4.9	4	20	4.6
2	1	5.5	4	7	5.7
2	13	4.9	4	6	4.4
2	3	6.3	4	12	4.1
2	18	4.2	5	16	6.3
2	9	3.3	5	15	7.5
2	14	4.8	5	8	6.1
3	16	6.1	5	14	6.4
3	12	5.3	5	3	5.1
3	8	5.2	5	20	5.7
3	2	4.3	5	5	3.8
3	11	4.0	5	1	3.8

10-7   (a) Open the Fabric1 Worksheet. Use the Minitab Assistant (**Hypothesis Tests ▶ One-Way ANOVA**) to redo the analysis shown in Figure 10.3, p. 303.
(b) Do the results come out the same?
(c) What else do you learn when doing the analysis with the Assistant?

Exercises for Section 10.2

10-8   Use Two-way analysis of variance to see whether there is a significant interaction in the Driving Performance data in Table 10.3, p. 305.

10-9   An experiment was done to see whether the damage to corn, wheat, dried milk, and other stored crops by a beetle, Trogoderma glabrum, could be reduced. The idea was to lure male adult beetles to inoculation sites where they would pick up disease spores, which they would then carry back to the remaining population for infection. The experiment consisted of two factors, "sex attractant or none" and "disease pellet present or not." The dependent variable was the fraction of larvae failing to reach adulthood.

Analyze the data presented in Table 10.15. Open the Beetle worksheet and get cell means; plot these means "by hand." Then do an analysis of variance. Interpret your results.

Table 10.15 Beetle data

Attractant	Disease	
	Yes	No
Yes	0.979	0.222
	0.743	0.189
	0.775	0.143
	0.885	0.262
No	0.312	0.239
	0.293	0.253
	0.188	0.159
	0.388	0.241

10-10   The Pendulum data of Exercise 10-3 involve two factors, length and bob. Do a two-way analysis of variance on the data.

(a) Display and interpret a Main Effects Plot.

(b) Does length seem to affect the estimate of g? Does the type of bob used affect the estimate? Is there a significant interaction between length and bob?

(c) Suppose you wanted to estimate the acceleration of gravity, using a pendulum. What do the results in part (a) tell you about designing such an experiment?

10-11   A substantial percentage of the potatoes raised in this country never have a chance to reach the grocery store. Instead, they fall victim to potato rot while being stored for later use. To find out what could be done to reduce this loss, an experiment was carried out at the University of Wisconsin. Potatoes were injected with bacteria known to cause rot, and were then stored under a variety of conditions. After five days the diameter of the rotted portion of each potato was measured. Three factors were varied in this experiment: (1) the amount of bacteria injected into the potato (1 = low amount, 2 = medium amount, 3 = high amount); (2) the temperature during storage (10°C, 16°C); (3) the amount of oxygen during storage (2%, 6%, 10%). Table 10.16 displays the data, which are stored in the Potato worksheet.

Table 10.16 Potato data

Bacteria	Temperature	Oxygen	Rot	Bacteria	Temperature	Oxygen	Rot
1	1	1	7	2	2	1	17
1	1	1	7	2	2	1	18
1	1	1	9	2	2	1	8
1	1	2	0	2	2	2	3
1	1	2	0	2	2	2	23
1	1	2	0	2	2	2	7
1	1	3	9	2	2	3	15
1	1	3	0	2	2	3	14
1	1	3	0	2	2	3	17
1	2	1	10	3	1	1	13
1	2	1	6	3	1	1	11
1	2	1	10	3	1	1	3
1	2	2	4	3	1	2	10
1	2	2	to	3	1	2	4
1	2	2	5	3	1	2	7
1	2	3	8	3	1	3	15
1	2	3	0	3	1	3	2
1	2	3	10	3	1	3	7
2	1	1	2	3	2	1	26
2	1	1	4	3	2	1	19
2	1	1	9	3	2	1	24
2	1	2	4	3	2	2	15
2	1	2	5	3	2	2	22
2	1	2	10	3	2	2	18
2	1	3	4	3	2	3	20
2	1	3	5	3	2	3	24
2	1	3	0	3	2	3	8

These data probably should be analyzed by using a three-way analysis of variance, but you can discover most of what is going on by using several two-way analyses and some plots and tables.

(a) Do a two-way analysis of variance on all the data, using bacteria and temperature as the two factors (ignore oxygen for the moment). Use the Intcraction Plot command to plot cell means.

(b) Next, do a two-way analysis of variance on all the data, using bacteria and oxygen as the two factors (ignore temperature in this analysis). Use the Interaction Plot command to plot cell means.

(c) How do each of the three factors influence potato rot? Do any of the factors appear to interact?

(d) The factors that can be controlled, to some extent, by the food supplier are temperature and oxygen. Analyze these two factors at each of the three levels of bacteria. Do temperature and oxygen seem to have the same effect at each level?

(e) What recommendations would you make to someone who was storing potatoes?

Exercises for Section 10.3

10-12   An experiment was performed at the University of Wisconsin to compare the yield of six varieties of alfalfa. Four fields were used. Each field was divided into six plots, one plot for each variety. The four fields can be considered to be four blocks. They are analogous to the 10 batches in the billiard ball experiment. The total yield from each combination was recorded. The data are stored in the file Alfalfa and are displayed in Table 10.17.

Use appropriate displays and tests to analyze the data.

Table 10.17 Alfalfa data

| | Field | | | |
Alfalfa Variety	1	2	3	4
Atlantic	3.22	3.31	3.26	3.25
Buffalo	3.04	2.99	3.27	3.20
Culver	3.06	3.17	2.93	3.09
Lohontar	2.64	2.75	2.59	2.62
Narragansett	3.19	3.40	3.11	3.23
Rambler	2.49	2.37	2.38	2.37

10-13   An experiment was done to compare different methods of freezing meat loaf. Meat loaf was to be baked, then frozen for a time, and finally compared by expert tasters. Eight loaves could be baked in the oven at one time. However, some parts of an oven are usually hotter or differ in some other important way from other parts. If this is the case, loaves baked in one part of the oven might taste better than those baked in another part, and differences in freezing methods might be masked. Therefore a preliminary test was conducted to see whether there were any noticeable differences among the eight oven positions used in the study.

The data shown in Table 10.18 came from this preliminary experiment. One batch of eight loaves was mixed; one loaf was assigned at random to each oven position. The loaves then were baked and analyzed. A second batch of eight loaves was mixed, assigned at random to the eight oven positions, baked, and analyzed. A third batch was tested in the same manner. Each batch is a block, and there are eight treatments (oven positions) within each block. Each loaf was analyzed by measuring the percentage of drip loss (that is, the amount of liquid that dripped out of the meat loaf during cooking divided by the original weight of the loaf).

Analyze the data, using appropriate tests and displays. Make sure you take into account all the information you are given about this experiment.

Table 10.18 Meat loaf data

	Batch		
Oven Position	1	2	3
1	7.33	8.11	8.06
2	3.22	3.72	4.28
3	3.28	5.11	4.56
4	6.44	5.78	8.61
5	3.83	6.50	7.72
6	3.28	5.11	5.56
7	5.06	5.11	7.83
8	4.44	4.28	6.33

Oven position on the shelf

5	6	7	8
4	3	2	1

Front of Shelf

(Note: Thermometers to measure temperature were inserted into loaves 1, 4, 5, 7, and 8.)

11

Regression and Correlation

Some of the most interesting problems in statistics occur when we try to find a model for the relationships among several variables. The data in Table 11.1 illustrate a common situation. Two tests were given to 31 individuals.

Table 11.1   Two test scores for 31 people

Subject	First score	Second score	Subject	First score	Second score
1	50	69	17	60	56
2	66	85	18	89	87
3	73	88	19	83	91
4	84	70	20	81	86
5	57	84	21	57	69
6	83	78	22	71	75
7	76	90	23	86	98
8	95	97	24	82	70
9	73	79	25	95	91
10	78	95	26	42	48
11	48	67	27	75	52
12	53	60	28	54	44
13	54	79	29	54	51
14	79	79	30	65	73
15	76	88	31	61	52
16	90	98			

We often want to answer questions such as the following: What is the correlation between the scores on the two tests? If you know someone's score on the first test, does that help you predict that person's score on the second test? What is the prediction of the second score for a person who scored 70 on the first test and how good is that prediction? In this chapter we will see how questions like these can be answered.

11.1 Correlation

There are several ways to measure the association between two variables, (x, y). The most common measure is the **Pearson product moment correlation coefficient**, or just the **correlation coefficient** or **correlation** for short. This is usually designated by the letter r and given by

$$r = \frac{1}{n-1}\sum_i \tilde{x}_i \tilde{y}_i$$

where $\tilde{x}_i$ and $\tilde{y}_i$ are standardized versions of x and y, namely, $\tilde{x}_i = (x_i - \bar{x})/s_x$ and $\tilde{y}_i = (y_i - \bar{y})/s_y$.

Table 11.1 presents the data and Figure 11.1 shows the scatterplot of the two test scores.

To compute the correlation between the two test scores,

1. Select **File ▸ New ▸ Project** to start a new Minitab session.
2. Open the **Scores** worksheet.
3. Choose **Graph ▸ Scatterplot ▸ Simple**, click **OK**, and plot **Second** versus **First**. Click **OK**. Minitab produces the graph.
4. Choose **Stat ▸ Basic Statistics ▸ Correlation**.
5. Enter **Second** and **First** as the variables. Click **OK**. Minitab reports the correlation in the Session Window as $r = 0.703$.

It can be shown mathematically that the a correlation coefficient is always between −1 and +1. We will denote the two variables by x and y. Here x would be the first score and y the second. The correlation coefficient is positive if y tends to increase as x increases—that is, if a plot of y versus x tends to slope upward. Conversely the correlation is negative if y tends to decrease as x increases—that is, if a plot of y versus x tends to slope downward. If the points fall exactly on some straight line, then $r = +1$ if the points slope upward and −1 if the trend is downward.[†] The closer the points are to forming a straight line, the closer r is to +1 or −1. The closer r is to +1 or to −1, the easier it is to predict y from x.

If there is almost no association between x and y, then r will be near 0. The converse, however, is not true. There are cases in which r is near 0 but there is still a clear association between x and y. We will see some examples in the exercises. The correlation coefficient measures one type of association—how closely the points fall on a straight line. Fortunately this is the most common type of association. But a correlation coefficient can be quite misleading at times, and plots may give you a better view of what is really going on.

[†] If the data fall exactly on a vertical or horizontal straight line, the correlation coefficient is not defined since then one of s_x or s_y is zero.

Figure 11.1   Scatterplot of test scores data

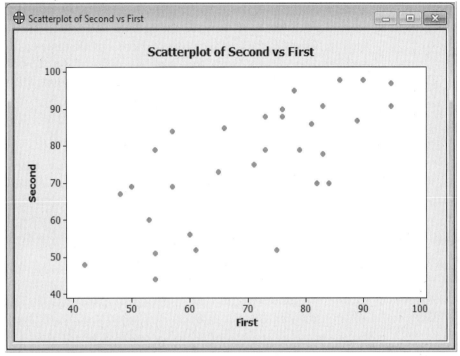

Figure 11.2 displays the scatterplot of standardized second and first exam scores. We added a horizontal line at $y = 0$ and a vertical line at $x = 0$.

Figure 11.2   Plot of standardized second score versus standardized first score

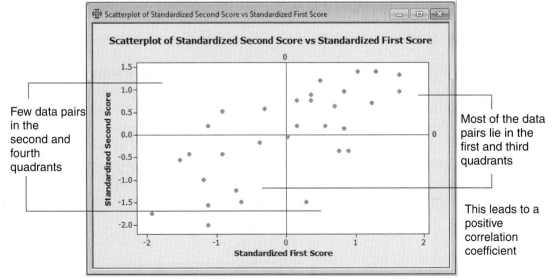

Since the correlation coefficient is the average of the products of these pairs of values, all pairs in the first or third quadrant of the graph produce positive values in this average. Pairs in the second and fourth quadrants contribute negative values to the average product. Here a predominance of pairs in the first and third quadrants leads to a positive correlation. In other situations a predominance of pairs in the second and fourth quadrants would produce many negative products and lead to a negative correlation.

11.2 Simple Regression: Fitting a Straight Line

Correlation tells us how much straight-line association there is between two variables; regression analysis goes much further. It gives us an equation that uses one or more variables to help explain the variation in another variable. Regression analysis also assesses the closeness of the data to that equation. In this section we will show how to use the Regression command to fit straight lines.

Look at the data in Table 11.1 again. Suppose we wanted an equation that, at least approximately, relates the second test score to the first test score. From Figure 11.1 we see that if a person scored 70 on the first test, we might expect a score of about 75 on the second test. If another person scored 90 on the first test, we might expect about a 90 on the second test. If still another scored 55 on the first test, we might expect a second score of about 65.

In fact it looks as if we might be able to draw a straight line on the plot and use it to help predict scores on the second test. Suppose we decide to do that. How would we draw the line? Any straight line we draw will miss some of the points. Figure 11.3 gives a version of this plot, with several lines drawn. Which one looks best?

A method known as **least squares** is often used to decide which line to choose. Suppose we look at the vertical differences or deviations between what a given line suggests and what actually happened for each person. If the line suggests a second score of 79 and the actual score was 88, then the deviation is $88 - 79 = 9$. Suppose we compute all the deviations for a given line, square them, and add them up. Notice that we use *vertical* deviations since we are interested in predicting the second score from the same person's first score.

Intuitively a line with large deviations is not as good as a line with smaller deviations. The least-squares criterion says we should use the line that gives the smallest sum of squared deviations. The surprising thing is that it is relatively easy to find the equation for that line.

Figure 11.3   Three possible lines to summarize the relationship

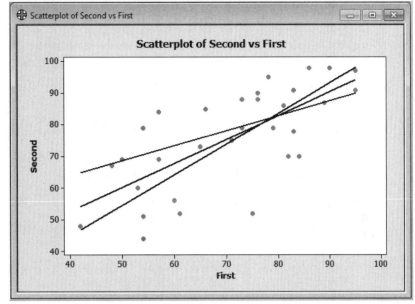

If we are trying to use x to determine y, the equation for any straight line can be written in the form $y = a + bx$, where a and b are two numbers that determine the particular line.[†] For example, $y = 3 + 2x$ and $y = -5 + 0.8x$ specify two different straight lines. It can be shown that for the least-squares line, a and b can be found with the formulas

$$b = r(s_y/s_x) \qquad a = \bar{y} - b$$

Minitab will give these least-squares values for a and b, and many other useful quantities, when the Regression command is used. The least squares regression line may also be added to any scatterplot with the click of a few buttons.

Suppose we want to add the least-squares regression line to the scatterplot shown in Figure 11.1, p. 329. To do so,

1. **Right-click** the scatterplot to make that window active and to open the graph extras menu. (Alternatively, use the Editor menu or Graph Editing toolbar.)
2. Select **Add ▸ Regression Fit** and a dialog box opens.
3. Select Model Order: **Linear** and check **Fit Intercept**. (These will likely already be selected.)

[†] b is the *slope* of the line and a is the *y-intercept*.

4. Click **OK** and the least-squares regression line is added to your scatterplot.

Now suppose we want to use the Regression command to obtain the full regression calculations. You enter a response variable and predictor variable (or variables) in the Regression dialog box, and Minitab generates the Regression output. You can control how much output you will receive by using the Results button. To regress Second score on First score,

1. Choose **Stat ▸ Regression ▸ Regression**.
2. Enter **Second** as the response variable and **First** as the predictor variable. See Figure 11.4.
3. For this illustration, click the **Results** button and click the final option button, shown in Figure 11.5, to display all possible results. This dialog box tells you what is included in the output for each topic; online Help can give you much more information.
4. Click **OK** in the Regression-Results dialog box and then click **OK** in the Regression dialog box. The Session Window, shown in Figure 11.6, displays the results.

Figure 11.4   Regression dialog box

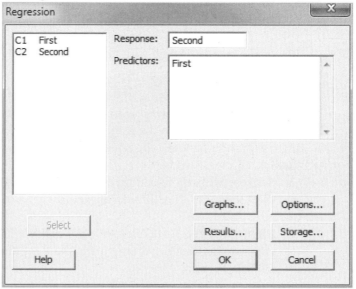

Figure 11.5   Regression results dialog box

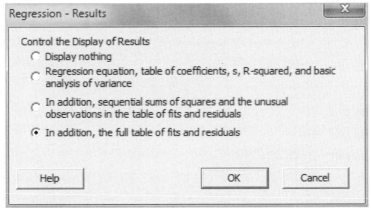

Figure 11.6 shows the full output that you can obtain from the Regression command. By changing the options in the Results dialog box you can receive an abbreviated version; we use the full output here to facilitate our explanation.

Look at some of the simpler parts of this output. The regression equation is Second = 22.5 + 0.755 First, or $y = 22.5 + 0.755x$. This equation corresponds to one of the lines in Figure 11.3. (Can you determine which one?) Since this is the least-squares line, it is impossible to find any other straight line that gives a smaller sum of squared deviations than this one. The sum of squared deviations for this line is 3844. It is printed in the Analysis of Variance table in Figure 11.6 as SS Residual Error. SS is an abbreviation for sum of squares and Error is another name for deviation or residual.

The next block of output again gives the values of $a = 22.47$ and $b = 0.7546$, with some additional information. Next comes the Analysis of Variance table, which gives, as we will see later, a breakdown of the variation in the data.

The last block of output lists each x value and corresponding y value, then gives the fitted y value from the least-squares equation. For example the first person scored 50 on the first test and 69 on the second test. The fitted score was 60.20, which was computed by calculating[†] 22.47 + 0.7546(50). The corresponding deviation or residual is thus $y - $ (fitted y) = 69 − 60.20, which gives 8.80.

[†] Minitab does this internally with many more digits for a and b. It then rounds the answer to the number of digits shown.

Figure 11.6   Regression results for second score regressed on first score

The equation of the least squares line

Estimate of σ

R-squared

Adjusted R^2

Fitted values

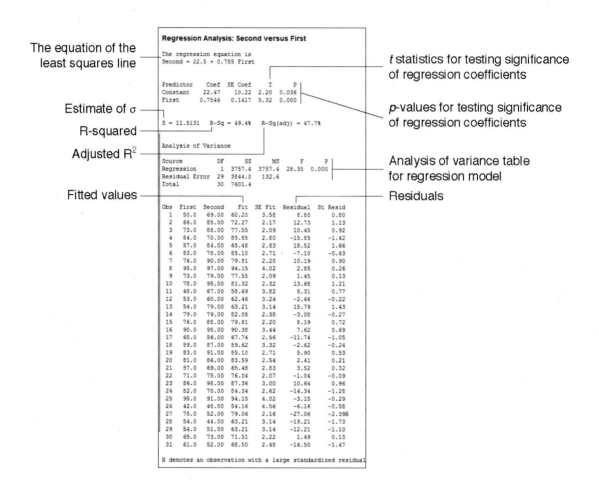

t statistics for testing significance of regression coefficients

p-values for testing significance of regression coefficients

Analysis of variance table for regression model

Residuals

Regression Analysis: Second versus First

The regression equation is
Second = 22.5 + 0.755 First

Predictor	Coef	SE Coef	T	P
Constant	22.47	10.22	2.20	0.036
First	0.7546	0.1417	5.32	0.000

S = 11.5131   R-Sq = 49.4%   R-Sq(adj) = 47.7%

Analysis of Variance

Source	DF	SS	MS	F	P
Regression	1	3757.4	3757.4	28.35	0.000
Residual Error	29	3844.0	132.6		
Total	30	7601.4			

Obs	First	Second	Fit	SE Fit	Residual	St Resid
1	50.0	69.00	60.20	3.58	8.80	0.80
2	66.0	85.00	72.27	2.17	12.73	1.13
3	73.0	88.00	77.55	2.09	10.45	0.92
4	84.0	70.00	85.85	2.80	-15.85	-1.42
5	57.0	84.00	65.48	2.83	18.52	1.66
6	83.0	78.00	85.10	2.71	-7.10	-0.63
7	76.0	90.00	79.81	2.20	10.19	0.90
8	95.0	97.00	94.15	4.02	2.85	0.26
9	73.0	79.00	77.55	2.09	1.45	0.13
10	78.0	95.00	81.32	2.32	13.68	1.21
11	48.0	67.00	58.69	3.82	8.31	0.77
12	53.0	60.00	62.46	3.24	-2.46	-0.22
13	54.0	79.00	63.21	3.14	15.79	1.43
14	79.0	79.00	82.08	2.38	-3.08	-0.27
15	76.0	88.00	79.81	2.20	8.19	0.72
16	90.0	98.00	90.38	3.44	7.62	0.69
17	60.0	56.00	67.74	2.56	-11.74	-1.05
18	89.0	87.00	89.62	3.32	-2.62	-0.24
19	83.0	91.00	85.10	2.71	5.90	0.53
20	81.0	86.00	83.59	2.54	2.41	0.21
21	57.0	69.00	65.48	2.83	3.52	0.32
22	71.0	75.00	76.04	2.07	-1.04	-0.09
23	86.0	98.00	87.36	3.00	10.64	0.96
24	82.0	70.00	84.34	2.62	-14.34	-1.28
25	95.0	91.00	94.15	4.02	-3.15	-0.29
26	42.0	48.00	54.16	4.56	-6.16	-0.58
27	75.0	52.00	79.06	2.16	-27.06	-2.39R
28	54.0	44.00	63.21	3.14	-19.21	-1.73
29	54.0	51.00	63.21	3.14	-12.21	-1.10
30	65.0	73.00	71.51	2.22	1.49	0.13
31	61.0	52.00	68.50	2.48	-16.50	-1.47

R denotes an observation with a large standardized residual

R-Squared: Coefficient of Determination

The output also gives R^2 = R-sq = 49.4%. Whenever a straight line is fit to a set of data, R^2 is just the square of the ordinary correlation coefficient we discussed in Section 11.1. There we found $r = 0.703$, which, when squared, gives 0.494, or 49.4%, the same value given here written as a percent in the Regression output.

R^2 also has two other, more general, interpretations. It is the square of the correlation between the observed y values and the fitted y values. It also is the fraction of the variation in y that is explained by the fitted equation. Look at the Analysis of Variance table in Figure 11.6. The total sum of squared deviations, written Total SS, is a measure of the variation of y about its mean. Here it is 7601.4. The Regression SS is the amount of this variation that is explained by the regression line. Here it is 3757.4. The fraction of variation explained is (3757.4)/(7601.4), which is again 0.494. It is sometimes more convenient to convert this to a percentage and to say the regression equation explains 49.4% of the variation in y.

11.3 Making Inferences from Straight-Line Fits

Sometimes we will be content to use Regression just to fit an equation to the set of data we happen to have. On other occasions we may want to generalize from the data at hand to some larger population. Sometimes our data will be a random sample from some population. In such cases we often will be more interested in the characteristics of the population than in the characteristics of the sample. At other times our data will have resulted from a carefully designed and executed experiment. Then we will usually be more interested in the underlying relationship between the variables than in the values we happened to get in this particular experiment.

In other cases our data will be produced more by happenstance. For example we may have good estimates of the automobile fatality rate and of the average highway speeds of vehicles in each of the 50 states. Sometimes we may want to pretend, for a moment, that such data are a random sample from some population. We will know that in reality there are only 50 states in all, but pretending our 50 are a random sample allows us to use the procedures of statistical inference for general guidance. Then we might seek to answer questions such as, "Is the relationship we observe between fatalities and speed one that could reasonably have occurred due to chance alone"?

Conditions for Inference

Several conditions must be met, at least approximately, before we can make reasonable statistical inferences:

1. In the underlying population the relationship between x and y should be a straight line. Suppose that for each value of x we find the mean of all the corresponding y values. Then these means must, at least approximately, fall on a straight line. We will denote this straight line by $A + Bx$, where A and B are some fixed values. For example suppose we could calculate the mean of all the y's in the population corresponding to $x = 1$, the mean of all the ys corresponding to $x = 2$, the mean of all the y's corresponding to $x = 3$, and so on. Then all these means should fall on a straight line. We will call this line the **population regression line**. The values a and b, discussed in Section 11.2, are our sample estimates of the population values A and B. We often say that, for each given x, the expected value of y is $A + Bx$.

2. For each x the amount of variation in the population of y's should be approximately the same. This variance is usually called the variance of y about the regression line, and is denoted by σ^2. Correspondingly σ is called the standard deviation of y about the regression line.

3. For each value of x, the distribution of y's in the population should be approximately normal.

4. The y values that are actually obtained should be approximately independent. In other words the amount by which a particular y value differs from its mean should not be related to the amount by which any other y value differs from its mean.

Later we will discuss some procedures for checking whether these conditions hold. Condition 4 is the most important and the most difficult to check. The best advice when doing surveys or experiments is to try to do them properly in the first place. For example try to make sure that no observations are unduly related to one another. Whenever practical, use randomization to determine the order in which measurements are made. Try to avoid biases or systematic errors.

Interpreting the Output

The following all require that the conditions for inference be at least approximately satisfied:

Estimate of σ. We begin by looking at s in Figure 11.6. The theoretical quantity σ is called the (population) standard deviation of y about the (population) regression line and it's estimate, s, is called the standard error of estimate. It is computed from the formula

$$s = \sqrt{\frac{\sum_i (y_i - \text{fitted } y_i)^2}{n - 2}}$$

The value of s can be thought of as a measure of how much the observed y-values differ from the corresponding average y-values as given by the least-squares line. It has $(n - 2)$ degrees of freedom and is used in all the formulas for standard deviations. All t tests and confidence intervals will be based on this s, and thus all will have $(n - 2)$ degrees of freedom. In our example the number of degrees of freedom is $(31 - 2) = 29$.

Standard Error of Coefficients. Under the conditions for inference the estimated coefficients a and b each have an approximate normal distribution. The estimated standard errors of these coefficients are given in the column headed "SE Coef" in Figure 11.6. The estimated standard deviation of a is 10.22, and the estimated standard deviation of b is 0.1417.

Confidence Intervals. The general formula for a t confidence interval is

(quantity) $\pm$ (value from t table)$\times$(estimated stdev of quantity).

The t_c value corresponding to 29 degrees of freedom and 95% confidence is 2.045. Thus a 95% confidence interval for A (the population value of a) is

$a \pm t_c \times$(SE of a)

or

$22.47 \pm 2.045(10.22)$

This gives the interval from 1.57 to 43.37. A 95% confidence interval for B (the population value of b) is similarly given by

$0.7546 \pm 2.045(0.1417)$

This gives the interval from 0.46 to 1.04.

Tests of Significance. We often want to know if there is any statistically significant evidence of an association between x and y. Thus a null hypothesis we frequently want to test is $B = 0$. We use the general formula

$$t = \frac{b - \text{(hypothesized value)}}{\text{SE of } b}$$

Here we find that

$$t = \frac{0.7546 - 0}{0.1417} = 5.32$$

This is given in the column headed T. With 29 degrees of freedom, this value of t is highly significant, giving us evidence that B is probably not 0. This in turn implies that the score on the first test is at least somewhat useful as a predictor of the score on the second test. (Note: This test is equivalent to testing whether the population correlation coefficient is 0.)

A similar t test of the null hypothesis that $A = 0$ gives

$$t = \frac{22.47 - 0}{10.22} = 2.20$$

This is also statistically significant at the usual levels. Thus from a statistical standpoint, both a and b have been shown to be useful in the equation.

Standard Deviation of a Fitted y Value. The estimated standard deviations of the fitted y values are given in the column headed "SE Fit" in Figure 11.6. These can be used to get confidence intervals for the population mean of all y values corresponding to a given value of x. For example the first line shows that the fitted value for $x = 50$ is 60.20. The 95% confidence interval for the mean of all ys corresponding to $x = 50$ is given by

$$(\text{fit}) \pm t_c \times (\text{stdev of fit})$$

or

$$60.2 \pm 2.045 \times (3.58)$$

This gives the interval 52.9 to 67.6. Thus we can be 95% confident that for all persons who would score 50 on the first test, the mean score on the second test is between 52.9 and 67.6.

This tells us how well persons with a first test score of 50 will do *on the average*. But how about one particular individual who scores 50 on the first test?

Prediction Interval for a Single y Value. Since individuals are never as predictable as averages, we must expect more uncertainty in this prediction. The theory that leads to the confidence interval for the mean, $A + Bx$, comes from consideration of the distribution of $a + bx$, namely that

$$\frac{(a + bx) - (A + Bx)}{SE_{(a + bx)}}$$

has a t distribution with $(n-2)$ degrees of freedom. However, to form a prediction interval for a single y value, say y_{new}, corresponding to a given x_{new}, we need to consider the distribution of

$$\frac{y - y_{new}}{SE_{(y - y_{new})}}$$

which is also a t distribution with $(n-2)$ degrees of freedom. But the standard error of $y-y_{new}$ is $\sqrt{(SE_y)^2 + (SE_{y_{new}})^2}$ and this leads to the prediction interval for y_{new} of

$$(a + bx_{new}) \pm t_c \sqrt{(SE_y)^2 + (SE_{y_{new}})^2}$$

The following calculations give us an interval that we can be 95% confident will contain the second test score of an *individual* who gets a 50 on the first test:

$$60.2 \pm 2.045 \sqrt{(11.51)^2 + (3.58)^2}$$

and this gives the interval from 35.5 to 84.9. The 3.58 is, again, the estimated standard deviation of the fitted value, and $s = 11.51$ is the estimated standard deviation of an individual. Note that the interval for an individual is, as expected, larger than the interval for the mean.

Most texts give the following, slightly different-looking, formula for a prediction interval for y_{new} corresponding to a given x_{new},

$$(a + bx_{new}) \pm t_c \sqrt{1 + \frac{1}{n} + \frac{(x_{new} - \bar{x})^2}{\sum(x - \bar{x})^2}}$$

but this formula can be rewritten as the one given above.

Predictions for New Values of x. The procedure we just used for $x = 50$ will work for any value of x that was in our set of data, such as $x = 66$, $x = 73$, or $x = 84$. But how about a prediction interval for a value of x that was not in our original set of data? For example could we find a prediction interval for $x = 68$? To get the predicted second score is not too difficult. All we have to do is substitute 68 into the regression equation. This gives

$$y = 22.47 + (0.7546)(68) = 73.78$$

However, the easiest way to get Prediction intervals with Minitab is to use the Regression Options dialog box. To predict a value for $x = 68$,

1. Click the **Edit Last Dialog** button 🔲 to reopen the Regression dialog box and then click the **Options** button. See Figure 11.4, p. 332.
2. Enter the value **68** in the Prediction intervals for new observations box, as shown in Figure 11.7.

Figure 11.7   Regression options dialog box

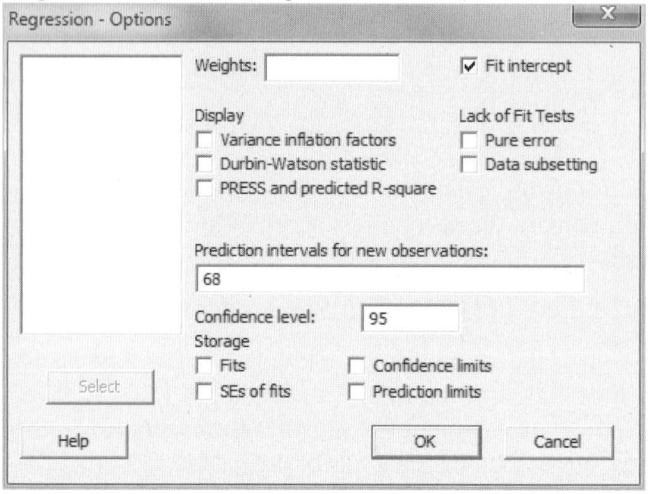

Regression - Options

Weights: [   ]   ☑ Fit intercept

Display   Lack of Fit Tests
☐ Variance inflation factors   ☐ Pure error
☐ Durbin-Watson statistic   ☐ Data subsetting
☐ PRESS and predicted R-square

Prediction intervals for new observations:
[68   ]

Confidence level: [95]
Storage
☐ Fits   ☐ Confidence limits
☐ SEs of fits   ☐ Prediction limits

[Select]

[Help]   [OK]   [Cancel]

3. Click **OK** twice. Minitab displays the additional results shown in Figure 11.8 in the Session Window.

Figure 11.8   Prediction interval results

```
Predicted Values for New Observations

New Obs    Fit    SE Fit      95% CI              95% PI
      1  73.78     2.10   (69.48, 78.08)   (49.84, 97.71)


Values of Predictors for New Observations

New Obs   First
      1    68.0
```

For someone who scores 68 on the first exam, the model predicts a second exam score of 73.78—the Fit given in Figure 11.8. Notice, however, that the prediction interval (PI in Figure 11.8) for the second exam score, goes from 49.84 to 97.71—quite a large range—especially when compared to the confidence interval (CI in Figure 11.8) for the *mean* of all those who score 68 on the first exam. Prediction intervals and confidence intervals answer different questions. It is always important to know which question is appropriate in which situation. Are you interested in predicting a new value of *y* or are you interested in estimating the *mean* of *y*? The prediction interval is appropriate for the first question while the confidence interval answers the second.

11.4 Multiple Regression

So far we have described how one variable can be used to help explain the variation in another—for example how the score on one test relates to that on another test. But what if you have two or three, or even more variables that could help with the explanation or prediction? One technique you can use, and the one we will describe in this section, is multiple regression. We begin with an example.

Many universities use multiple regression to estimate how well the various applicants would do if they were admitted. An equation used at one major university was:

$$(\text{Freshman GPA}) = 0.61813(\text{HS GPA}) + 0.00137(\text{SAT Verbal}) + 0.00063(\text{SAT Math}) - 0.19787$$

This equation shows that the estimated grade point average (GPA) at the end of the freshman year is equal to 0.61813 times the high school grade point average, plus 0.00137 times the Scholastic Aptitude Test verbal score, plus 0.00063 times the Scholastic Aptitude Test mathematics score, minus 0.19787. This equation was an important criterion in deciding whom to admit to that university. The variables HS GPA, SAT Verbal, and SAT Math are often called **predictor variables** or **explanatory variables**. Here they are used to predict the response variable, Freshman GPA.

This equation was obtained by using multiple regression on the records of previous students. The university had the freshman-year GPAs for some past students, as well as the high school GPA and the two SAT scores for each. They asked, "Which equation best explains freshman GPA from these other variables?"

The Grades example (in the Appendix, p. 527) gives some data from another university. Suppose we wanted to develop a similar equation for freshman GPAs at that university, using just the SAT verbal and SAT math scores. We used Minitab to find an equation based on a sample of 100 freshmen (sample A from the GRADES worksheet, stored in the GA worksheet). To find the regression equation for predicting GPA from SAT verbal and math scores,

1. Open the **Ga** worksheet and choose **Stat ▸ Regression ▸ Regression**.
2. Press **F3** to return the dialog box settings to their default values.
3. Enter **GPA** as the response variable and enter both **Verbal** and **Math** in the Predictors textbox.
4. Click **OK**. Minitab shows the results in Figure 11.9.

Figure 11.9 gives the default output from the Regression command: Only those observations which are "unusual" because of their x values (marked by X on the output) or because of their residuals (marked by R on the output) are printed. If you want all 100 observations printed, use the Results dialog box as previously indicated.

Interpreting the Output

The equation for Freshman GPA is

(fitted GPA) = 0.471 + (0.00356)(Verbal) + (0.000158)(Math)

We can use this equation to predict how well a student will do who scored 500 on both SATs. We compute

$$\text{(fitted GPA)} = 0.471 + (0.00356)(500) + (0.000158)(500)$$
$$= 0.471 + 1.78 + 0.08$$
$$= 2.33$$

Thus we predict that the student's GPA will be 2.33.

For a student who scored 500 on the verbal test and 800 on the math test we would predict

(fitted GPA) = 0.471 + (0.00356) (500) + (0.000158) (800) = 2.38

Thus a student with scores of 500 and 500 and another student with scores of 500 and 800 have predicted GPAs that differ by only 0.05. This seems to indicate that SAT math scores are not very useful estimates of GPA at that university.

Figure 11.9   Regression results for predicting freshman GPA from SAT scores

```
Regression Analysis: GPA versus Verbal, Math

The regression equation is
GPA = 0.471 + 0.00356 Verbal + 0.000158 Math


Predictor          Coef      SE Coef      T       P
Constant         0.4706       0.5433    0.87    0.388
Verbal        0.0035628    0.0007350    4.85    0.000
Math          0.0001576    0.0008514    0.19    0.854


S = 0.501777    R-Sq = 23.5%    R-Sq(adj) = 22.0%


Analysis of Variance

Source            DF       SS       MS       F       P
Regression         2   7.5137   3.7568   14.92   0.000
Residual Error    97  24.4227   0.2518
Total             99  31.9364


Source  DF  Seq SS
Verbal   1  7.5051
Math     1  0.0086


Unusual Observations

Obs  Verbal     GPA     Fit   SE Fit   Residual   St Resid
  2     454  2.3000  2.1624   0.1544     0.1376       0.29 X
 40     490  1.2000  2.3269   0.1149    -1.1269      -2.31R
 54     592  2.4000  2.6493   0.1863    -0.2493      -0.54 X
 89     361  2.4000  1.8517   0.1682     0.5483       1.16 X

R denotes an observation with a large standardized residual.
X denotes an observation whose X value gives it large leverage.
```

How good is this regression equation as a whole? Put another way, how much might the GPAs for individual students vary from what we predict? One way to answer this question is to look at the value of R^2. It is 23.5%, which means that our equation explains only 23.5% of the variation in GPAs. The remaining 76.5% of the variation in freshman GPAs is left unexplained.

From these results it appears that SAT scores have only limited value in forecasting who will succeed in college. We can give several possible reasons for this. First, we do not have results for a random sample of all students. All we have is students who applied to, were admitted to, and attended that university. Second, perhaps those students with low SAT scores were advised to take easier courses and thus received higher grades than they ordinarily would have, whereas those with higher SAT scores were encouraged to take more difficult courses. Third, it is possible that tests such as the Scholastic Aptitude Test simply do not do a very good job of measuring whatever it takes to get good grades in college.

Notation and Assumptions in Multiple Regression

Suppose there are two predictor or explanatory variables—call them x_1 and x_2 which are fixed and not random. We assume the following:

1. The underlying population regression line is approximately
   $y = B_0 + B_1 x_1 + B_2 x_2$
2. For all values of x_1 and x_2, the y's have approximately the same variance, σ^2.
3. For all values of x_1 and x_2, the y's have approximately normal distributions.
4. The y's are approximately independent.

In three-dimensional space, the equation $y = B_0 + B_1 x_1 + B_2 x_2$ represents a plane. Least-squares finds values for the parameters B_0, B_1, and B_2, to obtain the plane that best fits the 3D scatterplot—that is, minimizes the sum of squared deviations from the plane to the data. We use the notation b_0, b_1, b_2, and s for the estimated values of B_0, B_1, B_2, and σ, respectively.

An alternative, but equivalent, way of expressing the multiple regression model, is

1. $y = B_0 + B_1 x_1 + B_2 x_2 + e$ where the errors, e, satisfy
2. For all values of x_1 and x_2, the e's have approximately the same variance, σ^2.
3. For all values of x_1 and x_2, the e's have approximately normal distributions with mean zero.
4. The e's are approximately independent.

Confidence Intervals and Tests

All the confidence intervals we calculated for straight lines in Section 11.3 can be obtained here in essentially the same way. For example in Figure 11.9 the standard error of b_2 is 0.0008514. Since there are 97 degrees of freedom, a 95% confidence interval for B_2 is $0.000158 \pm (1.99)(0.0008514)$, which gives the interval from -0.00154 to 0.001185.

We can also do t tests on the coefficients just as we did in simple regression. For example to test H_0: $B_2 = 0.005$, we form the t-ratio $t = (b_2 - 0.005) /$ (SE of b_2). Under the null hypothesis, this ratio has a t distribution with 97 degrees of freedom.

The *t*-ratio for the null hypothesis H_0: $B_2 = 0$ is given on the output and is just 0.19 with a *p*-value of 0.854. Thus b_2 is not statistically different from zero. This test indicates that SAT math scores *in addition to* verbal scores were not statistically significant in explaining freshman GPAs in this sample of students. If you regress GPA on Verbal alone or on Math alone, you will find each is statistically significant by itself. Multiple regression attempts to assess the significance of the predictors jointly.

Suppose we want a 95% confidence interval for the population mean value of *y* corresponding to an SAT Verbal score of 454 and an SAT Math score of 471. This pair of scores happens to be in the data we used in the regression; it is the second observation and is shown in Figure 11.9. The fitted GPA is 2.1624, and the SE Fit is 0.1544. This means that 2.1624 ± (1.99)(0.1544), or 1.86 to 2.47, gives a 95% confidence interval for the mean GPA of all students in the population who had an SAT Verbal score of 454 and an SAT Math score of 471.

To obtain a confidence interval or prediction interval with Minitab, use the Prediction intervals box in the Options dialog box. To predict a GPA for a student with SAT scores of 600 and 750,

1. Click the **Edit Last Dialog** button ▣.
2. Click the **Options** button and enter **600 750** in the Prediction intervals box. (The values must be entered in the *same order* as the predictor variables in the model.)
3. Click **OK** twice and Minitab displays the predicted values shown in Figure 11.10.

We see that a student with an SAT Verbal score of 600 and an SAT Math score of 750 has an expected freshman GPA of 2.73, as we calculated. In addition we are 95% confident that the average GPA of all such freshmen at this university is between 2.5371 and 2.9159. The 95% prediction interval for an individual freshman is 1.7128 to 3.7402. Rotating a 3D scatterplot of these data may convince you that no plane is going to explain very much of the relationship between GPA and SAT Verbal and Math scores. A matrix plot will also help you see the limited relationships among these variables.

Figure 11.10   Prediction intervals in multiple regression

```
 Predicted Values for New Observations

 New Obs     Fit  SE Fit        95% CI              95% PI
       1  2.7265  0.0954  (2.5371, 2.9159)  (1.7128, 3.7402)


 Values of Predictors for New Observations

 New Obs  Verbal  Math
       1     600   750
```

11.5 Fitting Polynomials

So far we have talked about fitting straight lines (Sections 11.2 and 11.3) and about fitting equations with several variables (Section 11.4)[†]. Now we will show how to fit curved data such as those shown in Figure 11.11. In this section we will describe how to use polynomial models.[‡] In Section 11.6 we will show how transformations give an alternative, often preferable, method of analysis.

Polynomials are equations that involve powers of the x variable. The second-degree polynomial, or quadratic equation, is

$$y = B_0 + B_1 x + B_2 x^2$$

The data in Figure 11.11 were gathered in conjunction with an environmental impact study to find the relationship between stream depth and flow rate. Flow rate is the total amount of water that flows past a given point in a fixed amount of time. Data were collected on seven different streams. The data in Figure 11.11 are all from the same site on one stream. We are interested in estimating the flow rate from the depth of the stream. To plot the stream data,

1. Open the **Stream** worksheet.
2. Choose **Graph ▸ Scatterplot ▸ Simple** and click **OK**.
3. Enter **Flow** as the Y variable and **Depth** as the X variable. Click **OK**. Minitab produces the graph shown in Figure 11.11.

The plot shows a gap in the data. There are no observations of stream depth between 0.5 and 0.7. The relationship between flow and depth would be clearer if there were no gap. But these are all the data we have, so we will do our best

---

[†] That is, fitting planes or hyper-planes in higher dimensions.

[‡] See Section 11.9, p. 365, for an alternative approach to polynomial regression with the Minitab Assistant.

with the information on hand. The overall pattern in the plot is certainly not a straight line. There seems to be an upward curve. Let's try a quadratic polynomial. We continue to use the Least-squares principle to find the "best" values for the regression coefficients, B_0, B_1, and B_2. That is, we choose the values of B_0, B_1, and B_2 that minimize the sum of squares of the vertical deviations of the quadratic curve from our data.

Figure 11.11   Scatterplot of flow versus depth in stream data

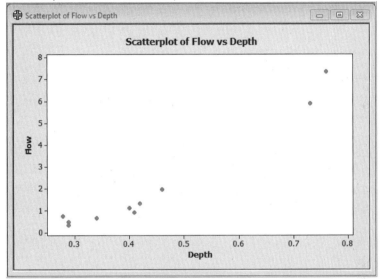

Minitab provides several different ways to obtain and plot the least-squares regression curve. If all we want to do is add the curve to the scatterplot, we can right-click the scatterplot, choose Add ▸ Regression Fit, Select Model Order: Quadratic, click OK, and the quadratic curve is added to your scatterplot. Alternatively, we could create the x^2 variable (using, for example, the Calculator). Then we would do a multiple regression with both x and x^2 as predictors. This would give us the required least-squares estimates of the various coefficients but no graph.

The most thorough way to fit a quadratic polynomial is somewhat hidden. To obtain the least-squares quadratic curve numerical results and plot,

1. Choose **Stat ▸ Regression ▸ Fitted Line Plot**.
2. Enter **Flow** as the response and **Depth** as the predictor. see Figure 11.12.

Figure 11.12   Fitted line plot dialog box

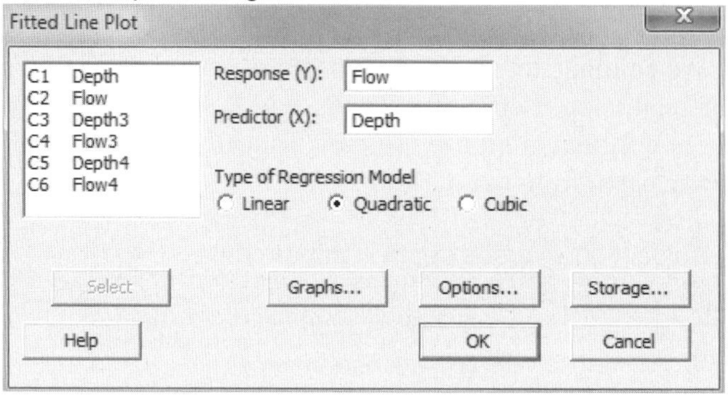

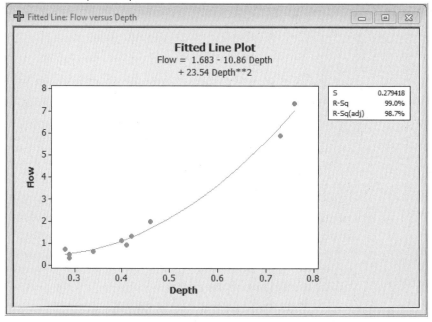

3. Select **Quadratic** for Type of Regression Model. Click **OK**. Figure 11.13
displays the quadratic fit and Figure 11.14 shows the numerical output.

Figure 11.13   Plot of least-squares quadratic fit to stream data

Figure 11.14   Session Window results of fitting a quadratic curve to the stream data

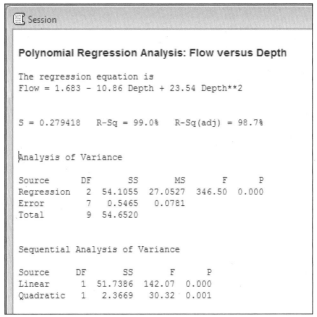

The fitted regression equation is given both in the Session Window and on the plot as

Flow = 1.683 - 10.86 Depth + 23.54(Depth)^2

The value of s, R^2 and adjusted R^2 are also given in the legend of the plot and in the Session Window results. The rest of the output in the Session Window is somewhat different from that for our previous regressions. When fitting a quadratic curve, the natural null hypothesis is $H_0: B_1 = B_2 = 0$ which asks whether the overall quadratic is "significant" or not. This hypothesis cannot be tested with a t-ratio—rather we use an F-ratio as shown in the output. The p-value of 0.000 shows the statistical significance of the quadratic curve.

As in many regression situations, one might ask, "Just what is the population in this case?" To answer this question, we need to imagine the amount of water in this stream fluctuating up and down over a long period of time, but without changing the basic flow pattern of the stream. Then all the measurements of flow rate and depth that might be made over this long period of time are our population. We assume (but don't really believe) that the errors in our model behave like a random sample. All we can really hope is that our measurements act just about the same as a random sample. If all our measurements were made in the spring, or early in the morning, or just after a storm, or within the same week, we could be fairly confident this would not be the case. It is hoped that the experimenter took such factors into consideration when the data were collected. If not, our inferences may be seriously compromised.

11.6 Interpreting Residuals in Simple and Polynomial Regression

Whenever we fit a model to a set of data, we always should plot the data and the residuals. A **residual** is the difference between the observed response and the fitted value determined by the regression equation. Thus residuals tell us how the model missed in fitting the data. For example in Figure 11.14, for the first observation, the stream depth is 0.340, the observed flow rate is 0.6360, and the estimated or predicted flow rate is 0.7107. Thus the first residual is $(0.6360 - 0.7107) = -0.0747$, and the second residual is -0.1933. Minitab offers a variety of options to calculate, store, and look at the residuals visually.

You can also obtain and graph **standardized residuals**, which are calculated by (residual)/(standard deviation of this residual). The standard deviations are different for each residual. One version of this standard deviation is given by the formula

$$(\text{Standard deviation of residual}) \ = \ \sqrt{\text{MSE} - (\text{StDev of fit})^2}$$

To illustrate residual calculation and plotting, let's use the straight-line model for Flow versus Depth.[†] To plot residuals against stream depth,

1. Choose **Stat ► Regression ► Regression** and press **F3** to return the dialog box to its defaults.
2. Enter **Flow** as the response variable and **Depth** as the predictor variable.
3. Click the **Graphs** button. In the Regression-Graph dialog box, click the **Regular** option button. See Figure 11.15.
4. Click the **Residuals versus fits** check box. Click **OK.**
5. Click **OK** in the Regression dialog box. Minitab stores the residuals in the next available empty column in the worksheet, naming the column RESI1. Figure 11.17 shows the Session Window output. Minitab also creates the plot of residuals versus fitted values, shown in Figure 11.18.

[†] We know this model is not the best we can do but we use it merely to illustrate calculating residuals and using them to diagnose models.

Figure 11.15   Regression graphs dialog box

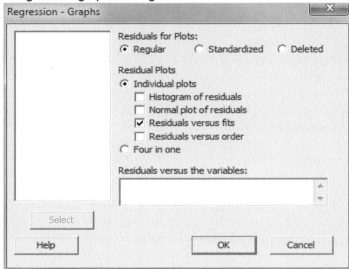

6. Click the **Storage** button and then click the **Residuals** check box so Minitab will store the residuals in an empty column.

Figure 11.16   Regression storage dialog box

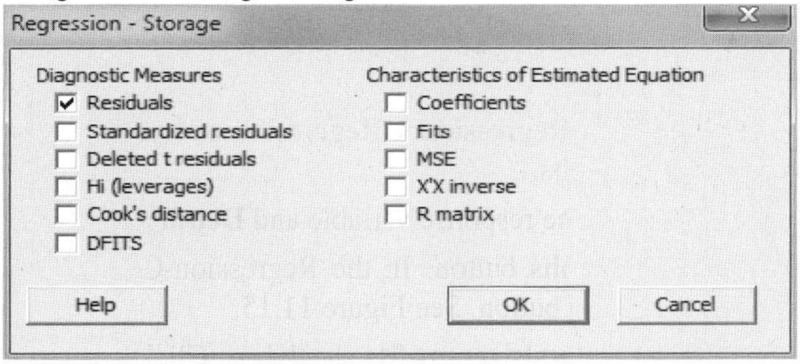

7. Click **OK**. See Figure 11.16.

Figure 11.17   Stream data with straight-line regression results

```
Regression Analysis: Flow versus Depth

The regression equation is
Flow = - 3.98 + 13.8 Depth


Predictor      Coef  SE Coef       T       P
Constant    -3.9821   0.5430   -7.33   0.000
Depth        13.834    1.161   11.92   0.000


S = 0.603470   R-Sq = 94.7%   R-Sq(adj) = 94.0%


Analysis of Variance

Source          DF      SS      MS       F       P
Regression       1  51.739  51.739  142.07   0.000
Residual Error   8   2.913   0.364
Total            9  54.652
```

Figure 11.18   Plot of residuals versus fitted values

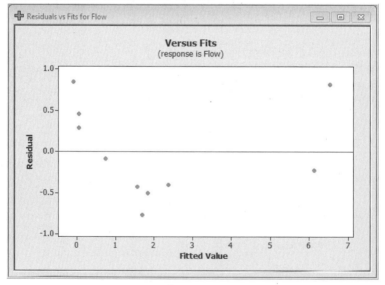

This plot shows what can happen when we fit an inadequate model to the stream-flow data. The plot of Flow versus Depth is curved, but we tried to represent it with a straight line anyway. Here it is easy to see that the regression equation is not a good fit. For low values of Depth the observed values of Flow are greater than the fitted values in the Fit column (above the horizontal line); for middle Depth values the observed Flow values are below the fitted values (below the line); and for the largest Depth value the observed Flow is above the fitted values (above the line). This means the residuals will be positive for low values of Depth, negative for middle values of Depth, and positive again for the high value of Depth. This pattern shows up nicely in the plot of Residuals versus Fitted Value shown in Figure 11.18.

Patterns such as the one in Figure 11.18 sometimes show up more clearly in plots of the residuals than in plots of the original data. The plots of the data and residuals help us see when the model we have fit is seriously wrong. A strong pattern in the residual plot indicates that we probably have a poor model. Often the pattern in the plot of residuals versus fitted values will suggest a better model.

What happens to the residuals when we fit a quadratic polynomial to the stream data? To fit and then plot the residuals from a quadratic polynomial,

1. Choose **Stat ▸ Regression ▸ Fit Line Plot**.
2. Enter **Flow** as the response and **Depth** as the predictor.
3. Click the **Graphs** button and, click the **Regular** option button.
4. Click the **Residuals versus fits** check box. Click **OK.**
5. Click the **Storage** button and then click the **Residuals** check box so Minitab will store the residuals. Click **OK.**
6. Click **OK** once more and Minitab produces the regression output as well as a new column, RESI2, that contains residuals from this quadratic regression. Minitab also produces the new residuals versus fits graph shown in Figure 11.19.

Figure 11.19   Residuals versus fitted values for quadratic model

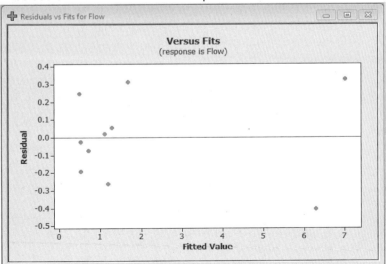

Is there any pattern here? For example do low fitted values tend to have low residuals? How about the residuals for middle and high values? Are there any patterns that would indicate that our model is not adequate? In this plot we do not see any. It appears that the model we fit was all right, at least as far as this plot is concerned.

If we fit a straight line to data when the basic relationship between x and y is a curve, a plot of the residuals versus x will be curved. The plot of y versus x also will be curved, but the curvature may not be as apparent. We might then fit a quadratic polynomial. If the residual plot after this fit is no longer curved (and no other problems are indicated), we may have a good fit. If the residual plot still shows curvature, we could try fitting a third-degree, or even a higher-degree, polynomial. As a general rule, however, it is better to use a transformation, as discussed in Section 11.7, than to use polynomials of degree higher than two.

Residual plots also help us spot outliers—observations that are far from the majority of the data or far from the fitted equation. Outliers should always be checked for possible errors or special causes. In some cases they should be temporarily set aside to see whether they have any effect on the practical interpretation of the results of the analysis.

11.7 Using Transformations

Rather than fit a polynomial to curved data, it is often preferable to transform them to see if a simpler model can be found. The use of transformations is new in some areas of research, but engineers and physical scientists have long used transformations to simplify relationships. In fact they have a saying: "Anything is a straight line after using a logarithmic transformation." Of course this statement is an exaggeration, but it does indicate the usefulness of transformations.

Suppose we have two variables, say x and y, and we want a simple way to describe the relationship between them. We start by plotting y versus x. If the plot is more or less a straight line, we have our simple description. But suppose the plot is a curve. Three types of trends frequently encountered in data analysis are shown in Figure 11.20.

Plots (a) and (b) both have an upward trend—that is, as x increases, y also increases. But there is a difference: plot (a) curves down, whereas (b) curves up. Plot (c) has a downward trend—as x increases, y decreases—and it curves up. Our objective is to transform x or y or both in order to get a straight line.

Figure 11.20
Three
common
trends in
relationships

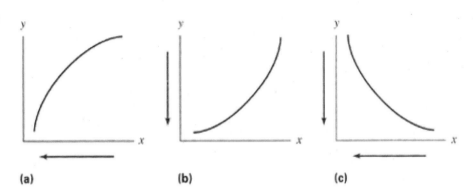

(a)   (b)   (c)

Three popular transformations are square root, log, and negative reciprocal—*negative* reciprocal to preserve the order in the relationship. These can be applied to *x* or *y* or both. The arrows in Figure 11.20 indicate which ones to try. In panel (a) the arrow points down on the *x* axis. This tells us that if we pull the upper end of the curve in the direction of the arrow, we will tend to straighten the curve. Put another way, we need to compress the high values of *x* to straighten the curve.

We often start with $\sqrt{x}$. If this is not strong enough, we try log(*x*), and then $-1/x$, until we find a transformation that works. In panel (b) we need to compress the high values of *y*. So we start with $\sqrt{y}$ and then, if necessary, try log(*y*), and then $-1/y$. In panel (c) there are downward arrows on both axes. This means we should try transforming either *x* or *y* or both. For example we might try *y* versus $\sqrt{x}$ or log(*y*) versus $\sqrt{x}$. Our goal in all cases is to get a straight line. Of course this doesn't always happen. But it does work surprisingly often.

We can use the stream data from Figure 11.11, p. 347, as an example. Plotting Flow versus Depth did not give us a straight line. The curvature in the plot of Flow versus Depth is similar to that in panel (b) of Figure 11.20. To achieve a straight line, we need to pull in the upper end of the Flow scale. The indicated transforms are thus $\sqrt{\text{Flow}}$, log(Flow), and $-1/\text{Flow}$. These are plotted versus Depth in Figure 11.21. Also shown is log (Flow) versus log (Depth).

Several of these plots seem to be reasonably close to a straight line. Let's try $\sqrt{\text{Flow}}$ versus Depth. If we fit a "straight line" to this plot, the resulting equation is really

$$\sqrt{\text{Flow}} = A + B(\text{Depth})$$

Figure 11.21   Various transformations of the stream data

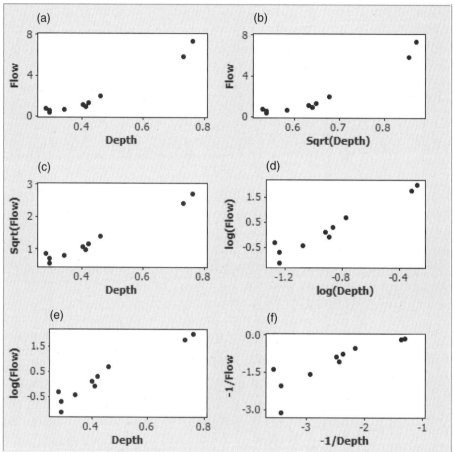

Let's try fitting this model.

1. Use **Calc ➤ Calculator** to create a new column, **Sqrt(Flow)**, that contains the square root of the values in the Flow column.
2. Choose **Stat ➤ Regression ➤ Regression**, enter **Sqrt(Flow)** as the response variable and **Depth** as the predictor variable. Click **OK**. Minitab produces the output shown in Figure 11.22.

Figure 11.22   Regression results for square root model

```
Regression Analysis: Sqrt(Flow) versus Depth

The regression equation is
Sqrt(Flow) = - 0.558 + 4.16 Depth


Predictor      Coef  SE Coef       T      P
Constant    -0.5579   0.1149   -4.86  0.001
Depth        4.1584   0.2456   16.93  0.000


S = 0.127693   R-Sq = 97.3%   R-Sq(adj) = 96.9%


Analysis of Variance

Source          DF       SS      MS       F      P
Regression       1   4.6751  4.6751  286.72  0.000
Residual Error   8   0.1304  0.0163
Total            9   4.8055


Unusual Observations

Obs  Depth  Sqrt(Flow)     Fit  SE Fit  Residual  St Resid
  3  0.280      0.8567  0.6065  0.0560    0.2503     2.18R

R denotes an observation with a large standardized residual.
```

Minitab tells us that the least-squares equation is

Sqrt(Flow) = −0.558 + (4.16)(Depth)

If Depth were 0.7, we would predict Sqrt(Flow) as

Sqrt(Flow) = −0.558 + 4.16(0.7) = 2.35

Then to predict Flow, at Depth = 0.7, we calculate

Flow = $(2.35)^2$ = 5.52

Another plot that looks fairly good is log(Flow) versus log(Depth). Fitting a straight line to this plot gives the equation

log(Flow) = A + Blog(Depth)

Either of these equations might do well enough for prediction, particularly within the range of depths for which we have data. In addition they might give us some idea of a good theoretical model for the relationship between stream-flow rate and depth. Using a polynomial might be equally good for estimation over the range of the data, but it would not work very well outside the range of the data, nor would it be likely to give us much theoretical insight.

Effect of Transformations on Assumptions

Section 11.3 lists the conditions necessary for inference in regression. Whenever we use a transformation, there will be some effect on the validity of the conditions. If we have an exactly straight line, and take the logarithm of x or y, we will not have an exactly straight line anymore. If y has exactly the same variance for all values of x, then $\log(y)$, $\sqrt{y}$, and $-1/y$ will not. If y is exactly normally distributed, then $\log(y)$, $\sqrt{y}$, and $-1/y$ will not be.

On the other hand if some of these conditions were not met before we transformed the data, they may be afterward. Sometimes a transformation will help with one of these conditions, but will cause problems with another. Surprisingly often, though, a transformation that helps with one condition also helps with others.

To summarize, our advice is to use transformations as you would use any other statistical technique—not blindly, but with a healthy skepticism that says, "I know none of these assumptions are met exactly, and I know that some are more important than others. I will check them all, particularly the most important ones, and always use the results of my analysis as guidance, not as gospel."

11.8 Plotting Regression Lines, Planes and Surfaces

In this section we will show you how to draw fitted regression lines, curves, planes, and more general surfaces on a data plot.

Straight Lines, Quadratics, and Cubics

As we have seen, if the regression equation is a straight line, quadratic, or cubic curve, Minitab has many ways to display or add the regression line or curve to a scatterplot. To add a regression line or curve, review pages 331 and 348. A regression line (not quadratic or cubic) may also be placed on a scatterplot when it is first created. Choose Graph ▸ Scatterplot and select **With Regression** from the Scatterplot Gallery shown in Figure 11.23, and continue as before.

Figure 11.23   Scatterplots gallery

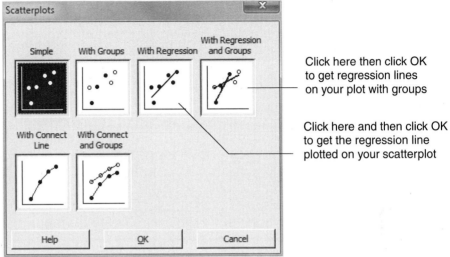

Click here then click OK
to get regression lines
on your plot with groups

Click here and then click OK
to get the regression line
plotted on your scatterplot

Regression Planes

Creating and plotting regression planes is not automated in Minitab but it can be accomplished without too much effort. We illustrate with the example of predicting GPA from SAT Verbal and Math scores—see p. 341.

The simplest way to get a crude plot of the fitted regression plane is to first store the fitted values from the regression.

1. Open the **Ga** data file and choose **Stat ▸ Regression ▸ Regression**.
2. Enter **GPA** as the **Response** and **Verbal** and **Math** as the predictors.
3. Click the **Storage** button and check the **Fits** checkbox.
4. Click **OK** twice to do the regression and store the fitted values into the first empty column. Minitab will name the column FIT1.
5. Choose **Graph ▸ 3D Scatterplot ▸ Simple** and click **OK**.
6. Enter **FIT1** for the Z variable, and **Verbal** and **Math** for the Y and X variables.
7. Click **OK** and Minitab produces the 3D plot shown in Figure 11.24. Also a 3D Graph Tools toolbar, as shown in Figure 11.25, opens automatically.

Figure 11.24   Initial 3D scatterplot of fitted values versus verbal and math scores

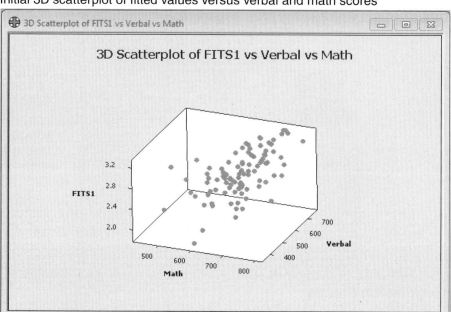

Figure 11.25   3D Graph Tools

It is difficult to see that the points in Figure 11.24 lie on a plane. However, using the 3D Graph Tools you can rotate the 3D plot around the z axis (the FITS1 axis) until it is easy to see the plane. Review p. 168 for 3D graph rotation. Figure 11.26 shows the FIT1 3D graph rotated until you view an edge of the plane.

Figure 11.26   Rotated 3D scatterplot of fitted values versus verbal and math scores

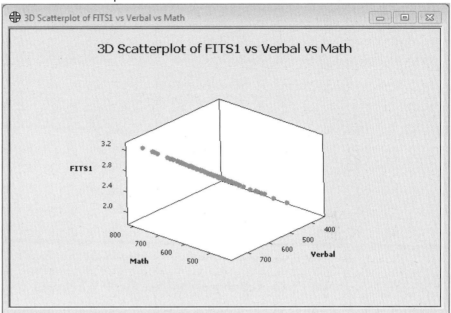

To plot the regression plane more precisely you first need to set up the Verbal and Math pairs that define the corners of the plane. Then you use the regression equation to calculate the height of the plane at those corners. Finally, you use Minitab's 3D surface plot command to do the plotting. In the data Verbal ranges from 361 to 752 and Math ranges from 441 to 800. Let's use the pairs (Verbal, Math) = (350, 440), (350, 800), (800, 440), and (800, 800) to define the corners. Here are the steps.

1. Enter the numbers **350 350 800 800** into a column named **V**.
2. Enter the numbers **440 800 440 800** into a column named **M**.
3. Choose **Stat ▸ Regression ▸ Regression** once more and enter **GPA** as the response and **Verbal** and **Math** as the predictors.
4. Choose **Options** and select **V** and **M** for Prediction interval for new observations: See Figure 11.27. Click **OK** twice and Minitab stores the predicted values in the next available column naming it PFIT1.

Figure 11.27   Regression Options dialog box

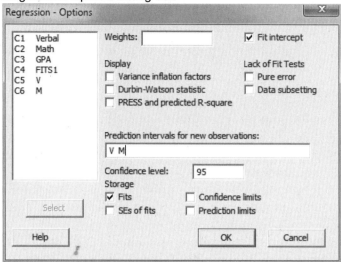

5. Now choose **Graph ▸ 3D Surface Plot ▸ Wireframe**.

6. Enter **PFIT1** as the Z variable and **V** and **M** as the Y and X variables.

7. Click **OK** and Minitab displays the 3D graph shown in Figure 11.28. (We edited the labeling on the graph after producing it.)

Figure 11.28   3D plot of regression plane for SAT versus Verbal and Math

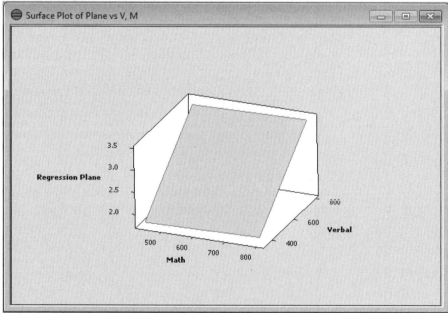

Once the regression plane is displayed it may be rotated to get more insight into the relationship between predicted SAT and Verbal and Math scores.

Regression Surfaces

Plotting a curved regression surface proceeds similarly but more (x, y) points are used to define the curves—the four corners are not sufficient as they are for a plane such as shown in Figure 11.28.

As an example let's use the property data first seen on p. 163. The matrix plot there suggested a curved relationship between Market and Sq.ft but perhaps linear between Market and Assessed. We will use a multiple regression model of the form Market $= B_0 + B_1$ Assessed $+ B_2$ Sq.ft $+ B_3 (\text{Sq.ft})^2$ and plot the fitted regression surface. You will need to create the variable Sq.ft.sq $= (\text{Sq.ft})^2$ then do the regression with Market as response and Assessed, Sq,ft, and $(\text{Sq.ft})^2$ as predictors.

Since, in the data set, Assessed value ranges from 4.6 to 32.2 and Sq.ft ranges from 521 to 1804, we will use a range of 4.5 to 33 for Assessed and 520 to 1820 for Sq.ft in defining the "mesh" for the surface plot. Here are the steps.

1. Open the **Property** data set.
2. Choose **Calc ► Make Mesh Data**.
3. Enter **Assess** for the Store in X variable. Enter **Sqft** for the Store in Y variable.
4. Enter **4.5** and **33** for the From and To values for X and **520** and **1820** for the From and To values for Y. See Figure 11.29.
5. Enter **41** for the Number of Positions in both cases and click **OK**. (This will ensure a reasonably smooth curve in the plot.) See Figure 11.29.
6. Choose **Calc ► Calculator** to make the new variable **Sqft2** where Sqft2 $= (\text{Sqft})^2$.
7. Choose **Calc ► Calculator** to make the new variable **Sq.ft2** where Sq.ft2 $= (\text{Sq.ft})^2$.
8. Choose **Stat ► Regression ► Regression** once more and enter **Market** as the response and **Assessed, Sq.ft**, and **Sq.ft2** as the predictors.
9. Choose **Options** and select **Assess, Sqft**, and **Sqft2** for Prediction interval for new observations: Select Fits under Storage. See Figure 11.27.
10. Click **OK** twice and Minitab stores the predicted values in the next available column naming it PFIT1.
11. Choose **Graph ► 3D Surface Plot ► Surface** and click **OK**.
12. Enter **PFIT1** as the Z variable, **Assess** as the Y variable, and **Sqft** as the X variable.

13. Click OK and Minitab produces the 3D surface plot shown in Figure 11.30. (again we edited the labels after the plot was produced.)

Figure 11.29   Make Mesh Data dialog box

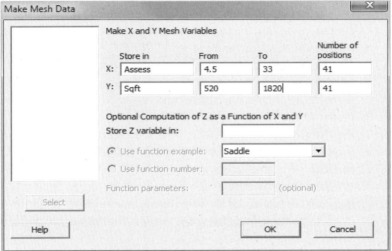

Minitab projects "lights" on its surface plots to help you see the shape of the surface. You may use the 3D graph toolbar shown in Figure 11.25 to rotate the lights to help further with the visualization.

Figure 11.30   Curved regression surface plot

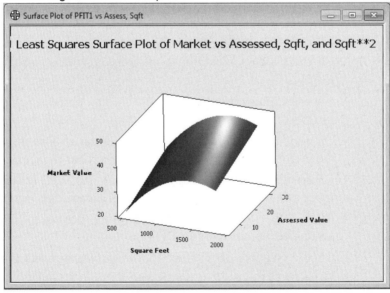

11.9 Regression with the Minitab Assistant

When you click on the Assistant menu and then select Regression, the following window appears.

Figure 11.31   Assistant Regression window

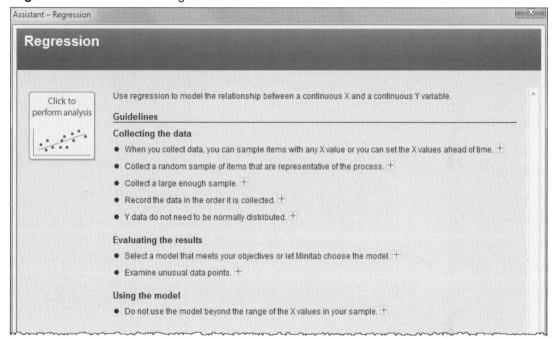

Notice that each of the bulleted items has a plus sign (+) in a small box at the end of the line of text. If you click on the plus sign, Minitab displays additional explanatory or cautionary information. As an example, Figure 11.32 shows what you see when you click on the line that reads "Record the data in the order it is collected."

Figure 11.32   Expanded bulleted item

- Collect a random sample of items that are representative of the process. +
- Collect a large enough sample. +
- Record the data in the order it is collected. −

    To detect time-related patterns in the data that may indicate problems with the regression model, make sure that you enter the data in the Minitab worksheet in the same order that you collect it.

- Y data do not need to be normally distributed. +

Evaluating the results

- Select a model that meets your objectives or let Minitab choose the model. +

When you are ready to proceed with the analysis you click on "Click to perform analysis" in the upper left hand corner shown in Figure 11.31. The dialog box displayed in Figure 11.33 appears. Notice that you are limited to one predictor variable (the X column) and linear, quadratic or cubic models. However, Minitab will choose the best model among these three if you select the default "Choose for me."

Figure 11.33   Assistant Regression dialog box

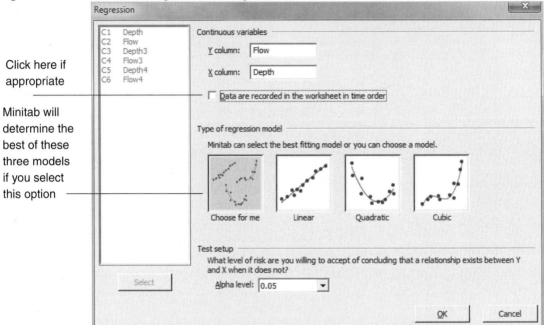

Click here if appropriate

Minitab will determine the best of these three models if you select this option

We'll use the stream flow data, p. 346, to illustrate. Enter **Flow** for the Y column and **Depth** for the X column and press OK. Minitab displays four screens entitled Summary Report, Model Selection Report, Diagnostic Report, and Report Card. The Summary Report is shown in Figure 11.34. Here you see the plot of the fitted model on the scatterplot of the data together with the equation of the best model and statistical significance results. Note that Minitab selected a quadratic curve or second degree polynomial as the best model.

Figure 11.34   Assistant Regression Summary Report

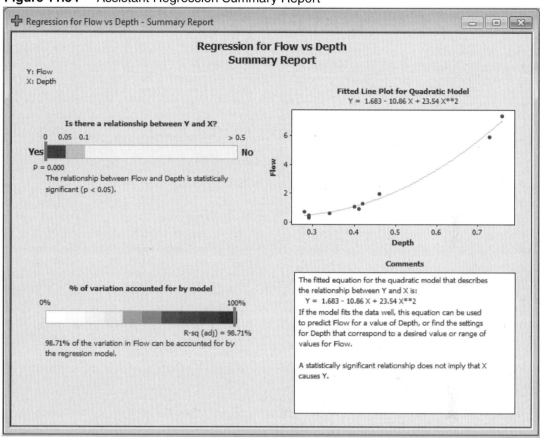

Figure 11.35 displays the Model Selection window. The fitted model is plotted once more with the raw data. The quadratic model was the best one based on various numerical values in the display.

Figure 11.35   Assistant Regression Model Selection Report

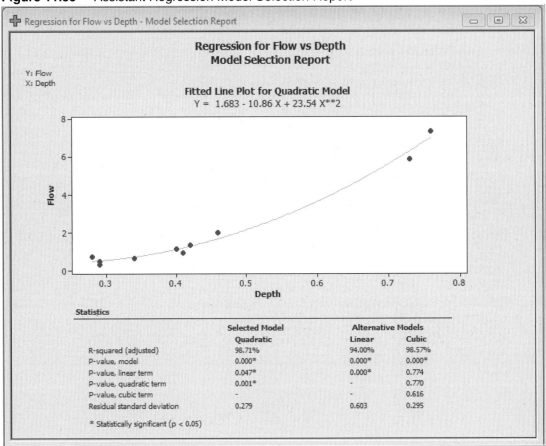

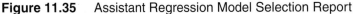

Figure 11.36 displays the plot of residuals versus fitted values along with advice about what patterns to look for when the model may be in doubt.

Figure 11.36   Assistant Regression Diagnostic Report

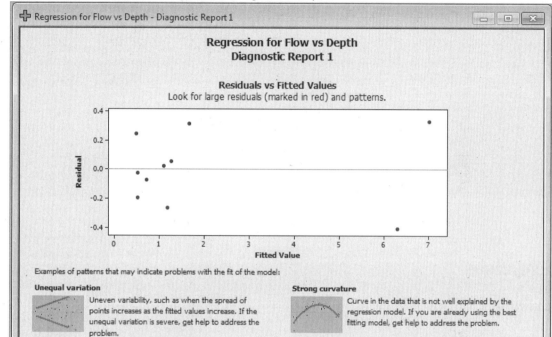

Finally, Figure 11.37 gives additional results, advice and caveats about the fitted model.

Figure 11.37   Assistant Regression Report Card

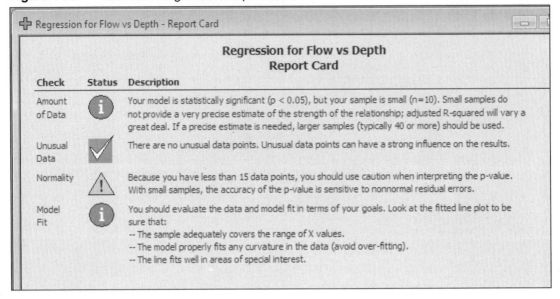

Exercises

Exercises for Section 11.1

11-1   Table 11.2 lists four variables:

Table 11.2   Four variables

X1	Y1	X2	Y2
1	4	1	1
2	5	2	4
3	6	3	7
4	7	4	7
5	8	5	4
6	2	6	1

   (a) First calculate the correlation coefficient between X1 and Y1.
   (b) Based on this result do you think there is any association between these two
        variables?
   (c) Plot Y1 versus X1. Now what do you think?
   (d) Repeat part (a) for the variables X2 and Y2.

11-2   There are three variables in the Trees data (described in the Appendix, p. 531): diameter, height, and volume. Open the Trees worksheet and use Matrix Plot to plot volume versus height, volume versus diameter, and height versus diameter. Find the correlation associated with each of the three plots. Which relationship seems strongest? Which variable would probably be the best predictor of volume? Is there any evidence to make you doubt the reasonableness of the linear (straight-line) correlation coefficients you computed?

11-3   Table 11.3 lists the price and number of pages for each of 15 books that were reviewed in the February 1982 issue of the journal *Technometrics*. Make a plot of price versus book length. Compute the correlation between book length and price.

Table 11.3   Books data

Pages	Price ($)	Pages	Price ($)
302	30	465	55
425	24	246	25
526	35	143	15
532	42	557	29
145	25	372	30
556	27	320	25
426	64	178	26
359	59		

11-4   In this exercise we simulate data under various conditions. We then plot the data, guess the correlations, and check our guesses. Use **Calc ▸ Random Data ▸ Normal** to simulate 50 rows of two independent normal random samples into columns C1 and C2. Plot C1 versus C3, guess the correlation, and check it. In each part below, use the LET Session command shown (or the equivalent Calculator menu command) to define C3.
(a) `LET C3 = C2`
(b) `LET C3 = C1 + C2`
(c) `LET C3 = C2 - C1`
(d) `LET C3 = 5*C1 + C2`
(e) `LET C3 = C2 - 5*C1`

11-5   Minitab has a built-in command to simulate multivariate normal samples. (Calc ▸ Random Data ▸ Multivariate Normal) which can be used to simulate pairs of bivariate normal data. However, it is a bit cumbersome to use since you must setup a mean vector and a full variance-covariance matrix. Instead we will use a Minitab macro based the theory that if X and Y are independent standard normal variables, then X and $Z = \rho X + \sqrt{1 - \rho^2}\, Y$ has a bivariate normal distribution with zero means, unit variances, and correlation coefficient ρ. The macro, named Corr, asks you to supply a value for ρ and the sample size. It then generates the data, stores it in C1 and C2, plots the scatterplot and calculates and displays the sample correlation coefficient.

(a) Use the Corr macro with ρ = 0.9 to generate and plot the data. Use a sample size of 30. Repeat at least four times to see the possible variation in the plots all generated from the same model. How does the sample correlation coefficient compare to the theoretical correlation coefficient that you specified?

(b) Repeat part (a) with ρ = −0.9.

(c) Repeat part (a) with ρ = 0.5.

(d) Repeat part (a) with ρ = 0.2.

(e) Repeat parts (a)-(d) with sample of size 100.

(f) Repeat parts (a)-(d) with samples of size 9.

11-6   (a) Figure 5.3, p. 148, plots natural gas consumption per hour versus the outside temperature. Guess the correlation of these data.

(b) Compute the correlation coefficient and check your guess.

(c) Repeat (a) and (b) for the plot of Cartoon scores versus Otis scores on p. 151.

11-7   Enter the integers 1 to 50 into C1. Now compute

```
C2 = C1*C1
C3 = SQRT(C1)
C4 = 3*C1
C5 = (C1-25.5)**2
```

In each case there is an exact relationship with C1. Plot each of C2, C3, C4, and C5 versus C1. Look at the relationships. Now compute the correlations among C1-C5. Explain why the correlation coefficients are not all 1.

Exercises for Section 11.2

11-8   Refer to the output from Regression in Figure 11.6, p. 334.

(a) Based on the fitted equation, if a person scored 54 on the first test, what score would you predict for the second test?

(b) How many individuals got 54 on the first test? What did each of them get on the second test? Find the residual (deviation) for each.

(c) If a person got 100 on the first test, what score would you predict on the second test? (Always check your answers for reasonableness.)

11-9   Table 11.4 shows the mean Scholastic Aptitude Test scores (SAT scores) for the years 1967-1981.

Table 11.4   SAT scores 1967-1981

Year	Verbal	Math	Year	Verbal	Math
1967	466	492	1975	434	472
1968	466	492	1976	431	472
1969	463	493	1977	429	470
1970	460	488	1978	429	468
1971	455	488	1979	427	467
1972	453	484	1980	424	466
1973	445	481	1981	424	466
1974	444	480			

(a) Open the SAT worksheet and plot the verbal SAT scores versus year. Repeat for math scores.
(b) Fit a regression line using year to predict the verbal SAT scores.
(c) Fit a regression line using year to predict the math SAT scores.
(d) Use Minitab to draw in the two regression lines.
(e) Do both verbal and math scores seem to be changing at about the same rate (same number of points per year)?
(f) What is the predicted average SAT score for math in the year 1970? In 1985? In 1990? In 2000? Which of these seem to make sense? Repeat for verbal scores.
(g) Table 11.5 shows the actual data for some of the years since 1981. What do you think of your predictions? It is generally considered risky to forecast much beyond the data you have. Was that a reasonable concern here?

Table 11.5   SAT scores 1984-1992

Year	Verbal	Math
1984	426	471
1985	431	475
1987	430	476
1988	428	476
1989	427	476
1990	424	476
1991	422	474
1992	423	476

11-10   Table 11.6 shows the winning times in seconds for the men's 1500-meter run in the Olympics from 1900 to 1992. (Note: No Olympics were held in 1916, 1940, and 1944 because of wars; in 1980 the United States and some other countries did not participate.)

(a) Open the Track15 worksheet and plot winning time versus year. Does winning time seem to be changing over the years? If so, is it changing according to a straight line?

(b) Fit a regression line for predicting winning time from year. Use Minitab to draw the regression line on the plot.

Table 11.6   Track data

Year	Time	Year	Time	Year	Time
1900	246.0	1932	231.2	1968	214.9
1904	245.4	1936	227.8	1972	216.3
1908	243.4	1948	229.8	1976	219.2
1912	236.8	1952	225.2	1980	218.4
1920	241.8	1956	221.2	1984	212.5
1924	233.6	1960	215.6	1988	216.0
1928	233.2	1964	218.1	1992	220.1

(c) On the average how much has the winning time decreased in the four years between Olympics? Use the fitted equation.

(d) Are there any appreciable departures from the straight line that you might have anticipated from outside information? Explain.

11-11   The data in Table 11.7 were collected to study the relationship between the temperature of a battery and its output voltage. The eight measurements were taken in the order shown.

Table 11.7   Battery data

	Reading							
	1	2	3	4	5	6	7	8
Temperature (°C)	10.0	10.0	23.1	23.1	34.0	34.0	45.6	45.6
Voltage	4290	4270	4470	4485	4723	4731	4920	4935

(a) Enter the data into a worksheet and then plot voltage versus temperature.

(b) Use the Regression command to fit a straight line to estimate voltage as a function of temperature. Does the regression line seem to fit the data well?

(c) Do you spot any weakness in the order in which the readings were taken? Can you think of some better orders in which to make the eight readings if you had to do this experiment again? Discuss.

11-12

(a) How well can you predict the volume of lumber in a tree from its diameter? Use the Trees data (described in the Appendix, p. 531) to find an equation for black cherry trees in Pennsylvania.

(b) Open the Trees worksheet and plot the data (volume versus diameter) and draw your regression line on it. How well does the line seem to fit the data? Do you see any problems?

(c) Repeat (a) and (b), but use height as the predictor of volume.

(d) Which is the better predictor, height or diameter? Which equation would be easier to use in the woods to predict the volume of a given tree? Why?

Exercises for Section 11.3

11-13

Refer to the output in Figure 11.6, p. 334.

(a) Calculate a 90% confidence interval for B, the slope of the underlying population regression line.

(b) Find a 90% confidence interval for A, the intercept of the underlying population regression line.

(c) Find a 90% confidence interval for the average of the second test scores of all persons in the population who scored 90 on the first test.

11-14

Refer to the output in Figure 11.6, p. 334. Suppose a person scored an 86 on the first test. What score should be expected on the second? Find an interval that you are 95% confident will cover the second score for that individual.

11-15

Refer to the output in Figure 11.4. Test the null hypothesis, H_0: $A = 0.5$ versus the alternative hypothesis, H_1: $A \neq 0.5$.

11-16

Refer to the test data in Table 11.1, p. 327.

(a) Find a 90% confidence interval for the mean of second test scores for all persons in the population who obtained a score of 77 on the first test. You will need to use the Prediction intervals option in the Options dialog box, because 77 is not in the data set.

(b) Find a 90% prediction interval for the second test score for an individual who achieved a score of 77 on the first test.

11-17

Refer to Exercise 11-10 for the 1500-meter race. Give 95% confidence intervals for the two regression coefficients.

11-18   Maple trees have winged fruit, called samara, which come spinning to the ground in the fall. A forest scientist was interested in the relationship between the velocity with which the samara fall and their "disk loading." The disk loading is a function of the size and weight of the fruit and is closely related to the aerodynamics of helicopters. Tests were run on samara from three trees. Table 11.8 shows the results, which are also stored in the worksheet Maple.

(a) Does velocity seem to be a straight-line function of loading? Examine this question separately for each tree.

(b) A scientist hypothesizes that the straight lines will go through the origin (the point $(x,y) = (0,0)$. Do they seem to do this, at least approximately?

(c) Test $H_0: A = 0$ and compare the result to your answer in (b).

(d) Is there any difference in the relationship between velocity and loading for the three different trees?

Table 11.8 Maple tree data

	Tree 1		Tree 2		Tree 3
Loading	Velocity	Loading	Velocity	Loading	Velocity
.239	1.34	.238	1.20	.192	0.91
.208	1.06	.206	1.06	.200	1.13
.223	1.14	.172	0.88	.175	1.00
.224	1.13	.235	1.24	.187	0.98
.246	1.35	.247	1.37	.181	0.96
.213	1.23	.239	1.37	.195	0.88
.198	1.23	.233	1.43	.155	0.81
.219	1.15	.234	1.32	.179	0.91
.241	1.25	.189	0.99	.184	1.00
.210	1.24	.192	1.00	.177	0.87
.224	1.34	.209	1.12	.177	1.02
.269	1.35			.186	0.94

11-19   (a) We can simulate data from a regression model as follows. To choose a model, we need to specify three things: A, B, and σ. Suppose we use $A = 3$, $B = 5$, and $\sigma = 0.5$. Next, we must specify values for x. Suppose we take two observations at each integer from 1 through 10. First, place the 20 values of x into a column. Then calculate $A + Bx$. Now simulate 20 observations from a normal distribution with $\mu = 0$ and $\sigma = 0.5$. Add these to $A + Bx$ to get the observed y's. Display a plot and do a regression for these simulated data. Record a, b, s, and R^2.

(b) Repeat part (a), using $\sigma = 2.00$. Compare the results with those of part (a).

(c) Repeat part (a), using $\sigma = 10.0$. Compare the results with those of parts (a) and (b).

Exercises for Section 11.4

11-20   Refer to the output in Figure 11.9, p. 343.
(a) Get a 95% confidence interval for the theoretical regression coefficient associated with the SAT verbal score.
(b) Is the coefficient of SAT verbal score significantly different from zero (use $\alpha = 0.05$)? What does this say about the relationship between SAT verbal score and GPA?
(c) Is your answer to (b) consistent with the low value of R^2? Explain.

11-21   (a) Use the data in sample B from the Grades data (described in the Appendix, p. 527, and saved in the Gb worksheet) to develop another equation for predicting GPA from SAT scores.
(b) How does this equation compare to the equation in Figure 11.7? Compare the estimates of the regression coefficients, σ, and R^2 for the two equations.

11-22   In Exercise 11-21 we fitted an equation for estimating volume of a black cherry tree from its diameter. Suppose we use height as a second explanatory variable.
(a) Find an equation for estimating volume from diameter and height. How much extra help does height seem to give you when you are predicting volume?
(b) Use the Prediction intervals option to calculate a 95% confidence interval for estimating the average volume of trees with diameter = 11 and height = 70.

11-23   In this exercise we will fit a model to predict systolic blood pressure, using the Peru data set. Read the description of this data set in the Appendix, p. 528.
(a) Open the Peru worksheet and regress systolic blood pressure on years since migration. What is the relationship?
(b) Add in a second predictor, weight. How does this model compare to the one in part (a)?
(c) What do the results in parts (a) and (b) tell you about fitting models to data?
(d) Try adding a third predictor to the model in part (b). Can you find one that improves the model?

Exercises for Section 11.5

11-24   In Exercise 11-12 we fitted the equation (volume) $= B_0 + B_1$(diameter) to the Trees data. In Exercise 11-22 we added height as a second explanatory variable and fitted the equation (volume) $= B_0 + B_1$(diameter) $+ B_2$(height). Now fit the quadratic equation (volume) $= B_0 + B_1$(diameter) $+ B_2$(diameter)^2.

(a) How well does this quadratic equation fit? Compare its fit to that of the straight line we fitted in Exercise 11-12.

(b) Does a quadratic equation seem a reasonable choice from a theoretical standpoint? Hint: Consider the geometry of diameter versus volume for a tree.

(c) Compare the quadratic equation in (a) to the equation in Exercise 11-22, which was based on height and diameter. How well does each fit?

(d) It is considerably more difficult to measure the height of a tree than its diameter. Based on this information, what equation would you most likely use in practice to estimate the volume of a tree?

Exercises for Section 11.6

11-25   Table 11.9 displays stream-flow data from another site, site 3.

Table 11.9   Stream-flow data from site 3

Flow Rate	.820	.500	.433	.215	.120	.172	.106	.094	.129	.240
Stream Depth	.96	.92	.90	.85	.84	.84	.82	.80	.83	.86

(a) These data are stored in the Stream worksheet in C3 and C4. Plot flow rate versus stream depth.

(b) Find a 95% confidence interval for the slope, B.

(c) Find a 95% confidence interval for the intercept, A.

11-26   (a) Plot residuals versus stream depth for the straight-line fit in Exercise 11-25. Does the plot indicate any lack of fit or any other problem? Explain.

(b) Fit a quadratic model to the data from site 3. Does it fit any better? Explain.

11-27   Stream-flow data from another site, site 4, are given in Table 11.10.

Table 11.10   Stream-flow data from site 4

Flow Rate	.352	.320	.219	.179	.160	.113	.043	.095	.278
Stream Depth	.71	.72	.64	.64	.67	.61	.56	.73	.72

(a) These data are stored in C4 and C6 of the Stream worksheet. Plot flow rate versus stream depth.

(b) Fit a straight line to explain flow rate based on stream depth. Where does this line fall on the plot?

(c) Plot the residuals against stream depth. Is there any indication that another model should be used? Is anything else indicated?

(d) Fit a quadratic model. Does it fit any better? Explain.

11-28   Compare the analyses for sites 3 and 4 on the following points:

(a) How well you can predict flow rate from depth.

(b) Whether or not a quadratic model fits better than a straight line.

(c) Whether similar relationships between flow rate and depth seem to hold for both sites.

11-29   A simple pendulum experiment from physics consists of releasing a pendulum of a given length (L), allowing it to swing back and forth for 50 cycles, and recording the time it takes to swing through these 50 cycles. Data from five trials of this experiment are given in Table 11.11.

Table 11.11 Pendulum experiment

Length	175.2	151.5	126.4	101.7	77.0
Time for 50 Cycles	132.5	123.4	112.8	101.2	88.2

(a) Let T be the average time per cycle. Compute T for each trial by dividing the time by 50. Plot T versus L and fit a straight line to explain T based on L. How well can you estimate time per cycle from pendulum length, using a straight line?

(b) Plot the residuals versus L. Are there any indications that a higher-degree polynomial should be used? Fit a better model if one seems needed.

11-30   In Exercise 11-19 we simulated data from a regression equation. Repeat those simulations, and each time plot the residuals versus x. This should give you some idea of how a residual plot looks when a correct model is fitted.

11-31   Frank Anscombe constructed the data listed in Table 11.12 to make an important point. The following steps should illustrate his point quite dramatically. (Note: y_1, y_2, and y_3 all use the same x values.) The data are shown in Table 11.12 and are saved in the worksheet Fa.

Table 11.12   Anscombe data

x_1, x_2, x_3	y_1	y_2	y_3	x_4	y_4
10	8.04	9.14	7.46	8	6.58
8	6.95	8.14	6.77	8	5.76
13	7.58	8.74	12.74	8	7.71
9	8.81	8.77	7.11	8	8.84
11	8.33	9.26	7.81	8	8.47
14	9.96	8.10	8.84	8	7.04
6	7.24	6.13	6.08	8	5.25
4	4.26	3.10	5.39	19	12.50
12	10.84	9.13	8.15	8	5.56
7	4.82	7.26	6.42	8	7.91
5	5.68	4.74	5.73	8	6.89

(a) Open the Fa worksheet and use the Regression command to fit a straight line to each pair of variables, y_i and x_i. Compare the regression output from the different data sets. Do they have anything in common? Based on the regression output, would you tend to think the data pairs were related pretty much alike?

(b) For each data set make a plot of y versus x. Now think back to the regression output in part (a). Can you guess what important point Anscombe was trying to make?

Exercises for Section 11.7

11-32   In Exercise 11-29 you fitted a line to some pendulum data. A residual plot indicates a curve in the data—a curve that may not have been apparent in a data plot. You then probably fitted a quadratic. This gives a residual plot with no apparent pattern (of course, with just five observations, it's difficult to do a very precise analysis). It is known from physics that the correct relationship is

$$\pi T = 2\pi\sqrt{L/q}$$

where $q = 981$ cm/ (at the latitude where the experiment was done). Thus if we fit a straight line of T versus $\sqrt{L}$, we should find $A = 0$ and

$$B = (2\pi)/\sqrt{981}.$$

(a) Fit the model $T = A + B\sqrt{L}$. How well does it fit compared to a quadratic model? If you didn't know any physics, could you decide between these two models based on these data?

(b) Do your estimates of A and B seem close to the theoretical values?

(c) Now test H_0: $A = 0$ and H_0: $B = 0.201$ (use $\alpha = 0.05$). Are the results of these tests consistent with your answers in (b)? Explain.

11-33   An experiment was run in which a tumor was induced in a laboratory animal. Table 11.13 shows the size of the tumor as it grew. The data are in the file named Tumor.

Table 11.13 Tumor growth

Number of days after induction	Size of tumor (cc)	Number of days after induction	Size of tumor
14	1.25	28	16.70
16	1.90	30	21.00
19	4.75	33	27.10
21	5.45	35	30.30
23	7.53	37	40.50
26	14.50	41	51.40

Investigate the relationship between time and tumor size. Is the relationship linear? Can it be "linearized" by an appropriate transformation?

11-34   In Exercise 11-23, we studied the Peru data (described in the Appendix, p. 528) with the aim of predicting systolic blood pressure from other variables in the data set. We will continue that study here.

(a) One model in Exercise 11-23 used Weight and Years to predict systolic blood pressure. Fit that model if you have not already done so.

(b) The researchers in this study created a new variable, fraction of life since migration, using Fraction = Years/Age. They reasoned that younger people adapt to new surroundings more quickly than older people. Therefore, Fraction might be a better measure of how long a person lived in the new environment than Years. Combining variables in this way is a type of transformation. Use this new predictor along with Weight to predict systolic blood pressure. How does this model compare to the one in part (a)?

(c) Use the three predictors Years, Fraction, and Weight to predict systolic blood pressure. How does this model compare to the ones in parts (a) and (b)?

(d) Can you find another predictor to add to the model in part (c) that improves things? Look at just the original predictors (there are seven remaining) in the data set.

(e) There is (at least) one major lesson in the analyses you have done using these data. What is it?

11-35   When a magnifying glass is used to view an object, there is a simple relationship between the apparent size of the object as seen through the lens (image size) and the distance of the object from the lens (object distance). Table 11.14 contains data that were obtained in an experiment.

Table 11.14 Magnifying glass data

Object distance (centimeters)	12	13	14	15	16	17	18	19	20	21	22
Image size (centimeters)	12.0	9.4	7.2	6.2	5.2	4.5	4.0	3.6	3.2	3.0	2.7

(a) Plot image size versus object distance. How does image size vary with object distance?

(b) Suppose you were to place an object at a distance of 12.5 centimeters from the lens. What would you estimate the image size to be? Suppose you were to place it at 23 centimeters. Approximately what image size would you expect?

(c) Now let's investigate the relationship between these two variables. It is obviously some sort of curve, but what curve?

(d) See if you can find a way to transform one of the variables so that the resulting plot is approximately a straight line. Look at the plots in Figure 11.20, p. 355, for ideas.

Exercises for Section 11.8

11-36   Continuation of Exercise 11-34 with the Peru data. Again consider prediction of Systolic with Weight and the calculated variable, Fraction.

(a) Run the regression and save the residuals and fitted values.

(b) Use the 3D scatterplot to display the fitted values versus the predictors Weight and Fraction. The fitted values should all lie on a plane. Rotate the plot until you can see the edge of the plane.

(c) Use the 3D scatterplot to display the residuals versus the predictors Weight and Fraction. With a good model there should be no pattern in the residuals. Rotate the plot to check for patterns.

(d) Follow the method on p. 364 and display a 3D graph of the regression plane. Rotate it to see how the predicted relationship varies with Weight and Fraction.

11-37   Use the Property data.
(a) Use Minitab to verify the fitted multiple regression model given by the curved relationship (Save the residuals and fitted values):

$$\text{Market} = -4.21 + 0.318 \text{ Assessed} + 0.0556 \text{ Sq.ft} - 0.000019 \ (\text{Sq.ft})^2$$

(b) Reproduce the 3D surface graph of the curved regression surface shown in Figure 11.30, p. 364.
(c) Rotate the graph in (part (b) to better visualize the curved surface.
(d) Rotate the "lights" shown in the graph of part (b) to better visualize the curved surface.
(e) Display a 3D scatterplot of the residuals versus the predictors Assessed and Sq.ft. Rotate the graph to look for problems with the regression model

Exercises for Section 11.8

11-38   Use the data displayed in Table 11.12, p. 380, and saved in the Worksheet named Fa. In each case let the Minitab Assistant choose the best polynomial model.
(a) Fit a regression model of y_1 versus x using the Assistant.
(b) Summarize the results displayed in the various reports.
(c) Repeat part (a) using y_2 versus x.
(d) Repeat part (a) using y_3 versus x.
(e) Repeat part (a) using y_4 versus x_4.

11-39   Use the data in Table 11.2, p. 370. In each case let the Minitab Assistant choose the best polynomial model.
(a) Fit a regression model of Y1 versus X1 using the Assistant.
(b) Summarize the results displayed in the various reports.
(a) Fit a regression model of Y2 versus X2 using the Assistant.
(b) Summarize the results displayed in the various reports for these pairs.

11-40   Use the Worksheet named Tree and let the Minitab Assistant choose the best polynomial model.
(a) Fit a regression model of Volume versus Diameter.
(b) Summarize the results displayed in the various reports.

11-41   The data file named Redwoods contains heights (Height), diameter at breast height (DBH) and bark thickness (Bark) of 21 California redwood trees. DBH of a tree is easy to measure but measuring Height is a bit more of a challenge. Bark thickness is also relatively easy to measure. It would be useful if foresters could predict Height from DBH and Bark with reasonable accuracy.

(a) Use the Minitab Assistant to approach this issue but select a linear (straight-line) model of Height versus DBH alone. Describe the results.

(b) Now use the Assistant but let Minitab choose the best model based only on DBH. How do the results differ from what you found in part (a)?

(c) Use the Minitab Assistant and select a linear (straight-line) model of Height versus Bark alone. Describe the results.

(d) Now use the Assistant and let Minitab choose the best model based only on Bark. How do the results differ from what you found in part (c)?

(e) Now consider models that use both DBH and Bark to explain Height. Fit a plane using DBH and Bark. How does it compare to the models chosen in parts (b) and (d)? (Note: You *cannot* do this with the Assistant.)

(f) Finally consider a model that uses DBH, DBH^2, Bark and $Bark^2$ to model Height. Discuss the results and compare these results to those obtained in parts (b), (c) and (e).

11-42   A moving company would like to be able to predict the man-hours required to complete a move. The data file named Moving gives the man-hours and volume (in cubic feet) of household goods from 36 past moves.

(a) Use the Minitab Assistant and select a linear (straight-line) model to see if man-hours can be reasonably predicted from volume. Discuss the results.

(b) Repeat the analysis of part (a) but now let Minitab select the best model. Discuss the results.

12

Chi-Square Tests and Contingency Tables

Many times we count the number of occurrences of an event. For example we count how many fatal accidents there are on a holiday weekend or how many people in different age groups support restrictions on international trade. We may then be interested in whether or not our counts are in agreement with some theory. Minitab may be used to test agreement in such cases.

12.1 Chi-Square Goodness-of-Fit Test

Several years ago an article in the Washington Post described a high-school boy named Edward who made 17,950 coin flips and "got 464 more heads than tails and so discovered that the United States Mint produces tail-heavy pennies." Is this result statistically significant? The statistician W. J. Youden called Edward and asked how he had done this experiment. Edward explained that he had tossed five pennies at a time and his younger brother had recorded the results as Edward called them out. Table 12.1 shows the results.

Table 12.1 Data from coin-tossing

Number of heads in five tosses	Number of times found
0	100
1	524
2	1080
3	1126
4	655
5	105
Total Tosses	3590

The standard model for coin tossing assumes that heads and tails are equally probable and the results from toss to toss are independent. Thus, under this model, if you toss five coins, the number of heads follows a binomial distribution with $n = 5$ and $p = 0.5$. This is our null hypothesis. To test this hypothesis, we use a chi-square goodness-of-fit test.

First, we enter the data into C1 and C2, and assign names to all the columns we plan to use. There are six cases, or "cells"; 0 heads in 5 tosses, 1 head in 5 tosses, 2 heads in 5 tosses, and so on. The Probability Distributions ▸ Binomial PDF command calculates the proportion of the time each case would occur if H_0 were true. Next, we multiply each proportion by the total number of observations (the number of times Edward tossed the five coins). This gives the number we would expect to get in each case. To test the hypothesis,

1. If necessary, open a new project or restart Minitab to set all dialog boxes to their defaults.

2. Enter the first six rows of data in Table 12.1 into the first two columns of the worksheet (don't include the Total Tosses row).

3. Name C1 **Heads** and C2 **Frequency**,

4. Choose **Calc ▸ Probability Distributions ▸ Binomial** and select the **Probability** button.

5. Enter **5** as the Number of trials and **0.5** as the Event probability.

6. Enter **Heads** as the input column and enter **Probability** as the column for Optional storage. See Figure 12.1.

7. Click **OK**.

Figure 12.1   Binomial Distribution dialog box

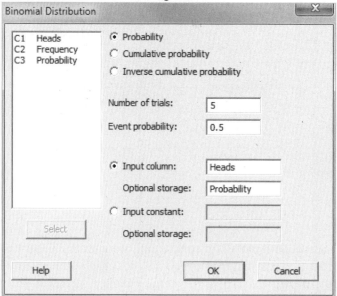

Now we can let Minitab perform the Chi-Square goodness-of-fit test.

1. Choose **Stat ▸ Tables ▸ Chi-Square Goodness-of-Fit Tests (One Variable)...**.
2. Click the **Observed counts** button and enter **Frequency**. See Figure 12.2.
3. Enter Category names of **0 1 2 3 4 5**. (Note the spaces in between.)
4. Click the **Specific proportions** button and enter **Probability**.
5. Click **OK**. The Session Window results are shown Figure 12.3.

Figure 12.2   Chi-Square goodness-of-fit dialog box

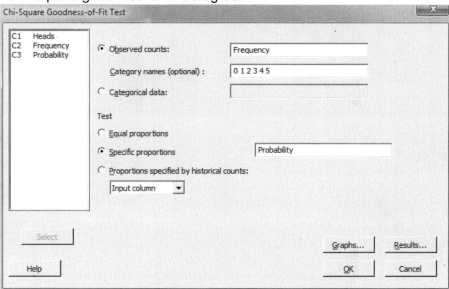

Figure 12.3   Chi-Square goodness-of-fit Session Window results

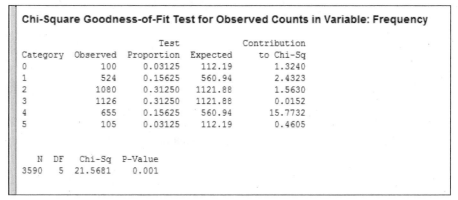

Chi-Square Goodness-of-Fit Test for Observed Counts in Variable: Frequency

Category	Observed	Test Proportion	Expected	Contribution to Chi-Sq
0	100	0.03125	112.19	1.3240
1	524	0.15625	560.94	2.4323
2	1080	0.31250	1121.88	1.5630
3	1126	0.31250	1121.88	0.0152
4	655	0.15625	560.94	15.7732
5	105	0.03125	112.19	0.4605

N	DF	Chi-Sq	P-Value
3590	5	21.5681	0.001

In addition to the numerical results, Minitab displays two charts. One, Figure 12.4, shows the discrepancy between the Observed and Expected numbers and the other, Figure 12.4, shows ordered individual contributions to the Chi-Square statistic. (Minitab displays both of these charts by default but either one or both may be omitted if you select the Graphs button in Figure 12.2.)

Figure 12.4   Chart of Observed and Expected Values

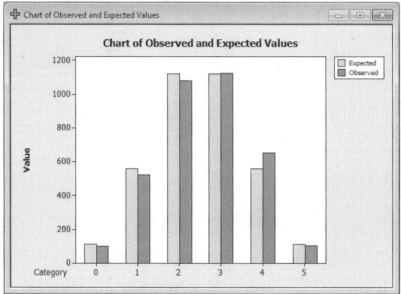

Of course the Observed and Expected numbers do not agree exactly. Even if our hypothesis were exactly true, we would not expect perfect agreement. Some amount of chance variation will always be present. On the other hand, if our hypothesis were true, we should not expect "too much" difference between the Observed and Expected counts. How much difference is "too much"? The chi-square test gives us a way of doing a formal test. The chi-square statistic is

$$\chi^2 = \sum [(\text{Observed} - \text{Expected})^2 / \text{Expected}]$$

If the Observed and Expected numbers are very different, the value of χ^2 will be large. In Figure 12.3 the Chi-Square statistic is reported as $\chi^2 = 21.581$ with 5 degrees of freedom and a p-value of 0.001. This is highly significant. We have good evidence that Edward's pennies do not follow a binomial distribution with $n = 5$ and $p = 0.5$.[†]

Let's look just a bit further. Look at the individual contributions to the chi-square statistic in Figure 12.3 and in Figure 12.4. One number stands out,

---

[†] Also note that we are not testing whether or not the individual coin flips are "fair" nor whether or not they are independent.

far out; the number 15.7732, corresponding to 4 heads. Look at the counts for this case. There are many more observed counts than expected. Of course, there may be some mistakes in the data gathering. For example, perhaps some of the tosses with 4 tails were recorded as 4 heads.

Figure 12.5   Chart of Ordered Contributions to the Chi-Square Statistic

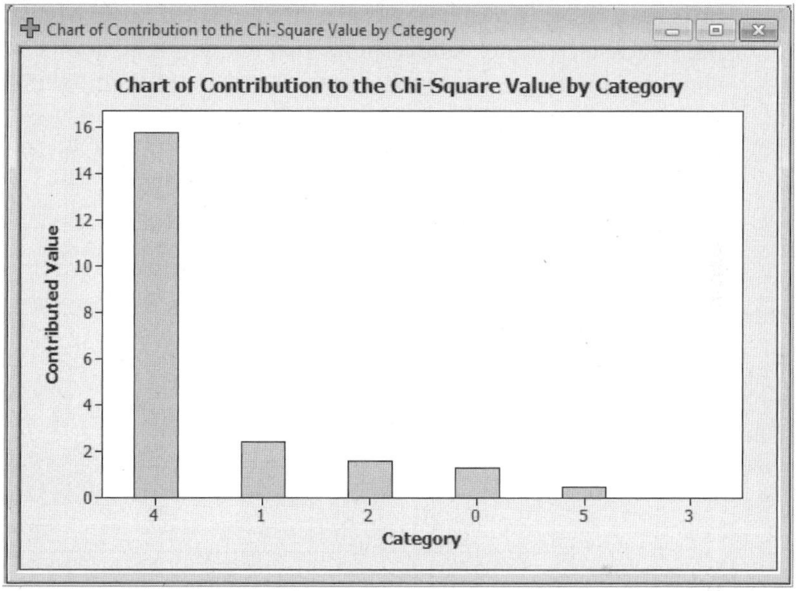

12.2 Contingency Tables

A researcher did a study to investigate the relationship between being an artist or not and believing in extrasensory perception (ESP). He asked 114 artists and 344 nonartists to classify themselves into one of three categories: (1) believe in ESP, (2) believe more-or-less, (3) do not believe. His results are given in a contingency table shown in Figure 12.2.

Table 12.2 Contingency table from a study of ESP

	Believe in ESP	*Believe more-or-less*	*Do not believe*	*Total*
Artists	67	41	6	114
Nonartists	129	183	32	344
Total	196	224	38	458

One question that is frequently asked about such data is ""Is there any association between being an artist and belief in ESP"? Another version of the same question is, "Do artists and nonartists seem to have about the same degree of belief in ESP?" The corresponding null hypothesis can be phrased as "The proportions in the three categories for belief in ESP are the same for artists as for nonartists."

Of the 458 people in the study, 196/458, or 42.79%, believe in ESP; 224/458, or 48.91%, believe more-or-less; and 38/458, or 8.30%, do not believe. There are 114 artists. Therefore, if the null hypothesis is true, 42.79% of the 114 artists, or 48.79 people, should believe in ESP. Similarly 48.91% of 114, or 55.76 artists, should believe more-or-less, and 8.30% of 114, or 9.46 artists, should not believe. The same proportions should hold for the nonartists. Thus 42.79% of 344 = 147.21 should believe, 48.91% of 344 = 168.26 should believe more-or-less, and 8.30% of 344 = 28.55 should not believe. These are the Expected counts if our null hypothesis is true.

Are the Observed and Expected counts close? To determine whether the differences are more than could reasonably be due to chance alone, we use a chi-square test. The test statistic is, as before,

$$\chi^2 = \sum [(\text{Observed} - \text{Expected})^2 / \text{Expected}]$$

For the ESP data

$$\chi^2 = \frac{(67 - 48.79)^2}{48.79} + \frac{(41 - 55.76)^2}{55.76} + \frac{(6 - 9.46)^2}{9.46}$$

$$+ \frac{(129 - 147.21)^2}{147.21} + \frac{(183 - 168.26)^2}{168.26} + \frac{(32 - 28.55)^2}{28.55}$$

$$= 15.94$$

The number of degrees of freedom is computed as follows: Let r be the number of rows in the contingency table. Here there are two: artists and nonartists. Let c = the number of columns. There are three: believe, believe more-or-less, and do not believe. The number of degrees of freedom is

$$(r - 1)(c - 1) = (2 - 1)(3 - 1) = 2$$

Let's do these calculations using Minitab's Chi-Square Test command. To perform the chi-square test,

1. Clear the worksheet.
2. Name C1 **Believers**, C2 **More-or-less Believers**, and C3 **Non-Believers.**
3. Enter the ESP data, placing **67** and **129** into C1, **41** and **183** into C2, and **6** and **32** into C3.
4. Choose **Stat ▸ Tables ▸ Chi-Square Test (Table in Worksheet).**
5. Enter **C1-C3** in the Columns containing the table box and then click **OK**. Minitab displays the output shown in Figure 12.6.

Figure 12.6   Chi-Square test results

```
┌─────────────────────────────────────────────────────────────────┐
│  Chi-Square Test: Believers, More-or-less Believers, Non-Believers │
│                                                                   │
│  Expected counts are printed below observed counts                │
│  Chi-Square contributions are printed below expected counts       │
│                                                                   │
│                    More-or-less                                   │
│          Believers   Believers  Non-Believers  Total              │
│    1         67          41            6        114               │
│            48.79       55.76         9.46                         │
│            6.800       3.905         1.265                        │
│                                                                   │
│    2        129         183           32        344               │
│           147.21      168.24        28.54                         │
│            2.254       1.294         0.419                        │
│                                                                   │
│  Total      196         224           38        458               │
│                                                                   │
│  Chi-Sq = 15.936, DF = 2, P-Value = 0.000                        │
└─────────────────────────────────────────────────────────────────┘
```

The *p*-value is 0.000. Therefore we have strong evidence that the distribution of ESP beliefs among artists is different from that among nonartists.

Remember that row 1 contains the artists data and row 2 the nonartists. Notice that the output gives the cell-by-cell contributions to the overall value. In this case the largest contribution, 6.800, comes from the "artists who believe in ESP" cell. The Observed count is 67, but the Expected count if the null hypothesis were true is only 48.79. Many more artists than expected believe in ESP.

12.3 Making the Table and Computing Chi-Square

In the preceding section we started with the contingency table and Minitab did the rest. In this section we show how Minitab can make the contingency table from raw data and then do the chi-square test all in one operation. The procedure is a simple extension of the one discussed in Chapter 4 for making tables.

Consider again the Restaurant data. In Chapter 4 we used the Cross Tabulation command to tabulate Owner by Size, which made a table that classified the restaurants by type of ownership and size. The Cross Tabulation command will also do a chi-square analysis if you click the Chi-Square analysis check box in the Cross Tabulation dialog box. See Figure 4.36, p. 135. To compute chi-square, using cross tabulation,

1. Open the **Restaurant** worksheet. (We have coded the numerical variables Owner and Size to text and Value-ordered them as we did in Section 4.7, p. 127.)

2. Choose **Stat ▸ Tables ▸ Cross Tabulation and Chi-Square**.

3. Enter **Owner** for rows and **Size** for columns as the Categorical variables.

4. For Display, select the **Counts** check box.

5. Select the **Chi-Square** button.

6. Select Display: **Chi-Square analysis** and **expected cell counts**.

7. Click **OK** twice. Minitab produces the output shown in Figure 12.7.

Minitab prints two values in each cell: the first is the Observed count while the second is the Expected count—the count we would expect if the null hypothesis of no association were true. The calculated value of chi-square is 67.917. This is quite large for a table with $(3 - 1)(3 - 1) = 4$ degrees of freedom. The p-value is listed as 0.000. We therefore have strong evidence that the size of a restaurant is related to the type of ownership. (Figure 12.7 also displays the value of the Likelihood Ratio Chi-Square statistic—an alternative to Pearson's chi-square. Using this statistic our conclusion would be the same.)

To see how size and ownership are related, we can compare the Observed and Expected counts in Figure 12.7. For sole proprietorships there are more small restaurants (under 10) than expected and fewer large (over 20) than expected. For corporations the reverse is true—there are fewer small restaurants than expected and more large ones. The overall conclusion is not surprising: Corporate-owned restaurants tend to be larger than those owned by a single person.

Figure 12.7   Chi-Square test results from restaurant data

```
Tabulated statistics: Owner, Size

Rows: Owner   Columns: Size

                    under 10  10 to 20  over 20  Missing     All

sole proprietorship       83        18        2        3     103
                       54.85     26.05    22.10        *  103.00

partnership               16         6        4        1      26
                       13.85      6.57     5.58        *   26.00

corporation               40        42       50        4     132
                       70.30     33.38    28.32        *  132.00

Missing                    1         0        1        8       *
                           *         *        *        *       *

All                      139        66       56        *     261
                      139.00     66.00    56.00        *  261.00

Cell Contents:      Count
                    Expected count


Pearson Chi-Square = 67.917, DF = 4, P-Value = 0.000
Likelihood Ratio Chi-Square = 77.732, DF = 4, P-Value = 0.000
```

12.4 Tables with Small Expected Counts

A survey was done in an introductory statistics class. Each student was asked to give his or her year in college and political preference. This gave the contingency table shown as Table 12.3.

Table 12.3 Data from a survey of political preference

	Freshman	Sophomore	Junior	Senior
Democrat	4	16	16	6
Republican	1	7	7	3
Other	3	4	12	5

Let's use the Chi-Square Test command on these data. To perform a chi-square test on the political party data,

1. Open a new, blank worksheet and enter the new data in C1-C4.
2. Name the columns **Freshmen**, **Sophomore**, **Junior**, **Senior**.
3. Choose **Stat ▸ Tables ▸ Chi-Square Test**.

4. Enter **C1-C4** in the Columns box and then click **OK**. The Minitab output is displayed in Figure 12.8.

Notice that the last line in the output contains the message:

```
5 cells with expected counts less than 5.
```

This indicates that the chi-square analysis done on this table may not be reliable. As with many statistical tests, the chi-square test is an approximate test. The approximation becomes better and better as the expected cell frequencies increase. Consequently if too many cells have expected frequencies that are small, a chi-square analysis may not be appropriate.

A good rule of thumb is that not more than 20% of the cells should have expected cell frequencies of less than 5, and no cell should have an expected frequency of less than 1. The table in Figure 12.8 has 12 cells, so no more than $0.2 \times 12 = 2.4$ cells should have expected frequencies of less than 5. But, as the message says, and as we can see ourselves from the table, there are five such cells.

Figure 12.8   Chi-Square test results for political preference data

```
Chi-Square Test: Freshman, Sophomore, Junior, Senior

Expected counts are printed below observed counts
Chi-Square contributions are printed below expected counts

         Freshman  Sophomore  Junior  Senior  Total
  1          4         16       16       6      42
           4.00      13.50    17.50    7.00
           0.000      0.463    0.129    0.143

  2          1          7        7       3      18
           1.71       5.79     7.50    3.00
           0.298      0.255    0.033    0.000

  3          3          4       12       5      24
           2.29       7.71    10.00    4.00
           0.223      1.788    0.400    0.250

Total        8         27       35      14      84

Chi-Sq = 3.982, DF = 6, P-Value = 0.679
5 cells with expected counts less than 5.
```

We do not have quite enough data here to analyze the original 3×4 table, so if we want to get something out of the data, we can try looking at a table with fewer levels.

There are several procedures we can follow with tables having too many small expected frequencies. We can try to combine cells so that we reduce the number of rows and/or columns. For example if we combine columns 1 and 2 into one level, called "lower class," and combine columns 3 and 4 into another level, called "upper class," we get the 3×2 table in Table 12.4.

Table 12.4 Combining columns

	Lower class	Upper class
Democrat	20	22
Republican	8	10
Other	7	17

Now if no more than one cell in this smaller table has expected frequency less than 5, we can be more comfortable with the chi-square analysis. The smaller table was read into the worksheet and a second analysis was done shown in Figure 12.9. In this case no cell had expected frequency less than 5, so the chi-square approximation is probably good enough. Notice that the p-value is 0.331 so there is no evidence that Lower or Upper classification is related to political preference.

Figure 12.9   Chi-Square test results for political preference data with reduced classes

```
Chi-Square Test: Lower class, Upper class

Expected counts are printed below observed counts
Chi-Square contributions are printed below expected counts

        Lower  Upper
        class  class  Total
   1      20     22     42
        17.50  24.50
        0.357  0.255

   2       8     10     18
         7.50  10.50
        0.033  0.024

   3       7     17     24
        10.00  14.00
        0.900  0.643

Total     35     49     84

Chi-Sq = 2.212, DF = 2, P-Value = 0.331
```

Of course we have changed our situation slightly. We are now testing a different null hypothesis: that there is no relationship between a student's political preference and whether the student is in the upper or the lower class. Before, we were testing the null hypothesis that there is no relationship between a student's political preference and whether the student is a freshman, sophomore, junior, or senior.

Another way to handle the problem is to remove one or more levels of a classification. In the political preference table we might omit the first column, corresponding to freshmen. This leads to Table 12.5.

Table 12.5 Removing a level of classification

	Sophomore	Junior	Senior
Democrat	16	16	6
Republican	7	7	3
Other	4	12	5

This table provides an analysis relevant to students above the freshman level. Here, however, the omission doesn't help much. There are still too many cells with small expected frequencies.

Exercises

Exercises for Section 12.1

12-1   Table 12.6 gives accidental deaths from falls, by month, for the year 1979.

Table 12.6 Accidental deaths

Month	Number of deaths from falls	Month	Number of deaths from falls
Jan	1150	Aug	1099
Feb	1034	Sep	1114
Mar	1080	Oct	1079
Apr	1126	Nov	999
May	1142	Dec	1181
Jun	1100	Total	113,216
Jul	1112		

(a) Open the Falls worksheet. Do accidental deaths due to falls seem to occur equally often in all 12 months? Do a chi-square goodness-of-fit test. What is an appropriate null hypothesis? Calculate the expected number of falls under this null hypothesis. Complete the test.

(b) Can you give some reasons for the result in part (a)? What patterns do you see in the data?

12-2   Suppose we count the number of days in a week on which at least 0.01 inch of precipitation falls. Can we model this by a binomial distribution? Table 12.7 displays data collected in State College, Pennsylvania, during the years 1950-1969. All the observations are from the same month, February. This gives us (4 weeks in Feb.)×(20 years) = (80 weeks in all).

Table 12.7 Precipitation data

Number of precipitation days in a week	0	1	2	3	4	5	6	7
Number of weeks in which this occurred	3	12	17	25	14	5	4	0

(a) Enter these data. Use these data to estimate $p =$ the probability of precipitation on a given day.
(b) Test to see whether a binomial distribution with $n = 7$ and p as estimated in part (a) fits the data. You can use the Probability density option to help you get the expected counts for each group.
(c) Does your answer to part (b) agree with what you would expect of rainfall data? Explain.

12-3   The Missouri Department of Conservation wanted to learn about the migration habits of Canadian geese. They banded more than 18,000 geese from one flock and classified each goose into one of four groups according to its age and sex. Information on the bands asked all hunters who shot one of these geese to send the band to a central office and tell where they shot the goose. The number of geese that were banded in each of the four groups and the number of bands that were returned from geese that were shot outside the normal migration route for this flock are given in Table 12.8.

Table 12.8 Geese

Group	Number of geese banded	Number of bands returned
Adult Male	4144	17
Adult Female	3597	21
Yearling Male	5034	38
Yearling Female	5531	36
Totals	18,306	112

(a) Suppose the four groups were equally likely to stray from the normal migration routes. Calculate the number of bands you would expect to be returned in each of the four groups. Notice that you must take into account the fact that the number of geese banded in each group was not the same.

(b) Are the four groups equally likely to stray? Do an appropriate test.

Exercises for Section 12.2

12-4   Two researchers at Penn State University studied the relationship between infant mortality and environmental conditions in Dauphin County, Pennsylvania. This county has one large city, Harrisburg. The researchers recorded the season in which each baby was born and whether or not it died before one year of age. Data on infant deaths and births for Dauphin County, 1970, are given in Table 12.9.

Table 12.9 Infant data for Dauphin county

	Season of birth			
	Jan Feb Mar	Apr May Jun	Jul Aug Sep	Oct Nov Dec
Died before one year	14	10	35	7
Lived one year	848	877	958	990

(a) Is an infant more likely to die if it is born in one season than in another? Which season has the highest risk? The lowest risk?

(b) Newspapers reported severe air pollution, covering the entire East, during the end of July. Air pollution, especially during the end of pregnancy and in the first few days after birth, is suspected of increasing the risk of an infant's death. Is this theory consistent with the data and the analysis in part (a)?

12-5   The data in Exercise 12-4 are for all of Dauphin County. However, environmental conditions as well as socioeconomic conditions are different for the city of Harrisburg and for the surrounding countryside. Table 12.10 and Table 12.11 give the data separately for these two regions. Do the analysis in Exercise 12-4 separately for each region. What are your conclusions now?

Table 12.10 Infant data for Harrisburg

	Season of birth			
	Jan Feb Mar	Apr May Jun	Jul Aug Sep	Oct Nov Dec
Died before one year	6	6	25	3
Lived one year	306	334	347	369

Table 12.11 Infant data for Dauphin County, excluding Harrisburg

	Season of birth			
	Jan Feb Mar	Apr May Jun	Jul Aug Sep	Oct Nov Dec
Died before one year	8	4	10	4
Lived one year	542	543	611	621

12-6   A survey on car defects was done in Dane County, Wisconsin, based on a random sample of people who had just purchased used cars. Each owner was sent an invitation to bring the car in for a free safety inspection. Car owners who did not respond to the invitation were sent postcard reminders. Eventually about 56% of the owners had their cars inspected, giving a total of 8842 cars. Here we will look at four types of defects: brakes, suspension, tires, and lights. There is one contingency table for each type of defect. Each car was classified as to whether it was purchased from a car dealer or from a private owner. Table 12.12 shows the tables.

Table 12.12 Safety inspection data

	Brakes				Suspension	
	Defective	Not defective			Defective	Not defective
Dealer	931	2723		Dealer	1120	2534
Private	1690	3498		Private	1171	3477

	Tires				Lights	
	Defective	Not defective			Defective	Not defective
Dealer	1147	2507		Dealer	602	3052
Private	1765	3423		Private	1098	4090

(a) Is there a relationship between a car's having defective brakes and whether it was purchased from a dealer or a private owner? Do an appropriate test. What percent of dealer-sold cars have defective brakes? What percent of privately sold cars? (Calculate these percents by hand.) How do the percents compare?

(b) Repeat part (a) for defects in the suspension system.

(c) Repeat part (a) for defective tires.

(d) Repeat part (a) for defective lights.

(e) Compare the results from parts (a)-(d). Is there a common pattern?

(f) Try to think of some possible reasons for the results in parts (a)-(e).

12-7   Table 12.13 gives more data from the Wisconsin car survey described in Exercise 12-6. The age of each car was also recorded. These ages were then grouped into three categories: cars under three years old, cars from three to six years old, inclusive, and cars over six years old.

Table 12.13 Safety inspection data including car age

	Under three years	Three-six years	Over six years
Dealer	570	1792	1292
Private	513	1859	2816

Is there a relationship between the age of a car and whether it was purchased from a dealer or from a private owner? Do an appropriate test. Describe the relationship. Calculate (by hand) row percents to use in your description.

12-8   Mark Twain has been credited in numerous places with the authorship of 10 letters published in 1861 in the New Orleans Daily Crescent. The letters were signed "Quintus Curtius Snodgrass." Did Twain really write these letters? In a 1963 paper Claude S. Brinegar used statistics to compare the Snodgrass letters to works known to have been written by Mark Twain. We present some of his very interesting analyses in this exercise. The data are stored in the worksheet Twain.

There are many statistical tests of authorship. The one Brinegar used is quite simple in concept—just compare the distributions of word lengths for the two authors. If these distributions are very different, we will have some evidence that the authors are probably different people. In using this type of test, we must assume that the distribution of word lengths is about the same in all works written by the same author. Parts (a), (b), and (c), below, attempt to provide some checks on this assumption.

The 10 Snodgrass letters were first divided into three groups. Then the number of two-letter words in each group was recorded, followed by the number of three-letter words in each group, the number of four-letter words, and so on. One-letter words were omitted. There are only two such words, "I" and "a," and the use of "I" tends to characterize content (work written in first person or not) more than an unconscious style. Data for Mark Twain were obtained from letters he wrote to friends at about the same time and from two works written at a later time. These will be used to see whether the word-length distribution remained about the same throughout Twain's writings.

(a) Compare the three groups of Snodgrass letters to see whether they are consistent in word-length distribution. First, compare them graphically. To make the numbers comparable, change the frequencies into relative frequencies (proportions). Then plot each column versus word length. Put all three plots on the same axes.

Do the distributions look similar? Next, do a chi-square test of the null hypothesis that all three sets of letters have the same word-length distribution. Is there any evidence that they differ?

(b) Next, compare the three groups of Mark Twain letters. Do both a plot and a chi-square test, as in part (a). Is there any evidence that these three collections of writings differ in word-length distribution?

(c) Now compare Mark Twain's writings over a large span of years. The samples from Roughing It and Following the Equator were taken for this purpose. First, combine the three columns of Mark Twain letters into one column of "early works." Compare the "early," "middle," and "late" samples for Mark Twain. Do both a plot and a chi-square test, as in part (a). Is there any evidence that the word-length distribution changed over the years?

(d) Finally, now that we've checked the authors for consistency, let's compare the Twain and Snodgrass works. As in part (c) combine the three columns of Twain's early works into one column. Also combine the three columns of Snodgrass letters into one column. Finally compare Twain's letters with the Snodgrass letters. Do both a plot and a chi-square test. Do you think Mark Twain wrote the Snodgrass letters?

(e) In what ways do the two authors differ, as far as word-length distribution is concerned? Examine both the plot and chi-square output from part (d) to find out. Does the chi-square output tell you anything the plot does not, or vice versa?

Table 12.14 displays word counts for the Quintus Curtius Snodgrass letters; Table 12.15 for other known Mark Twain writings.

Table 12.14 Word counts for Quintus Curtius Snodgrass letters

Word length	First three letters	Second three letters	Last four letters
2	997	831	857
3	1026	828	898
4	856	669	777
5	565	420	446
6	366	326	300
7	318	293	285
8	258	183	197
9	186	150	129

Table 12.14 Word counts for Quintus Curtius Snodgrass letters (continued)

Word length	First three letters	Second three letters	Last four letters
10	96	94	86
11	63	49	40
12	42	30	29
13 and over	25	25	11
Totals	4798	3998	4055

Table 12.15 Word counts for known Mark Twain writings

Word length	Two letters from 1858 and 1861	Four letters from 1863	Letter from 1867	Sample from Roughing It, 1872	Sample from Following the Equator, 1897
2	349	1146	496	532	466
3	456	1394	673	741	653
4	374	1177	565	591	517
5	212	661	381	357	343
6	127	442	249	258	207
7	107	367	185	215	152
8	84	231	125	150	103
9	45	181	94	83	92
10	27	109	51	55	45
11	13	50	23	30	18
12	8	24	8	10	12
13 and over	9	12	8	9	9
Totals	1811	5794	2858	3031	2617

Exercises for Section 12.3

12-9   In Section 12.4, p. 393 it was suggested that deleting the Freshman column from the Chi-Square analysis does not solve the problem of small expected values for all the cells. Use the data in Table 12.5 on page 396 to verify that this is the case. How many cells are there with expected counts less than 5?

12-10   In the Restaurant survey described in the Appendix, p. 530, is there a relationship between the type of food sold and the size of the restaurant? Do an appropriate chi-square test.

12-11   In the Restaurant survey (described in the Appendix, p. 530) is there a relationship between the type of ownership and the overall outlook of the owner? Do an appropriate test. Describe this relationship, using appropriate row and/or column percents. (Recall that the Cross Tabulation dialog box has check boxes you can click to calculate these.)

12-12   Use the Pulse data (described in the Appendix, p. 529), to test to see whether there is a relationship between:

(a) sex and activity;

(b) sex and smokers;

(c) smokers and activity.

12-13   Use the Cartoon data (described in the Appendix, p. 524) to see whether there is a relationship between education and whether or not a person took the delayed test. You will have to create a variable for the second classification. Choose Data ▶ Code ▶ Numeric to Numeric and create a column, C20, in which those who took the test are assigned 1 and those who did not (those who have * for their score) are assigned 0.

(a) Do an appropriate test.

(b) Calculate appropriate percents and describe the relationship. Can you give some possible reasons for this relationship?

13

Nonparametric Statistics

The tests and confidence intervals described in Chapters 8, 9, and 10 assume we have either large random samples or random samples from normal populations. The methods they use are for inferences about population means. There are several nonparametric methods that also do tests and confidence intervals, but do not assume large sample sizes nor normal populations. These methods are based on medians rather than means.

There are several reasons why it may be more appropriate to use nonparametric procedures instead of the normal-theory (parametric) procedures. Among them are the following:

1. *Some populations are not even approximately normal.* To be strictly valid, the procedures in Chapters 8, 9, and 10 require that we have random samples from normal populations. If the samples do not come from normal populations, these procedures may give misleading results. For example if we construct a 95% t confidence interval from nonnormal data, the real confidence level may be only 91%, not 95% as we had planned. Another, and usually more serious, problem is loss of efficiency. Loosely speaking, a more efficient procedure makes better use of the data and enables us to get a better estimate or test with a smaller sample size. Some nonparametric procedures are more efficient than normal-theory procedures if we are sampling from a nonnormal population.

2. *Outliers can distort results.* Normal-theory methods are quite sensitive to even a few extreme, or outlying, observations. As a simple example consider the five numbers 3, 7, 9, 10, and 11. Normal theory uses their mean, whereas many of the methods presented in this chapter use the median. The median of these numbers is 9 and the mean is 8. But what would happen if one of the observations had been measured or recorded incorrectly? Suppose the 11 had been recorded as a 61 and we failed to notice this. Then the median would remain unchanged but the mean would more than double, to 18. The median, and the procedures described in this chapter, are more *resistant* than normal-theory methods to distortion by a few gross errors.

3. *Sometimes the median is a more informative measure of the center of the population.* If a population is very skewed, the mean can be much larger (or smaller) than many of the observations. A very common example is income in a small town where most people have moderate incomes but a few people are very wealthy. Those few can pull the town mean far upward, so that for some purposes the median would be a more informative measure of town income.

13.1 Sign Procedures

Let's take a look at some of the data from the Wisconsin Restaurant survey (described in the Appendix, p. 530). We want to analyze the market values of the fast-food restaurants owned by one person. There are two observations of zero and three asterisks (missing data) in the original data set. We will assume that zero market value really means missing and we replace those zeros by the missing value asterisk. Now we use the subsetting methods introduced on p. 74.

1. Open the **Restaurant** worksheet. (Restrnt.mtw recoded and value ordered)
2. Choose **Data ▸ Subset Worksheet**. See Figure 13.1.

Figure 13.1 Subset Worksheet dialog box

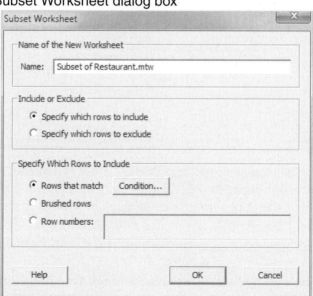

3. Click the **Condition** button and specify 'rows that match' and entering 'Typefood'="fast food" and 'Owner'="sole proprietorship" in the text box.[†] See Figure 13.2.

4. Click **OK** twice to obtain a new worksheet (by default named Subset of Restaurant.mtw) which contains the required subset of the **Values** variable. That is, market values for just the fast-food restaurants owned by a single owner.

5. Be sure the new worksheet is active and choose **Graph ▸ Stem-and-Leaf**,

6. Select **Value** as the Graph variable (be sure **Trim outliers** is deselected) and click **OK** to create a stem-and-leaf plot. Figure 13.3 shows the plot.

Figure 13.2 Subset Worksheet Condition dialog box

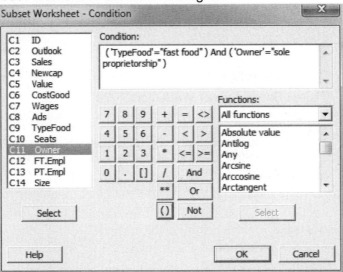

Notice that the stem-and-leaf display is not symmetric, but is somewhat skewed toward high values. (The N* = 5 means that 5 of the fast food restaurants owned by sole proprietorships had missing data for the Value variable.)

[†] The column names Typefood and Owner may be entered by selecting them from the list of available columns and Minitab will supply the single quotes enclosing them. The parentheses and the And operator may be selected by clicking the appropriate buttons. Unfortunately the "fast food" and "sole proprietorship" items will need to be typed in including the double quotes.

Figure 13.3

Stem-and-Leaf plot of Market Value (in thousands) of 37 fast-food restaurants owned by one person

```
Stem-and-Leaf Display: Value

Stem-and-leaf of Value   N  = 37
Leaf Unit = 10
N* = 5


  6   0   234444
 13   0   6777789
 (9)  1   002222334
 15   1   55677
 10   2   022
  7   2   577
  4   3   0
  3   3   6
  2   4   0
  1   4
  1   5   0
```

Sign Tests

Suppose we want to do a hypothesis test, using these values. It seems unreasonable to assume we have a sample from a normal distribution. A sign test does not make any assumptions about the shape of the population in our hypothesis: It could be very skewed, it could have "fat tails," (like the t distribution) and it could even be bimodal (that is, have two peaks).

As an example suppose the median market value for all fast-food restaurants in Michigan that are owned by one person is \$105,000. Are fast-food restaurants in Wisconsin generally worth more than those in Michigan? Let η be the (unknown) population median for Wisconsin. Then we test H_0: $\eta = 105,000$ versus H_1: $\eta > 105,000$.

Minitab's 1-Sample Sign command does the calculations for a sign test. For each column it tests the null hypothesis H_0: $\eta = K$, where η is the population median. The default alternative hypothesis is H_1: $\eta \neq K$. If you don't specify a value for η, Minitab tests H_0: $\eta = 0$. You can do a one-sided test with the Alternative list arrow.

In the Restaurant example Minitab first counts the number of values that are above $\eta = 105,000$ and the number that are below. If η were 105,000, we would expect, on the average, about half our observations to be above 105,000 and half to be below. The Restaurant example is skewed toward high values. Is this surprising? To perform a sign test,

1. Choose **Stat ▸ Nonparametrics ▸ 1-Sample Sign**.
2. Enter **Value** as the variable.

3. Click the **Test median** option button and enter **105** in the corresponding text box. (Remember the values are listed in *thousands*.)

4. Click the **Alternative** list arrow and then click **greater than**.

5. Click **OK** to obtain the output shown in Figure 13.4.

Figure 13.4
Sign test of
$H_0: \eta = 105$
versus
$H_1: \eta > 105$

```
Sign Test for Median: Value

Sign test of median =  105.0 versus > 105.0

          N  N*  Below  Equal  Above      P  Median
Value    37   5     15      0     22  0.1620   125.0
```

The output from the 1-Sample Sign test in Figure 13.4 tells us that the Restaurant data is not skewed toward high values compared to the Michigan median value. The *p*-value is 0.162. That is, more than 16% of the time we will get 22 or more observations above the median even if $\eta = 105,000$.

How Minitab Does a Sign Test. The sign test is based on the binomial distribution. Suppose it is true that the population median $\eta = 105,000$. If we pick one fast-food restaurant in Wisconsin at random, there is a 50-50 chance that its market value is above $105,000. If we pick a second fast-food restaurant, again there is a 50-50 chance that its market value is above $105,000, and so on, for all 37 observations. On each trial (sampling a fast-food restaurant at random), the probability of a success (a value over $105,000) is 0.5. Let X be the total number of successes. Then X has a binomial distribution with $p = 0.5$ and $n = 37$.

Sometimes there are several values equal to the hypothesized median value. This would be the case if our hypothesis were $\eta = 100,000$. There are two market values of $100,000. In such cases it is conventional practice to set aside those ties and apply the sign test to the remaining data. Then X would have a binomial distribution with $p = 0.5$ and $n = 35$.

Suppose we test $H_0: \eta = 105,000$ versus $H_1: \eta > 105,000$. In our sample $X = 22$. The probability of getting this many, or even more, observations over 105,000 is 0.1620 (you could also use Minitab's PDF option to get this value). This is the *p*-value given by the 1-Sample Sign test in Figure 13.2.

Suppose we test $H_0: \eta = 105,000$ versus $H_1: \eta < 105,000$. Then we let X be the number of observations below 105,000. If H_0 is true, X again has a binomial distribution with $p = 0.5$ and $n = 37$. In our sample $X = 15$. The probability under H_0 of getting this many, or even more, observations under 105,000 is 0.838, certainly not significant.

Suppose we do a two-sided test—for example H_0: $\eta = 105,000$ versus H_1: $\eta \neq 105,000$. Here we count the number of observations below the null value of η and the number above, and let X be the larger of the two counts. (If they are equal, either will do.) Here there are 22 observations below 105,000 and 15 above. The number below is the larger, so we calculate $P(X \geq 22$ under $H_0) = 0.1620$. The p-value of the two-sided test is twice this value, or $2(0.1620) = 0.3240$. Thus over 32% of the time we can expect to get a split this extreme or more so—that is, with 22 or more observations on one side or the other of 105,000.

Sign Confidence Intervals

Now let's try using the 1-Sample Sign confidence interval option to get a 95% confidence interval for the median market value for fast-food restaurants owned by one person. To calculate the confidence interval,

1. Choose **Stat ▸ Nonparametrics ▸ 1-Sample Sign** and choose **Value** as the variable.
2. Click the **Confidence interval** option button. Click **OK**. Figure 13.5 shows the output.

Figure 13.5
Sign confidence interval for the median market value

```
Sign CI: Value

Sign confidence interval for median

                                 Confidence
                      Achieved    Interval
        N  N*  Median  Confidence  Lower  Upper  Position
Value  37   5  125.0    0.9011    100.0  150.0      14
                        0.9500     91.0  159.0     NLI
                        0.9530     90.0  160.0      13
```

First, notice that Minitab prints out three confidence intervals. The top and bottom intervals are exact; that is, they have the achieved confidence given in the output. The middle interval is an approximate interval calculated by a nonlinear interpolation, abbreviated NLI, and corresponds to the confidence specified in the Confidence interval option button. Here it is the default confidence of 95%. With nonparametric methods it is rare that we can get the exact confidence level we ask for.

Let's look at the 90.11% interval. The 14 under Position tells us that this interval goes from the fourteenth smallest observation to the fourteenth largest observation. If we order the 37 observations and count 14 observations up from the smallest, we get 100,000; if we count 14 observations down from the largest, we get 150,000. This gives the interval ($100,000, $150,000) shown on the output.

Paired Data

One natural use of sign tests and confidence intervals is for paired data. In this case the null hypothesis is often that the median difference is zero. To test this, we first compute the differences, then use the Test median option in the 1-Sample Sign dialog box. Let's examine the Cholesterol data: the change from the second to the fourth day. We'll calculate those changes and put them into C5. Then we'll test H_0: $\eta = 0$ versus H_1: $\eta = 0$, where η is the median of the population of all changes. Let's also find a 90% confidence interval for η. To test the median and find the confidence interval,

1. Open the **Cholest** worksheet.
2. Choose **Calc ▸ Calculator** and create a new column, **Differ**, that contains the difference between 2-Day and 4-Day: **2-Day – 4-Day**.
3. Choose **Graph ▸ Stem-and-Leaf** and create a stem-and-leaf plot of the Differ column.
4. Choose **Stat ▸ Nonparametrics ▸ 1-Sample Sign**.
5. Press **F3** to return all settings to their defaults. Enter **Differ** as the variable. Click the **Test median** option button. Click **OK**.
6. Finally click the **Edit Last Dialog** button and click the **Confidence interval** button. Enter a level of **90**. Click **OK**. Figure 13.6 shows the output.

Figure 13.6
Sign procedures for paired data using the Cholesterol study

```
Stem-and-Leaf Display: Differ

Stem-and-leaf of Differ  N  = 28
Leaf Unit = 10


  1    -0   4
  3    -0   32
  9    -0   110000
 14    0    00111
 14    0    22333
  9    0    4555
  5    0    66
  3    0    89
  1    1    0


Sign Test for Median: Differ

Sign test of median =  0.00000 versus not = 0.00000

            N  Below  Equal  Above       P  Median
Differ     28      9      0     19  0.0872   19.00


Sign CI: Differ

Sign confidence interval for median

                                  Confidence
                       Achieved    Interval
            N  Median  Confidence  Lower  Upper  Position
Differ     28   19.00     0.8151    8.00  30.00        11
                          0.9000    3.28  36.29       NLI
                          0.9128    2.00  38.00        10
```

13.2 Wilcoxon Signed Rank Procedures

With sign procedures the population can have any shape; with Student's *t* procedures the population should be approximately normal. Wilcoxon procedures are somewhat in between: the population should be approximately symmetric but it need not be normal or have any other specific shape. Recall that for any symmetric population, the mean and median are equal. The Wilcoxon test can then be viewed as a test for either the population mean μ or the population median η. Similarly a Wilcoxon confidence interval can be viewed as an interval for either μ or η.

Wilcoxon Signed Rank Test

The Wilcoxon test, like the sign test, can be used to do a test on the median of one sample, but in most studies these methods are used to do paired tests. The Stat ▸ Nonparametrics ▸ 1-Sample Wilcoxon command does Wilcoxon signed rank tests. For each column this command tests the null hypothesis, $H_0: \eta = K$, where η is the population median and K is a specified null value for η. The default alternative hypothesis is $H_1: \eta \neq K$. If you don't specify a value for η, the command tests $H_0: \eta = 0$. We will use a paired example, the same example used in Figure 13.6. We want to see if there is a significant change in cholesterol levels from the second to the fourth day. Thus we test

$$H_0: \eta_1 = \eta_2 \text{ versus } H_1: \eta_1 > \eta_2$$

or equivalently we test

$$H_0: \eta = 0 \text{ versus } H_1: \eta > 0, \text{ where } \eta = \eta_1 - \eta_2$$

Table 13.1 shows how Minitab calculates the test statistic, W, for these data. We take the absolute value of each difference in the Differ column. Next, we assign ranks. The smallest absolute value is 2; it gets rank 1. The second smallest absolute value is 4, and there are three of these. Their ranks would be 2, 3, and 4. When there are ties, we average the ranks and assign the average rank to all of them. Thus they all get rank 3. The next rank is 5. The fifth smallest value is the last observation, so it gets rank 5. After assigning all ranks, we compute W. This is the sum of all the ranks that correspond to differences that are positive. Here W is 324.

If the null hypothesis is true, W has mean $= n(n + 1)/4$ and variance $= n(n + 1)(2n + 1)/24$, where n is the number of observations in our sample. There are 28 observations in our example, so the mean of $W = 28(29)/4 = 203$, and the variance $28(29)(57)/24 = 1928.5$. The standard deviation, then, is $\sqrt{1928.5} = 43.9$.

The exact distribution of W is given in special tables.[†] If we don't have such tables, we can use the fact that W has approximately a normal distribution. Minitab uses the normal approximation (with a continuity correction). In our example $W = 324$, which is more than two standard deviations above its mean..

[†] For example, see M. Hollander and D. Wolfe, *Nonparametric Statistical Methods*. New York: Wiley, 1973.

Table 13.1 Example of calculating the Wilcoxon statistics W

2-Day	4-Day	Differences	Absolute values	Ranks	Ranks of positive differences
270	218	52	52	21	21
236	234	2	2	1	1
210	214	−4	4	3	
142	116	26	26	13.5	13.5
280	200	80	80	2	26
272	276	−4	4	3	
160	146	14	14	10.5	10.5
220	182	38	38	17.5	17.5
226	238	−12	12	8.5	
242	288	−46	46	20	
186	190	−4	4	3	
266	236	30	30	10.5	15.5
206	244	−38	38	17.5	
318	258	60	60	24	24
294	240	54	54	22	22
282	294	−12	12	8.5	
234	220	14	14	10.5	10.5
224	200	24	24	12	12
276	220	56	56	23	23
282	186	96	96	27	27
360	352	8	8	6	6
310	202	108	108	28	28
280	218	62	62	25	25
278	248	30	30	15.5	15.5
288	278	10	10	7	7
288	248	40	40	19	19
244	270	-26	26	13.5	
236	242	-6	6	5	
					Sum = 324

Suppose we test H_0: $\eta = 0$ versus H_1: $\eta > 0$. The p-value for this test is $P(W \geq 324)$. To test the hypothesis,

1. Choose **Stat ▸ Nonparametrics ▸ 1-Sample Wilcoxon**.
2. Enter **Differ** as the variable. Click the **Test median** option button.
3. Click the **Alternative** list arrow and then click **greater than**. Figure 13.7 shows the output.

Figure 13.7
Wilcoxon test
for paired data
using the
Cholesterol
data

```
Wilcoxon Signed Rank Test: Differ

Test of median = 0.000000 versus median > 0.000000

                  N for   Wilcoxon              Estimated
           N      Test    Statistic      P       Median
Differ    28       28        324.0    0.003       22.00
```

Output from the Wilcoxon test, shown in Figure 13.7, gives the p-value of 0.003. Thus there is little chance of W being as large as 324 by chance alone. Therefore there is a statistically significant decrease in cholesterol from the second to the fourth day.

Now suppose we test H_0: $\eta = 0$ versus the alternative H_1: $\eta \neq 0$. Since this is a two-sided alternative hypothesis, we need two probabilities—one for the lower tail and one for the upper tail. The distribution of the test statistic is symmetric, so all we need to do is double the one-sided p-value, giving 2(.003) = 0.006 as the two-sided p-value.

Wilcoxon Confidence Interval

There is also a confidence interval associated with the Wilcoxon test. This is calculated by clicking the Confidence interval option button in the 1-Sample Wilcoxon dialog box. The confidence interval is essentially the set of all values d for which the test H_0: $\eta = d$ versus H_1: $\eta \neq d$ is not rejected, using $a = 1 -$ (specified confidence)/100.

As with most nonparametric procedures, it is seldom possible to achieve the specified confidence, so the closest value is printed in the output. Figure 13.6 gives a confidence interval for the change in cholesterol scores from the second to the fourth day. To calculate the confidence interval,

1. Choose **Stat ▸ Nonparametrics ▸ 1-Sample Wilcoxon**.
2. Select **Differ** as the variable.
3. Click the **Confidence interval** option button and then click **OK**. Figure 13.8 shows the output.

Figure 13.8
Confidence
interval
associated
with Wilcoxon
signed rank
test

```
Wilcoxon Signed Rank CI: Differ

                                    Confidence
               Estimated   Achieved   Interval
           N     Median  Confidence  Lower  Upper
Differ    28       22.0       95.1     7.0   38.0
```

13.3 Two-Sample Rank Procedures

In this section we describe a nonparametric procedure for comparing two populations. We assume that (1) we have a random sample from each population, (2) the samples were taken independently of each other, and (3) the populations have approximately the same shape (this means the variances must be approximately equal). We use to represent the median of the first population and for that of the second population.

Our null hypothesis is that the medians of the two populations are equal. Our alternative is that one population is shifted from the other; that is, it has a different median. Since we assume the two populations have the same shape, this procedure is analogous to the pooled t procedures discussed in Chapter 9.

This two-sample rank test was introduced by Wilcoxon, and is often called the Wilcoxon rank sum test. It was further developed by Mann and Whitney. To avoid confusion with the Wilcoxon test described in Section 13.2, the Minitab command is named Mann-Whitney.

An Example

In Chapter 9 we used two-sample t procedures to analyze data for patients with Parkinson's disease. Recall that Parkinson's disease, among other things, affects a person's ability to speak. Eight of the people in this study had received an operation to treat the disease. This operation seemed to improve the patients' condition overall, but it may also have affected their ability to speak. Each patient was given a test for speaking ability. We will now use nonparametric methods to analyze these data.

Table 13.2 shows how to do the two-sample rank test by hand. The first two columns give the original data. The second two columns show each sample ordered. Next, we combine the two ordered samples and assign ranks. The smallest observation in the combined sample is 1.2, and we give it rank 1. The next three observations are all equal to 1.3. Whenever we have two or more observations that are tied, we assign the average rank to each. The three observations of 1.3 are tied for ranks 2, 3, and 4. We therefore give each the average rank of $(2 + 3 + 4)/3 = 3$. It is not until we look at the eighth rank that we get an observation from the first sample. When we have assigned all ranks, we sum the ranks corresponding to the observations in the first sample. This sum is usually denoted by W. Here $W = 126.5$.

Table 13.2 Two-sample rank test, done by hand

Original data		Ordered data		Ranks of ordered data	
Operation	No operation	Operation	No operation	Operation	No operation
2.6	1.2	1.7	1.2	8.0	1.0
2.0	1.8	2.0	1.3	11.5	3.0
1.7	1.8	2.5	1.3	15.5	3.0
2.7	2.3	2.5	1.3	15.5	3.0
2.5	1.3	2.6	1.5	17.5	5.5
2.6	3.0	2.6	1.5	17.5	5.5
2.5	2.2	2.7	1.6	19.5	7.0
3.0	1.3	3.0	1.8	21.5	9.5
	1.5		1.8		9.5
	1.6		2.0		11.5
	1.3		2.2		1.3
	1.5		2.3		1.4
	2.7		2.7		19.5
	2.0		3.0		21.5
				Sum = 126.5	

The value of W reflects the relative locations of the two samples. If the values in the first sample tend to be larger than those in the second sample, W will be large; if the values in the first sample tend to be smaller, W will be small. Minitab's Mann-Whitney command does all this work for us. To use the Mann-Whitney procedure,

1. Open the **Parkinsn** worksheet.
2. Choose **Stat ▶ Nonparametrics ▶ Mann-Whitney**.
3. Enter **Op** as the first sample and **NoOp** as the second sample. Click **OK**. Figure 13.9 shows the output.

The output gives the value of $W = 126.5$ and tells us that the attained significance of the two-sided test is 0.0203. This means that the chance of observing two samples as separated as these, when in fact the two populations have the same median, is only 0.0203. We therefore have statistical evidence that the two populations differ.

The output also contains a confidence interval and a point estimate for $(\eta_1-\eta_2)$. Both of these are calculated by using procedures developed from the two-sample rank test.

Figure 13.9
Mann-Whitney test and confidence interval

```
Mann-Whitney Test and CI: Op, NoOp

         N   Median
Op       8   2.5500
NoOp    14   1.7000


Point estimate for ETA1-ETA2 is 0.7000
95.6 Percent CI for ETA1-ETA2 is (0.2002,1.2001)
W = 126.5
Test of ETA1 = ETA2 vs ETA1 not = ETA2 is significant at 0.0203
The test is significant at 0.0199 (adjusted for ties)
```

13.4 Kruskal-Wallis Test

The Kruskal-Wallis test is a generalization of the procedure used by Mann-Whitney, and offers a nonparametric alternative to one-way analysis of variance for comparing several populations.

We assume that (1) we have a random sample from each population, (2) the samples were taken independently of each other, and (3) the populations have approximately the same shape (this means the variances must be approximately equal). We use η_1 to represent the median of the first population, η_2 for the second population, η_3 for the third population, and so on.

Our null hypothesis is that the medians of all the populations are equal. Our alternative is that they are not.

In Chapter 10 we used analysis of variance to analyze data from a study of fabric flammability. Recall that there were 11 observations from each of five labs. Thus there are five populations.

Table 13.3 shows how to do the Kruskal-Wallis test by hand. The five columns headed 'Data' show the original data, with the numbers for each lab ordered from smallest to largest. The next step is to combine all the data and rank all 55 observations. If two or more observations are tied, we assign the average rank to each. The columns in Table 13.3 headed 'Ranks' give these ranks. Underneath each column of ranks we show the average of the ranks in that column.

Table 13.3 Kruskal-Wallis test done by hand

	Laboratory								
1		2		3		4		5	
Data	Ranks	Data	Ranks	Data	Ranks	Data	Ranks	Data	Ranks
2.9	11.5	2.7	3.0	2.8	7.0	2.6	1.0	2.8	7.0
2.9	11.5	3.2	21.5	2.8	7.0	2.7	3.0	3.1	17.0
3.0	14.0	3.3	26.0	2.8	7.0	2.7	3.0	3.5	34.5
3.1	17.0	3.4	30.0	3.2	21.5	2.8	7.0	3.5	34.5
3.1	17.0	3.4	30.0	3.3	26.0	2.9	11.5	3.5	34.5
3.1	17.0	3.6	38.0	3.3	26.0	2.9	11.5	3.7	40.5
3.1	17.0	3.8	44.5	3.5	34.5	3.2	21.5	3.7	40.5
3.7	40.5	3.8	44.5	3.5	34.5	3.2	21.5	3.9	47.5
3.7	40.5	4.0	49.0	3.5	34.5	3.3	26.0	4.1	51.0
3.9	47.5	4.1	51.0	3.8	44.5	3.3	26.0	4.1	51.0
4.2	53.5	4.3	55.0	3.8	44.5	3.4	30.0	4.2	53.5
				Average rank					
26.1		35.7		26.1		4.7		37.4	

The test statistic is

$$H = \frac{12\sum_i n_i [\bar{R}_i - \bar{R}]^2}{N(N+1)}$$

where n_i is the number of observations in group i, N is the total number of observations, $\bar{R}_i$ is the average of the ranks in group i, and $\bar{R}$ is the average of all the ranks. Using the fabric data, we get

$$H = \frac{12}{55(55+1)}[11(26.1-28)^2 + 11(35.7-28)^2 + 11(26.1-28)^2 +$$

$$11(14.7-28)^2 + 11(37.4-28)^2] = 14.9$$

Large values of H suggest that there are some differences among the populations. Under the null hypothesis the distribution of H can be approximated by a chi-square distribution with $k-1$ degrees of freedom, where k = the number of populations. Minitab's Kruskal-Wallis command does all this work for us. To use the Kruskal-Wallis procedure,

1. Open the **Fabric1** worksheet.
2. Choose **Stat ▸ Nonparametrics ▸ Kruskal-Wallis**.
3. Enter **Charred** as the response and **Lab** as the factor. Click **OK**. Figure 13.10 shows the output.

Figure 13.10
Kruskal-Wallis
test

```
Kruskal-Wallis Test: Charred versus Lab

Kruskal-Wallis Test on Charred

Lab        N  Median  Ave Rank      Z
1          11  3.100      26.1   -0.44
2          11  3.600      35.7    1.78
3          11  3.300      26.1   -0.44
4          11  2.900      14.7   -3.07
5          11  3.700      37.4    2.18
Overall    55             28.0

H = 14.19  DF = 4  P = 0.007
H = 14.26  DF = 4  P = 0.007  (adjusted for ties)
```

The output gives the p-value for the test, 0.007. This means that the chance of observing five samples as separated as these, when in fact the populations have the same median, is only 0.007. We therefore have strong statistical evidence that the two population medians differ.

Some authors (e.g., see E. L. Lehman, *Nonparametrics: Statistical Methods Based on Ranks*. New York: Wiley, 1975) suggest adjusting H when there are ties in the data, so that the chi-square approximation is better. Minitab prints this adjusted value. Minitab also prints a z value, z_i, for each group i. Under the null hypothesis, each z_i is approximately normal, with mean 0 and variance 1. The value of z_i indicates how the mean rank, $\overline{R}_i$, for group i differs from the mean rank, $\overline{R}_i$, for all N observations.

Exercises

Exercises for Section 13.1

13-1   Consider the Cartoon experiment (described in the Appendix, p. 524). The national median of all Otis scores is 100. Do the Otis scores for the participants in this study differ significantly from the national median? Do a sign test.

13-2   Suppose we wanted to test H_0: $\eta = \$105,000$ versus H_1: $\eta \neq \$105,000$, using the Value data you entered to create the stem-and-leaf plot shown in Figure 13.3. Use the output in Figure 13.4 to find the corresponding p-value.

13-3   Table 13.4 shows the results of a titration to determine the acidity of a solution in a chemistry class. These data are stored in C1 of the file Acid.

Table 13.4 Titration data (in one column in the worksheet)

.123	.109	.110	.109	.112	.109	.110	.110
.110	.112	.110	.101	.110	.110	.110	.110
.106	.115	.111	.110	.107	.111	.110	.113
.109	.108	.109	.111	.104	.114	.110	.110
.110	.113	.114	.110	.110	.110	.110	.110
.090	.109	.111	.098	.109	.109	.109	.109
.111	.109	.108	.110	.112	.111	.110	.111
.111	.107	.111	.112	.105	.109	.109	.110
.110	.109	.110	.104	.111	.110	.111	.109
.110	.111	.112	.123	.110	.109	.110	.109
.110	.109	.110	.110	.111	.111	.109	.107
.120	.133	.107	.103	.111	.110	.122	.109
.108	.109	.109	.114	.107	.104	.110	.114
.107	.101	.111	.109	.110	.111	.110	.126
.110	.109	.114	.110	.110	.110	.110	.110
.111	.107	.110	.107				

(a) The instructor knew the correct value for this solution was 0.110. Do a two-sided sign test of the null hypothesis H_0: $\eta = 0.110$. (This is a check to see whether the class is 'biased'—that is, to see whether it tends to be systematically too high or too low.)

(b) Make a histogram of the data. Do you think the population is symmetric? If it is, then $\eta = \mu$.

(c) A distribution is called heavy-tailed if it has a higher probability of very extreme values than in a normal distribution. Does the histogram for part (b) give any indication that the distribution of titration results from this class is heavy-tailed? (It may help to compare your histogram with those of normal distributions.)

13-4   The data from a second titration experiment are given in Table 13.5. These data are in C2 of the worksheet Acid.

Table 13.5 Second titration experiment (in one column in the worksheet)

.109	.111	.110	.110	.105
.110	.111	.110	.110	.111
.109	.111	.109	.112	.109
.109	.111	.110	.112	.112
.109	.110	.110	.109	.113
.108	.105	.110	.109	.109
.110	.110	.110	.104	.109
.110	.111			

(a) Make a histogram. Do you think $\eta = \mu$? It will be if the population is symmetric. Exercise 13-3 gives us some more information about the shape of the distribution for titration data. What did you conclude there?
(b) Find (as closely as possible) a 95% sign confidence interval for η.
(c) Find a 95% confidence interval for η, using normal-theory methods, and compare it to the sign confidence interval in part (a).
(d) Repeat parts (a) and (b), using 90% confidence intervals.

13-5   The following is a sample (which we got by simulation) from a normal distribution:

62, 60, 65, 70, 60, 67, 61, 66, 64, 64, 62, 63

(a) Calculate an approximately 95% sign confidence interval for η (the closest confidence level will be 96.1).
(b) Find a 96.1% t confidence interval using the 1-Sample Sign ▶ Confidence interval option. Compare this with the sign interval of part (a).
(c) Now suppose a mistake was made in recording the last observation and the number 36 was recorded instead of 63. Repeat parts (a) and (b).
(d) Repeat part (c), only now suppose the last observation was mistakenly recorded as 630.

13-6   We can use simulation to get a feel for how well sign confidence intervals work in various cases. If your data came from a normal population, a t confidence interval is the best way to estimate μ (= η). But how much worse might a sign confidence interval be? Use the Calc ▶ Random Data ▶ Normal command sequence to simulate 10 samples, each containing 18 observations, from a normal distribution with $\mu = 50$ and $\eta = 8$, and then use 1-Sample Sign to get a 90% sign confidence interval and 1-Sample t to get a 90% t confidence interval for each sample. Compare the two confidence intervals for each sample. Do both cover? Which is narrower?

13-7   (a) Repeat Exercise 13-6, but simulate data from a Cauchy distribution, using 0 and 1 as the location and scale. This gives an example of data from a very 'heavy-tailed' distribution.

(b) Repeat Exercise 13-6, but simulate data from a uniform distribution, using 0 and 1 as the lower and upper endpoints. A uniform distribution is an example of a distribution that has very 'skinny' tails.

13-8   The job of President of the United States is a very demanding, high-pressure job. This might cause premature deaths of presidents. On the other hand only vigorous people would run for president, so presidents might tend to live longer than other people. Table 13.6 lists the 'modern' (since Lincoln) presidents who died before December 1993, the number of years they lived after inauguration, and the life expectancy of a man whose age is that of the president at the time of his first inauguration. These data are in the worksheet Pres.

Table 13.6 President data

President	Life expectancy after first inauguration	Actual years lived after first inauguration
Andrew Johnson	17.2	10.3
Ulysses S. Grant	22.8	16.4
Rutherford B. Hayes	18.0	15.9
James A. Garfield	21.2	.5
Chester A. Arthur	20.1	5.2
Grover Cleveland	22.1	23.3
Benjamin Harrison	17.2	12.0
William McKinley	18.2	4.5
Theodore Roosevelt	26.1	17.3
William H. Taft	20.3	21.2
Woodrow Wilson	17.1	10.9
Warren G. Harding	18.1	2.4
Calvin Coolidge	21.4	9.4
Herbert C. Hoover	19.0	35.6
Franklin D. Roosevelt	21.7	12.1
Harry S. Truman	15.3	27.7
Dwight D. Eisenhower	14.7	16.2
John F. Kennedy	28.5	2.8
Lyndon B. Johnson	19.3	9.2

(a) Do a paired sign test of the null hypothesis that being president has no effect on length of life.

(b) Find an approximately 95% sign confidence interval for the median difference between the expected and attained life spans of presidents.

(c) If our main interest is the effect of stress on length of life, then perhaps we should not include the presidents who were assassinated (Garfield, McKinley, and Kennedy). Carry out the analysis of (a) and (b) without these three presidents.

(d) We have, perhaps, stretched the use of statistics rather far here, as we often must in real problems. Comment on this statement, paying particular attention to the assumptions needed for a sign test.

Exercises for Section 13.2

13-9   Do Exercise 13-1, using a Wilcoxon test in place of the sign test. Do you think a Wilcoxon test is appropriate for these data?

13-10   Do Exercise 13-3, using a Wilcoxon test. Do you think a Wilcoxon test is appropriate for these data?

13-11   (a) Do Exercise 13-4, using a Wilcoxon confidence interval.
    (b) Compare this interval to the sign interval and to the t interval of Exercise 13-4.

13-12   Do Exercise 13-5, using Wilcoxon confidence intervals in place of sign confidence intervals.

13-13   Do Exercise 13-6, using Wilcoxon confidence intervals in addition to sign confidence intervals. Compare all three intervals for each sample.

Exercises for Section 13.3

13-14   (a) By hand, compute the W statistic of a two-sample rank test, using the following data:
    Sample A: 10 7 6 12 14
    Sample B: 8 4 6 11
    (b) Use Minitab's Mann-Whitney command to check your answer.

13-15　Do migratory birds store, then gradually use up, a layer of fat as they migrate? To investigate this question, two samples of migratory song sparrows were caught, one sample on April 5 and one sample on April 6. The amount of stored fat on each bird was subjectively estimated by an expert. The higher the fat class, the more fat on the bird. Table 13.7 displays the data.

Table 13.7 Bird data

| | Number of birds in class | |
Fat Class	Found on April 5	Found on April 6
1	0	3
2	0	1
3	0	11
4	2	6
5	0	1
6	10	9
7	9	2
8	7	2
9	6	1
10	4	0
11	1	0
12	2	0

(a) Use a dotplot to compare these two groups visually.

(b) Do an appropriate test to see whether there is any evidence that birds use up a layer of fat as they migrate.

13-16　A study was done at Penn State University to see how much one type of air pollution, ozone, damages azalea plants. Eleven varieties of azaleas were included in the study. We will look at data from just two varieties.

During week 1, 10 plants of each variety were fumigated with ozone. A short time later, each plant was measured for leaf damage. The procedure was repeated four more times, each time using new plants. These data are in C1-C5 (variety A) and C6-C10 (variety B) of the worksheet Azalea and displayed in Table 13.8.

(a) Compare the first week's data for the two varieties, using a two-sample rank test. Does there appear to be any difference between the two varieties' susceptibilities to ozone damage?

(b) Repeat the test of part (a) for each of the other four weeks. Overall, how do the two varieties compare?

(c) Susceptibility to ozone varies with weather conditions, and weather conditions vary from week to week. Does this seem to show up in the data?

One simple way to look for a 'week effect' is to calculate the median leaf damage for all 20 plants sprayed during week 1, the median for week 2, and so on. Do these five medians seem to be very different?

(d) Another way we could test for a 'week effect' would be to use a contingency table analysis (discussed in Chapter 12). First form a contingency table as follows: For each week count the number of damaged plants and the number of undamaged plants (a plant is undamaged if its leaf damage is 0.00); form a table that has two rows and five columns. Then do a chi-square test for association between damage and week. Do the results agree with what was indicated in part (c)?

Table 13.8 Azalea data

Leaf damage for variety A				
Week 1	*Week 2*	*Week 3*	*Week 4*	*Week 5*
1.58	1.09	0.00	2.22	0.20
1.62	1.03	0.00	2.40	0.40
2.04	0.00	0.07	2.47	0.34
1.28	0.46	0.18	1.85	0.00
1.43	0.46	0.40	2.50	0.00
1.93	0.85	0.20	1.20	0.00
2.20	0.30	0.63	1.33	0.10
1.96	0.90	0.63	2.40	0.06
2.23	0.00	0.56	2.23	0.17
1.54	0.00	0.26	2.57	0.25

Leaf damage for variety B				
Week 1	*Week 2*	*Week 3*	*Week 4*	*Week 5*
1.29	0.00	0.78	0.40	0.00
0.70	0.20	0.64	0.00	0.40
1.93	0.00	1.00	0.20	0.47
0.98	0.98	0.42	0.40	0.00
0.94	0.00	0.97	0.14	0.00
1.06	0.62	2.43	0.44	0.00
0.94	0.67	0.65	1.23	0.00
1.65	0.00	0.00	0.35	0.00
0.70	0.00	0.30	0.17	0.00
0.35	0.00	0.00	0.20	0.00

Exercises for Section 13.4

13-17   **(a)** Using the azalea data for variety A, do a Kruskal-Wallis test to see whether ozone damage varied over the five weeks.
(b) Do the same test, using the data for variety B. How do the results compare to those for variety A?

13-18   A study was done to compare Pinot Noir wine made in three different regions. Wine samples from each region were taken and scored by a panel of judges on a number of characteristics. The scores for three characteristics are given in Table 13.9 and stored in the Wine worksheet, organized as unstacked. The higher the number, the more favorable the rating.

Table 13.9 Wine sample data

	Region 1			Region 2			Region 3	
Aroma	Body	Flavor	Aroma	Body	Flavor	Aroma	Body	Flavor
3.3	2.8	3.1	3.3	5.4	4.3	4.3	4.3	3.9
4.4	4.9	3.5	3.4	5.0	3.4	5.1	4.3	4.5
3.9	5.3	4.8	4.7	4.1	5.0	5.9	5.7	7.0
3.9	2.6	3.1	4.1	4.0	4.1	7.7	6.6	6.7
5.6	5.1	5.5	4.3	4.6	4.7	7.1	4.4	5.8
4.6	4.7	5.0	5.1	4.9	5.0	5.5	5.6	5.6
4.8	4.8	4.8	3.9	4.4	5.0	6.3	5.4	4.8
5.3	4.5	4.3	4.5	3.7	2.9	5.0	5.5	5.5
4.3	3.9	4.7	5.2	4.3	5.0	6.4	5.4	6.6
4.6	4.1	4.3				5.5	5.3	5.3
3.9	4.0	5.1				6.0	5.4	5.7
4.2	3.8	3.0				6.8	5.0	6.0
3.3	3.5	4.3						
5.0	5.7	5.5						
3.5	4.7	4.2						
4.3	5.5	3.5						
5.2	4.8	5.7						

(a) Use appropriate displays to compare the aromas of wines made in the different regions. Also do a Kruskal-Wallis test.
(b) Repeat part (a), using the other two characteristics, body and flavor.
(c) Compare the results for all three characteristics.

14

Control Charts

14.1 Control Charts and Process Variation

The concept of statistical process control was developed by Walter Shewhart in the 1920s to help managers at Western Electric (part of the Bell Telephone System) understand and detect changes in manufacturing processes. Initially, his ideas were used only on a limited basis in the United States. After World War II, Japan adopted them and started the quality revolution that eventually enabled Japan to produce products of very high quality and at reasonable cost. Now manufacturing and service industries worldwide are using process-control ideas to improve quality and reduce waste. Control charts, discussed in this chapter, are one of the techniques used to study a process.

A **process** is simply a sequence of steps that results in an outcome. A receptionist answers the phone and directs the call to the appropriate person. A laboratory measures the amount of cholesterol in a sample of blood. A machine fills bottles with 12 ounces of soda. In each case the specific activity is repeated again and again over time. Ideally each activity would be done "perfectly." The receptionist would answer each call on the first ring and direct the call to the correct person. The laboratory would determine the exact amount of cholesterol in each sample of blood. The machine would fill each bottle with precisely 12 ounces of soda. But this is the real world, and processes vary.

There are two basic types of process variation. **Common-cause variation** results from chance or the inherent variability in the system. The only way to reduce this type of variation is to change the process itself. For example, in the case of the receptionist, we might install a different type of phone system, create a chart so that the receptionist can more quickly and accurately determine to whom the calling party should be referred, or reduce distractions and noise in the office so the receptionist can work more efficiently.

The second type of variation, **special-cause variation**, results from special causes. In the receptionist example perhaps the company placed some extra ads one month, causing more calls than the receptionist could handle quickly, so response time went up. A process is considered to be in control if there are no special causes affecting it, only common causes.

Control charts are used to study a process over time. There are two basic types: charts for measurement data and charts for attribute data (count data). If we were to record the time from when a call first comes into a company until the calling party reaches the appropriate person within the company, we would have measurement data. If we were to record for each day, the total number of calls and the number that were incorrectly referred, we would have attribute data.

14.2 Control Charts for Measurement Data

Charts for Means

One part of the process of making ignition keys for automobiles consists of cutting grooves into raw key blanks. Some of the groove dimensions are critical to the proper functioning of the keys. Suppose we wish to study one critical groove. Keys are produced at a high volume, so we take a sample of five keys every twenty minutes and carefully measure the dimension of this groove. Each sample is often called a **subgroup**. Table 14.1 shows key measurements.

Table 14.1 Data for ignition-key study (in thousandths of an inch)

Sample Number	Samples					Means	Standard Deviations	Ranges
	x_1	x_2	x_3	x_4	x_5			
1	6.1	8.4	7.6	7.6	4.4	6.82	1.588	4.0
2	8.8	8.3	7.6	7.4	5.9	7.60	1.102	2.9
3	8.0	8.0	9.4	7.5	7.0	7.98	0.896	2.4
4	6.7	7.6	6.4	7.1	8.8	7.32	0.942	2.4
5	8.7	8.4	8.8	9.4	8.6	8.78	0.377	1.0
6	7.1	5.2	7.2	8.8	5.2	6.70	1.526	3.6
7	7.8	8.9	8.7	6.5	6.8	7.74	1.083	2.4
8	8.7	9.4	8.6	7.3	7.1	8.22	0.983	2.3
9	7.4	8.1	8.6	8.3	8.7	8.22	0.517	1.3
10	8.1	6.5	7.5	8.9	9.7	8.14	1.236	3.2
11	7.8	9.8	8.1	6.2	8.4	8.06	1.292	3.6
12	8.9	9.0	7.9	8.7	9.0	8.70	0.464	1.1
13	8.7	7.5	8.9	7.6	8.1	8.16	0.631	1.4
14	8.4	8.3	7.2	10.0	6.9	8.16	1.222	3.1
15	7.4	9.1	8.3	7.8	7.7	8.06	0.666	1.7
16	6.9	9.3	6.4	6.0	6.4	7.00	1.325	3.3
17	7.7	8.9	9.1	6.8	9.4	8.38	1.094	2.6
18	8.9	8.1	7.3	9.1	7.9	8.26	0.740	1.8
19	8.1	9.0	8.6	8.7	8.0	8.48	0.421	1.0
20	7.4	8.4	9.2	7.4	10.3	8.54	1.240	2.9

Here there are $k = 20$ subgroups and each subgroup has $n = 5$ observations. We use x_{ij} to denote the jth observation in subgroup i. We calculated three quantities for each subgroup: the mean (denoted $\bar{x}_i$), the standard deviation (denoted s_i), and the range (denoted r_i). We will use these to create control charts. A control chart for means first plots the sample means versus the sample numbers. This gives us an overall picture of how the sample means vary over time.

If the process is in control, the individual observations will all have come from the same population, and thus will all have the same population mean, μ, and standard deviation, σ. The sample means will then have mean μ and standard deviation $\sigma/\sqrt{n}$. In a normal distribution almost all observations (99.7%) are within three standard deviations of the mean. We add three lines to the plot to help us judge the variability in the data: The **center line** is located at the mean; the upper control limit (**UCL**) is three standard deviations above the mean; and the lower control limit (**LCL**) is three standard deviations below the mean.

In order to draw these lines, we must estimate μ and σ. We estimate μ by $\bar{\bar{x}}$, often called the grand mean:

$$\bar{\bar{x}} = \frac{1}{k}\sum_i \bar{x}_i = \frac{1}{kn}\sum_i\sum_j x_{ij}$$

Several different methods are commonly used to estimate σ. Minitab gives you a choice of two. By default Minitab uses the pooled standard deviation, s_p, defined by

$$s_p = \sqrt{[\sum_i\sum_j(x_{ij}-\bar{x}_i)^2]/(kn)}$$

Then the control limits are

$$UCL = \bar{\bar{x}} + 3s_p/\sqrt{n}$$

$$LCL = \bar{\bar{x}} - 3s_p/\sqrt{n}$$

Minitab can also estimate σ with s_r, defined by

$s_r = \bar{r}/d_n$, where $\bar{r} = \sum r_i/k$, and d_n is a constant that makes s_r an unbiased estimate of σ.

Then the control limits are

$$UCL = \bar{\bar{x}} + 3s_r/\sqrt{n}$$

$$LCL = \bar{\bar{x}} - 3s_r/\sqrt{n}$$

Figure 14.3 shows an Xbar chart for the ignition-key data. First we enter all the data, in order, into one column, C1; then we use the Xbar command to create a chart for the subgroup sample means. We enter the column and the subgroup size. If all subgroups have the same size, you can enter a constant (in this case, 5). We'll also request the eight tests for special causes, which we will discuss later.

To create the Xbar chart,

1. Open a new project or restart Minitab to set all dialog box options to their defaults.

2. Open the **Key** worksheet.

3. Choose **Stat ▸ Control Charts ▸ Variables Charts for Subgroups ▸ Xbar**.

4. Since the data are arranged as a single column, enter **Grove** as the single column.

5. Enter a subgroup size of **5** and click the **Xbar** options button. See Figure 14.1.

6. Click the **Tests** tab and then select **Perform all eight tests** from the drop-down list. See Figure 14.1.

7. Click **OK** in the Tests dialog box. You return to the Xbar Chart dialog box, shown in Figure 14.2.

8. Click **OK**. Minitab produces the Xbar chart shown in Figure 14.3.

Figure 14.1   Xbar chart dialog box

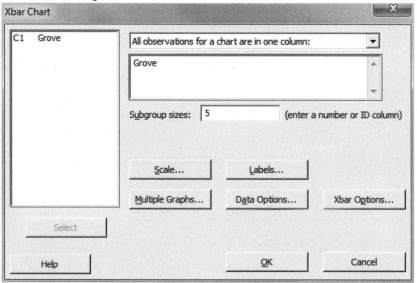

Figure 14.2   Xbar chart xbar options dialog box

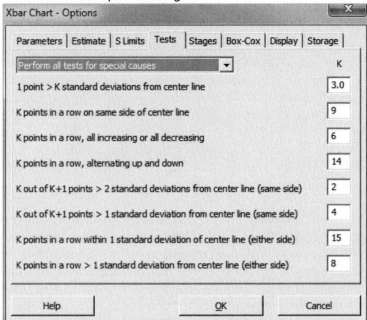

Figure 14.3   Xbar control chart for the ignition key data

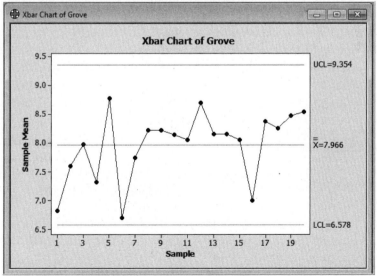

   The overall mean is equal to 7.966. The control limits are horizontal straight lines at 6.578 and 9.354. All the sample means are within these limits and none of the eight tests signal an out-of-control process.

The Xbar Chart dialog box offers many options. If you have historical values for μ and σ, you can specify these in the Xbar options dialog box under the Parameters tab. By default Minitab uses the pooled standard deviation to estimate the process standard deviation, but you can click the Estimate tab to choose several other options. You can click the Rbar option button, for example, to use an estimate based on the sample ranges. By default Minitab uses all the samples to calculate the center line and control limits. You can tell Minitab to use only certain samples by listing the ones you want to omit in the Omit box at the top of the Estimation dialog box. By default Minitab calculates the *UCL* and *LCL* based on three standard deviations. You can click the S Limits tab to enter different limits. You also have many of the same graphics options available to you as for the other core graphs.

Finding Problems. A process is considered to be in control if no special causes are influencing the variability in the process, only common causes. Only when a process is in control can we hope to make improvements in that process. A process can be out of control for many different reasons. One sample could be unusually large or small. The mean level could change so that from some point on, the mean is larger (or smaller) than it was in the past. There could be a trend such that the mean increases (or decreases) as time goes on. There could be cycles, in which the mean decreases for, say, five samples, then returns to its former level, then decreases for another five samples, and so on.

Control charts help us find unusual patterns, and hence, special causes. The next step is to determine what caused the patterns, by carefully examining all aspects of the process. For example we might discover that one sample was unusually low because the raw material used to produce that sample was mistakenly taken from an old batch. There could be a downward trend in the mean because a chemical used to produce the samples gradually deteriorated over time owing to improper storage conditions. We attempt to remove as many of the special causes of variation as we can in order to bring the process into control.

We included all eight tests for unusual patterns. In all cases these are patterns that are extremely unlikely to occur if the process is under control. The patterns are summarized in Figure 14.4.

In Figure 14.2 we requested all eight tests. Nothing was indicated in the chart, so the chart passed all the tests. Therefore, as far as we can tell, this process is in control.

Important!

The fact that a process is in control does not mean the process is acceptable. For example suppose that in the ignition-key study the value of the grand mean was 10.00, not 7.966. Suppose also that all points were within the control limits and no unusual patterns were detected. From a statistical standpoint the process is in control. From a practical standpoint keys produced by this company are probably not very good.

Figure 14.4   Eight tests for special causes

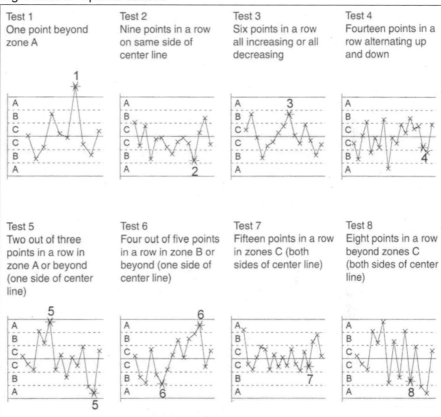

Note: The solid line in the center is the center line. The dotted lines are at one sigma limit and two sigma limits beyond the center line. The solid lines at the top and bottom are at three sigma limits beyond the center line.

Charts for Dispersion

Two charts are used to study the spread, or dispersion, of a process. An s-chart plots the standard deviation of each sample; an r-chart plots the range of each sample. These values were calculated in Table 14.1. To create an s-chart,

1. Choose **Stat ▸ Control Charts ▸ Variables Charts for Subgroups ▸ S**.
2. Enter **Grove** as the single column and enter a subgroup size of **5**. Click **OK**.

Figure 14.5 shows an s-chart for the data from Table 14.1. As with an chart, there is a center line, a *UCL*, and an *LCL*. To draw these lines, we need to know the mean and the standard deviation of s. This requires some mathematical derivations and special formulas. Fortunately Minitab does all the work for you. Formulas for the three lines are given below. The value of $c_4(n)$ and $c_5(n)$ are taken from special tables and depend on the sample size.

center line at $c_4(n)\sigma$

$$UCL = c_4(n)\sigma + 3c_5(n)\sigma$$

$$LCL = c_4(n)\sigma - 3c_5(n)\sigma \quad \text{(if } LCL < 0 \text{, then set } LCL = 0\text{)}$$

Figure 14.5   S-chart for the ignition key data

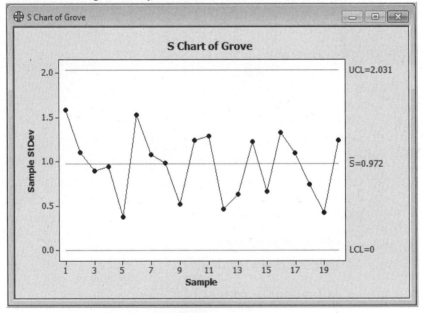

In order to calculate these three lines from data, we must estimate σ. Recall that Minitab gives you two choices, s_p and s_r, defined in the section on Xbar-charts. The formulas for the *UCL* and *LCL* are based on various approximations. Therefore it is possible for the *LCL* to be less than zero. If this happens, Minitab sets the *LCL* equal to zero.

Now let's create an r-chart, a chart for ranges, of the data from Table 14.1. To create the r-chart,

1. Choose **Stat ▸ Control Charts ▸ Variables Charts for Subgroups ▸ R**.

2. Enter **Grove** as the single column and **5** as the subgroup size. Click **OK**. Figure 14.6 shows the chart.

Figure 14.6   R-chart for ignition key data

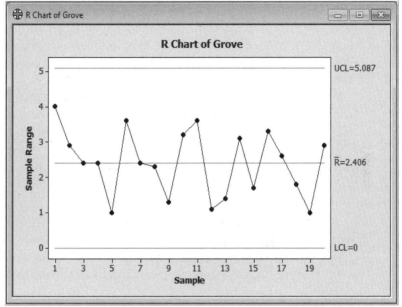

As in an s-chart, there is a center line, a *UCL*, and an *LCL*. These three lines are based on the formulas given below. The value of $d_2(n)$ and $d_3(n)$ are taken from special tables and depend on the sample size.

center line at $d_2(n)\sigma$

$$UCL = d_2(n)\sigma + 3d_3(n)\sigma$$

$$LCL = d_2(n)\sigma - 3d_3(n)\sigma \quad \text{(if } LCL < 0, \text{ then set } LCL = 0)$$

As always, in order to calculate these three lines from data, we must estimate σ. Minitab offers two choices, s_p and s_r, defined in the section on Xbar-charts. As before, the formulas for the *UCL* and *LCL* are approximations. Therefore it is possible for the *LCL* to be less than zero. If this happens, Minitab sets the *LCL* equal to zero.

Finding Problems. If a process is in control, the values plotted in an s-chart (or an r-chart) should vary about the center line, be within the control limits, and show no special patterns. What problems would we typically detect in a chart for spread? One value of s (or r) could be unusually large. The variability could change so that from some point on, s (or r) is larger than it was in the past. It is also possible for spread to decrease, but this is usually not a problem—rather a desirable goal. We would still investigate what was going on, but in the hope of using what we might find to improve the process.

As we mentioned for the Xbar-chart, a process that is in control is not necessarily a process that is acceptable. For example suppose that in the ignition-key study the center line was at 7.966, all points were within the control limits, and no unusual patterns were detected. Thus, from a statistical standpoint, the process is in control. Suppose, however, the groove dimension for individual keys varied from 3 to 13. This much variability may not be acceptable.

When we first study a process, we usually do a chart for dispersion before an Xbar-chart. If the chart for dispersion shows the process is not in control, we cannot properly interpret an Xbar-chart. In order to draw the *UCL* and *LCL* on an Xbar-chart, we need an estimate of *s*. If the dispersion of the process is not in control, the limits on the Xbar-chart will be affected.

Unequal Subgroup Sizes

The examples we have discussed so far are for the case in which all subgroups have the same number of observations. This is the most common situation. Minitab will also produce charts when the subgroups are not all equal. In this case you must supply a column containing integers to tell Minitab which subgroup each observation is in.

Now let's do an Xbar-chart using the data from Table 14.1 with some observations removed, so that three subgroups—10, 11, and 12—have only their first three observations. Minitab will handle missing data quite well.

1. Activate the Data Worksheet and navigate to row 49.
2. Replace the data value of 8.9 in C1R49 with an asterisk, *. (Recall that the asterisk is Minitab's missing value code.)
3. Similarly, enter asterisks in column C1 of rows 50, 54, 55, 59, and 60. You now have missing values for the last two data points of subgroups 10, 11, and 12.
4. Choose **Stat ► Control Charts ► Variables Charts for Subgroups ► Xbar**.
5. Enter **Grove** as the single column and **5** in the Subgroup size text box. Click **OK**. Figure 14.7 shows the resulting chart.

The control limits for the Xbar-chart are no longer straight lines. The limits are wider for the subgroups with only three observations. This is not surprising: when a sample mean is based on fewer observations, the standard deviation of that sample mean is larger.

Below, we give the formulas for the center line and control limits when s_p is used to estimate σ. We use n_i for the number of observations in subgroup i. Consult online Help for the formulas using s_r.

$$\text{center line at } \bar{\bar{x}} = \sum_i \sum_j x_{ij} / \sum_i n_i$$

$$s_p = \sqrt{\sum_i \sum_j x_{ij} - \bar{x}_i^2 / \sum_i (n_i - 1)}$$

$$UCL_i = \bar{\bar{x}} + 3s_p / \sqrt{n_i}$$

$$LCL_i = \bar{\bar{x}} - 3s_p / \sqrt{n_i}$$

Figure 14.7   Xbar chart with unequal subgroup sizes

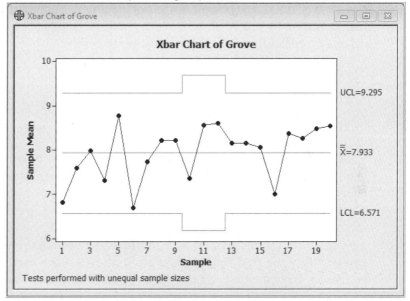

14.3 Control Charts for Attribute Data

We use a chart for attribute data when the variable of interest is counts: for example the number of purchase orders per day that contain one or more errors, or the number of incoming phone calls per day that are misdirected.

Let n_i denote the number of observations in the ith subgroup, and y_i the number of defectives in the ith subgroup. Then $p_i = y_i/n_i$ is the proportion of defectives in the ith subgroup. There are two charts you can do. A p-chart plots the proportions, p_i, and an np-chart plots the counts, y_i. The Stat ▸ Control Charts ▸ Attribute Charts ▸ P command produces a p-chart for the proportion of defectives. Minitab can do four different tests to determine whether there are any problems. If you have a historical value for p, you can specify it, and Minitab will use this value to calculate the center line and control limits. The Stat ▶ Control Charts ▶ Attribute Charts ▶ NP command generates an *np*-chart for the number of defectives. We will discuss the *p*-chart.

If the process is in control, then the probability of getting a defective will be the same for all trials. We denote this value by p. We estimate p by $\bar{p}$, using the formula

$$\bar{p} = \frac{\text{total number of defectives}}{\text{total number of observations}} = \frac{\sum_i y_i}{\sum_i n_i}$$

The distribution of each y_i is binomial with mean $n_i p$ and variance $n_i p(1-p)$. If the sample sizes, n_i, are not too small, we can approximate the distribution of each y_i by a normal distribution with the same mean and variance. It then follows, using some basic probability theory, that we can approximate the distribution of each p_i by a normal, with mean p and variance $p(1-p)/n_i$. Now we can calculate a center line and control limits for the chart.

center line at $\bar{p}$

$$UCL_i = \bar{p} + 3\sqrt{\bar{p}(1-\bar{p})/n_i}$$

$$LCL_i = \bar{p} - 3\sqrt{\bar{p}(1-\bar{p})/n_i}$$

Because we are using a normal approximation, it is possible for the calculated value for *LCL* to be less than zero. If this happens, Minitab sets the *LCL* equal to zero.

Thirty samples, each containing 50 assembled parts, were taken. For each sample the number of parts that had a plating defect was recorded. Try creating a *p*-chart with sample size 50. Use the Tests button to do the tests for special causes. Only the first four tests from Figure 14.4 are used by Minitab's P command. To create the *p*-chart,

1. Open the **Plating** worksheet.
2. Choose **Stat ▶ Control Charts ▶ Attribute Charts ▶ P**.
3. Enter **Plating** as the variable and **50** as the subgroup size.
4. Click the **P Charts Options** button and click the **Tests** tab.
5. Select **Perform all tests for special causes** from the drop-down list.
6. Click **OK** twice. Figure 14.8 shows the *p*-chart.

Points 20 and 21 failed test 1. They are both beyond the upper control limit. These samples had 15 and 12 defective parts, respectively. The next step would be to try to determine why these two samples had so many defective parts. We can redo the *p*-chart, calculating the center line and control limits with samples 20 and 21 omitted. This would give a better picture of the process.

Figure 14.8   P chart for plating data

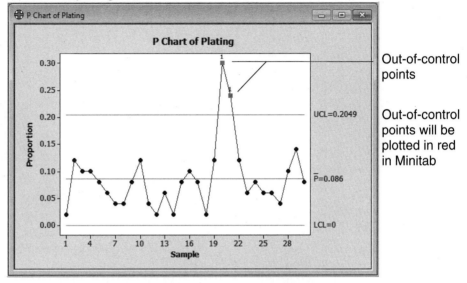

To redo the *p*-chart with the omitted samples,

1. Click the **Edit Last Dialog** button and click the **P Charts Options** button.
2. Click the **Data Options** button.
3. Select **Specify Which Rows to Exclude**.
4. Select **Row numbers** and enter **20 21** in the text box. (Note the space between 20 and 21.)
5. Check **Leave gaps for excluded points**.
6. Click **OK** twice to generate a new *p*-chart. See Figure 14.9.

Figure 14.9   P chart for plating data after samples 20 and 21 removed from calculations

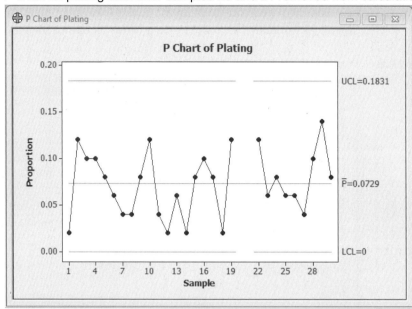

Now the center line is slightly lower and the control limits are closer together. All points, except the missing points at 20 and 21, are well within the control limits.

14.4 Control Charts with the Minitab Assistant

If you follow the command sequence **Assistant ▸ Control Charts**, the decision tree shown in Figure 14.10 will appear. Let's illustrate constructing a P chart by revisiting the plating data.

Notice that there are several choices and branches in the tree. If you let your cursor hover over various locations on the tree Minitab will display tooltips to help explain the item. Figure 14.11 shows an example. Since we are working with attribute (good/bad) data, click on the rectangle labeled P chart. If you hover the pointer over that rectangle, a tooltip will say "Click to create this chart". Tooltips are available for most of the items in the decision tree. Figure 14.11 displays the tooltip when you hover over the Data type rectangle. Since we have attribute data we can proceed and click the P Chart rectangle in the decision tree.

Figure 14.10   Assistant Control Charts Decision Tree

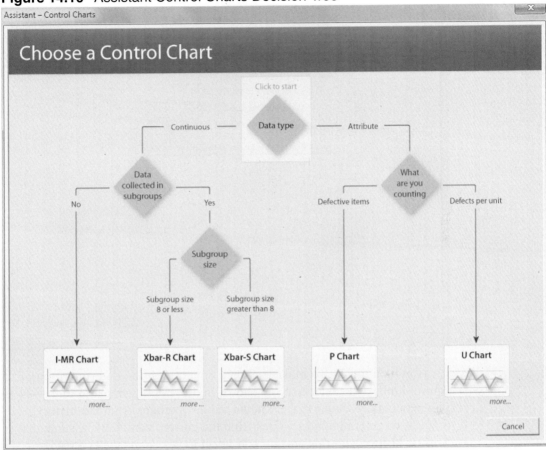

Figure 14.11   Tooltip help for Data type

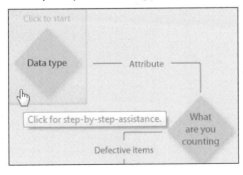

Figure 14.12 shows the dialog box for the Assistant P chart. We entered **Plating** as the column that gives the number of defectives and **50** for the constant subgroup size.

Figure 14.12 Assistant P chart dialog box

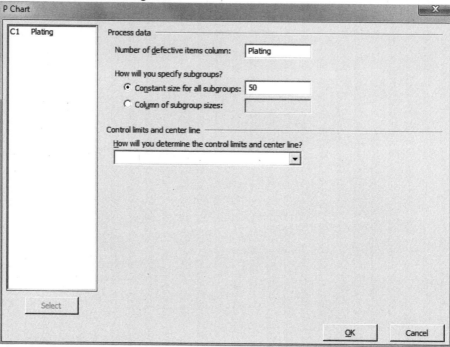

Now you must also make a choice concerning the determination of the control limits and center line. As shown in the drop-down list in Figure 14.13, you may choose either to use known values, perhaps from the history of the process, or determine new values from the current data. Let's assume that we have no history with this process and select to determine the control limits and center line from the current Plating data.

Figure 14.13 Choices for determining control limits

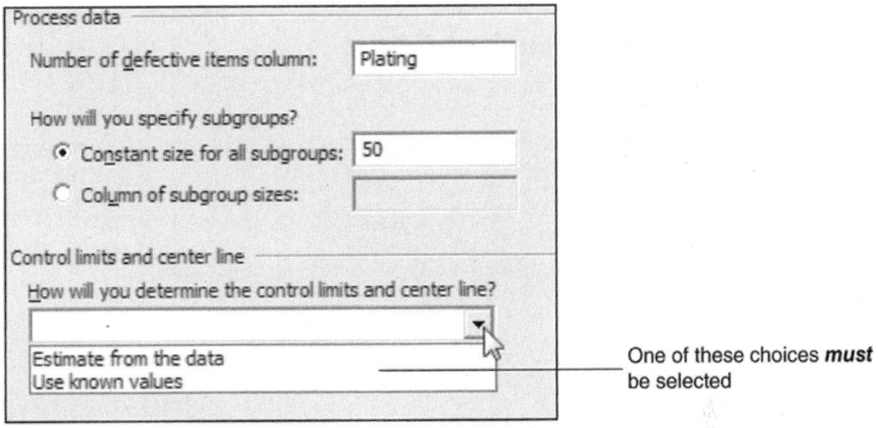

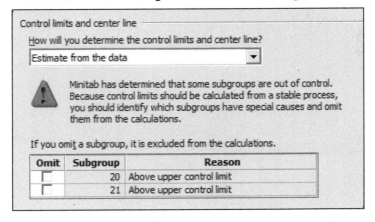

As soon as you select "Estimate from the data" Minitab computes the control limits in the background and checks the process for lack of control. In the case of the Plating data there are two subgroups that fall above the upper control limit (UCL). Before showing any further results, Minitab adds warning information to the dialog box as shown in Figure 14.14.[†]

Figure 14.14 Assistant P chart dialog box after determining estimation method

You can now either click OK to proceed with the control chart results or, preferably, select the check boxes as shown in Figure 14.15, then click OK and instruct Minitab to omit the out-of-control data in its calculation of the control limits.

[†] This is the only instance where Minitab alters a dialog box prior to showing you further results.

Figure 14.15 Assistant P chart dialog box after choosing to omit out-of-control data

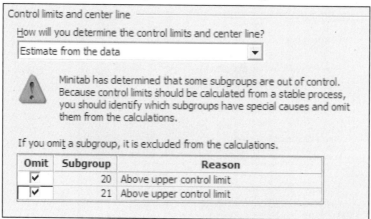

When you click OK, Minitab produces the three screens shown in Figures 14.16, 14.17 and 14.18. In the Summary Report we see that the process may not be stable. There are 6.7% defectives and the upper control limits are exceeded in subgroups 20 and 21. (The plotting symbols for these subgroups will be red on your screen.) The Stability Report displays the usual control chart but with control limits based on all but subgroups 20 and 21. Finally, the Report Card in Figure 14.18 warns one more time that the process may not be stable.

Figure 14.16 Assistant P chart Summary Report

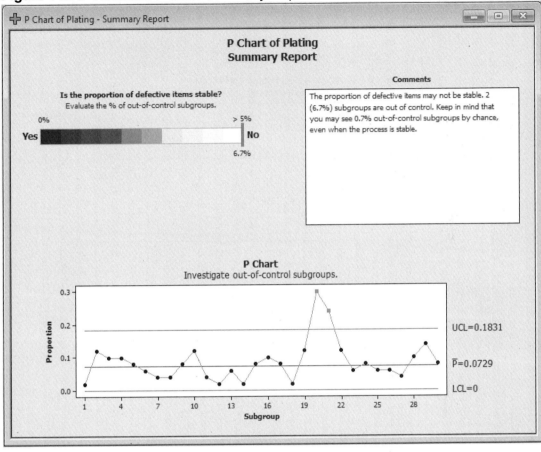

Figure 14.17 Assistant P chart Stability Report

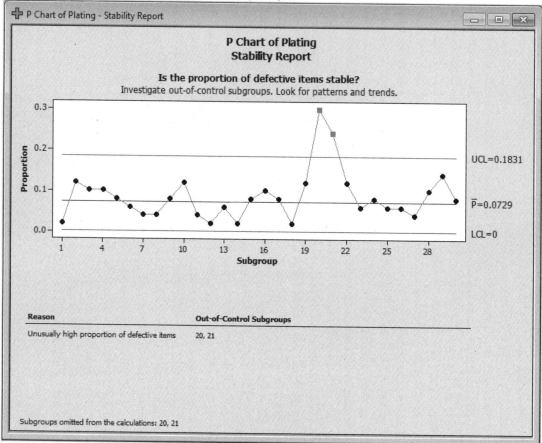

Figure 14.18 Assistant P chart Report Card

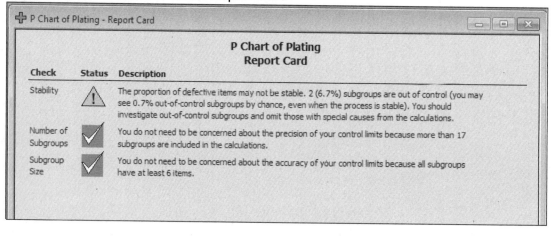

Exercises

Exercises for Section 14.2

14-1   Procter & Gamble was involved in studying a process that tightens the caps on containers of a hair conditioner. Cap torque, a measure of how tightly the caps are screwed onto the containers, was recorded for 17 samples, each of size 4. The data are given in Table 14.2 and are available as file CapTorque in the Data Library at www.CengageBrain.com associated with this book.

Table 14.2 Cap torque data

Sample Number	Samples			
	1	2	3	4
1	24	14	18	27
2	17	32	31	27
3	21	27	24	21
4	24	26	31	34
5	28	32	24	16
6	22	37	36	21
7	16	17	22	34
8	20	19	16	16
9	18	30	21	16
10	14	15	14	14
11	25	15	16	15
12	19	15	15	19
13	19	30	24	10
14	15	17	17	21
15	34	22	17	15
16	17	20	17	20
17	15	17	24	20

(a) Do an r-chart to check the stability in dispersion of cap torque. Does it appear to be in control?

(b) Do an Xbar-chart to check the mean level. Does it appear to be in control?

(c) Generate the eight tests to see whether Minitab flags any points as possible problems. How do these results compare to the results of your visual inspection in part (b)?

(d) Display a histogram and a normal probability plot of the individual observations. If the process is in control and data come from a normal distribution, then your displays should look "normal." Do they?

14-2   The Tests options in the control chart command includes tests that help you find unusual patterns in a control chart. There is a problem, however—the problem you always have when you do many tests. You may find a "problem" when nothing is actually wrong.

(a) Test 1 detects points that are more than 3 sigma limits from the mean. Suppose a process is generated from a normal distribution with a constant mean and standard deviation, and all observations are independent. That is, suppose there are no problems with the process. What is the probability that a sample will fail test 1? Does this depend on the true process mean, standard deviation, sample size, or number of subgroups?

(b) Suppose you were to chart a process like the one described in part (a). Each day you collect 50 samples, each of size 5, and do an Xbar-chart. You create a separate chart each day for 10 days. Do you think it is likely that at least one chart will fail test 1? On the average how many of the samples you collect over the 10 days would fail test 1?

(c) Use simulation to create the 10 charts described in part (b). Use the Tests button for all eight tests. Did any chart fail test 1? How many charts failed one or more of the eight tests for special causes?

14-3   The data file BreadLoaves contains data on the weight of bread loaves that leave a large bakery. Five loaves are selected every ten minutes and their weights (in decagrams) determined. The data file is available form the Data Library at www.CengageBrain.com associated with this book.

(a) Display an s-chart for these data and determine whether or not dispersion seems to be in control.

(b) Confirm your results in part (a) using an R-chart.

(c) Display an Xbar-chart to see if the mean weight is in control. Click the Xbar options button, then the Tests tab, and choose all tests from the drop-down list. (An investigation into the lack of control discovered that a scale was improperly calibrated at some point in time.)

14-4   (a) Control charts are very useful displays. However, they do not show all the important patterns that may occur in a data set. What other Minitab displays might be useful for investigating the control chart data?

(b) Table 14.3 contains a small data set. (file name Samples). Use control charts and the displays you listed in part (a) to investigate these data. What are some of the important features?

Table 14.3 Sample data

Sample Number	Samples				
	1	2	3	4	5
1	701	630	687	691	681
2	699	691	676	689	677
3	708	707	694	690	672
4	708	696	692	682	687
5	719	699	681	700	659
6	711	692	673	690	661
7	715	699	679	682	669
8	715	683	687	679	679
9	698	674	675	671	684
10	691	690	679	664	679

Exercises for Section 14.3

14-5   Documents used in a manufacturing system were monitored for correctness. Each day a sample of 100 documents was taken. The number of documents that contained one or more errors was recorded. Table 14.4 shows the data that are stored in the Docs worksheet.

Table 14.4 Documents data

	Number of Documents with Errors									
Days 1-10	10	12	10	11	6	7	12	10	6	11
Days 11-20	9	14	16	21	20	12	11	6	10	10
Days 21-25	11	11	11	6	9					

(a) What percentage of the documents (overall) have errors?
(b) Calculate, by hand, the center line and the upper and lower 3-sigma control limits for a *p*-chart.
(c) Use Minitab to display a *p*-chart. Compare your answers in part (b) to the values that Minitab prints on the chart.
(d) Does the process appear to be in control?

14-6   Past data indicate that a company's absenteeism rate is about 12% among its 987 workers. Managers instituted a plan to attempt to reduce that rate. Table 14.5 lists the total numbers of absent workers after the plan had been implemented. The data are available in file named Absent at the Data Library at www.CengageBrain.com associated with this book.

Table 14.5 Absenteeism data

	Number Absent			
Day	Week 1	Week 2	Week 3	Week 4
Monday	129	119	99	115
Tuesday	121	103	92	101
Wednesday	117	103	83	106
Thursday	109	89	92	83
Friday	122	105	92	98

(a) Display a p-chart of these data. Click the P-Chart options button, the Parameters tab, and enter 0.12 as the historical value for $\bar{p}$.

(b) Does the p-chart indicate that the process is in control?

(c) Does the p-chart give good evidence that the plan to reduce absenteeism is effective?

(d) Display a p-chart of these data now using the current data to estimate $\bar{p}$. (You will need to click the P-Chart options button, the Parameters tab, and delete the historical value of 0.12 that you entered for part (a)).

(e) Does the new p-chart indicate that the process is in control (but at a new, lower, level)?

14-7   Repeat Exercise 14-6 using np-charts instead of p-charts. Are the conclusions the same?

Exercises for Section 14.4

14-8   Use the Assistant to repeat the control chart analysis shown in Section 14.4 but this time do not omit the out-of-control subgroups in the calculations. What changes? Do any conclusions change?

14-9   Repeat Exercise 14-5 but now using the Assistant.

14-10   Repeat Exercise 14-6 using the Assistant for the analysis.

15

Additional Topics in Regression

15.1 Variable Selection in Regression

The Peru data described in the Appendix, p. 528, were collected to study the long-term effects of a change in environment on blood pressure. Some questions we might ask are, "What happens to the Indians' blood pressure as they live longer periods in the new environment?" and "What factors are related to blood pressure?" In this section we will show how variable selection procedures can be used to help answer these questions.

Both systolic and diastolic blood pressure were measured. Since systolic blood pressure is often a more sensitive indicator, we will use it as our explanatory or predictor variable. There are eight predictors: age, number of years since migration, weight, height, three measures of body fat (chin skin fold, forearm skin fold, and calf skin fold), and pulse rate. To these we will add one derived predictor, fraction of life since migration. This new variable might be a better measure of "how long" the Indians have lived in their new environment than just the number of years since migration. Younger people adapt to new surroundings more quickly than older people. A 25-year-old might be able to adapt as well in one year as a 50-year-old could in two years. There may be other derived variables that could be useful predictors, but we will stop here for now. To derive the new variable,

1. Open the **Peru** worksheet.
2. Choose **Calc ▸ Calculator** and create a new variable, **Fraction**, that equals **Years/Age**.

Now the problem is how to find a "good" regression equation, one that has few predictors, but still has good predictive ability. With nine predictors there are many possible subset models.[†] Of course, if we consider transformations we would have many more than nine predictors!

Minitab has two "automatic" methods that will help you in this search: stepwise regression and best subsets regression. Each sorts through subsets of the variables, using numerical criteria to choose good subsets. Automatic procedures such as these can be a valuable tool, particularly in the early stages

---

[†] $2^9 = 512$, to be exact or one less if you don't count the model with *no* predictors.

of building a model and when you have many predictors. At the same time they present certain dangers.

- Since the procedures automatically "snoop" through many models, the model selected may fit the data "too well." That is, the procedure may select variables that by chance alone fit well.

- Automatic procedures cannot take into account special knowledge the investigator may have about the situation, how important a variable is from a scientific standpoint, how difficult a variable is to measure, and so on. Therefore the model may not be the best from a practical point of view.

- The automatic procedures are heuristic algorithms, which often work very well but which may not select the model with the highest R^2 value for the given number of predictors.

Stepwise Regression

In its simplest form the Stepwise command starts by fitting all models with just one predictor variable. For each model it checks the p-value associated with that variable. The variable with the lowest p-value is added to the equation, provided its p-value is less than the Alpha to Enter, the value that controls which variables enter the equations (the default is $\alpha = 0.15$). If no variable has a p-value smaller than Alpha to Enter, Stepwise terminates. Let's assume variable X_3 entered the equation on step 1.

Next, Stepwise fits all two-variable models in which one of the variables is X_3. The variable with the smallest p-value is added to the equation, provided its p-value is less than Alpha to Enter. If no variable has a p-value smaller than Alpha to Enter, Stepwise terminates. Let's assume variable X_5 entered the equation on step 2.

If you used Regression to fit all models containing two predictors, X_5 and X_i, then chose the variable X_i with the smallest p-value, this would be the variable Stepwise chooses. (Assuming the p-value is smaller than the Alpha to Enter.) Some people use an F-statistic to guide Stepwise. Minitab will do this if you change the options.

At this point in the procedure, Stepwise looks to see if any of the other variables in the equation (other than the one that was just entered) can be removed. At this stage there is just one possible variable to remove, X_5. Later, when the equation contains more variables, there will be many candidates. The variable with the largest p-value is removed, provided its p-value is larger than the value specified by Alpha to Remove, which controls which variables are removed (again the default is $\alpha = 0.15$, but Alpha to Enter must be $\geq$ Alpha to

Remove). If no variable has a *p*-value larger than Alpha to Remove, Stepwise goes to the next step, where it attempts to enter another predictor.

Stepwise continues in this manner, first adding the variable with the smallest *p*-value (provided its value is larger than Alpha to Enter), then removing the variable with the largest *p*-value (provided its value is greater than the Alpha to Remove). Stepwise terminates when it can no longer add or remove any variables.

At each step Minitab prints the coefficient and *t*-ratio for each variable in the equation, and *s* (the square root of MSE) and R^2 for the equation.

Stepwise has several advantages:

- You can have many predictors, you can have more predictors than you have observations
- You can have predictors that are highly correlated. Stepwise can still look for good subsets. However, Stepwise uses a heuristic to add and remove variables, and it may not find models with the highest R^2 value.

Let's use Stepwise on the nine predictors in the Peru data set. To perform a stepwise regression on the Peru data set,

1. Choose **Stat ▸ Regression ▸ Stepwise**.
2. Enter **Systol** as the response variable.
3. Enter **C1-C8** and **Fraction** as the predictor variables. (We will use the default value of 0.15 for both Alpha to Enter and Alpha to Remove. This may be changed if you click the Methods button. See Figure 15.1.
4. Click **OK**. Minitab produces the output shown in Figure 15.2.

   TIP If you have command language enabled and run Stepwise, Minitab will ask if you want more output in the Session Window. Stepwise allows you to intervene in the algorithm at various points, and you can consult online Help if you want more information about this. You can answer "yes" to see more steps, but Minitab will tell you no variables could be entered or removed (using the current values for Enter and Remove). Answer "no" because for now there is no need to intervene.

Stepwise added Weight and then Fraction, and then terminated. Therefore no other variable had a *p*-value less than 0.15. Note that like the standard Regression command, Stepwise does not print the value of R^2 if you fit a model without a constant.

Figure 15.1 Stepwise regression dialog box

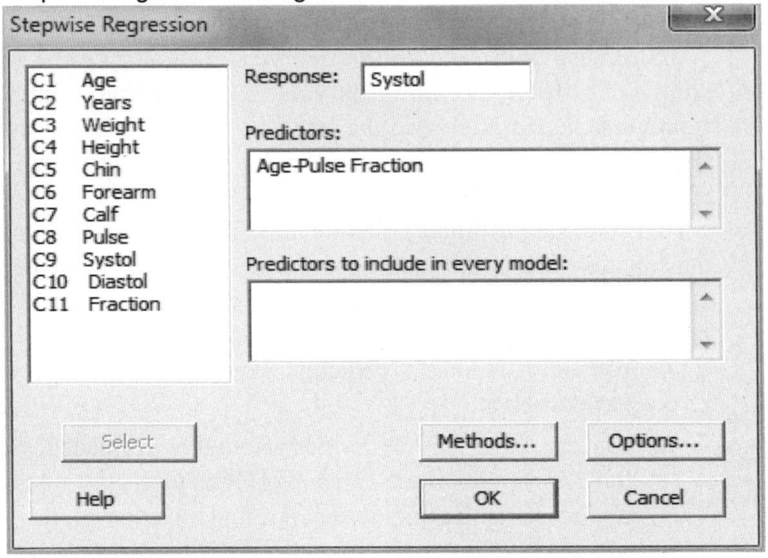

Figure 15.2 Stepwise regression results for systolic blood pressure

Stepwise Regression: Systol versus Age, Years, ...

```
  Alpha-to-Enter: 0.15  Alpha-to-Remove: 0.15


Response is Systol on 9 predictors, with N = 39


Step             1       2
Constant     66.60   60.90

Weight        0.96    1.22
T-Value       3.72    5.21
P-Value      0.001   0.000

Fraction             -26.8
T-Value              -3.71
P-Value              0.001

S             11.3    9.78
R-Sq         27.18   47.31
R-Sq(adj)    25.21   44.38
Mallows Cp    28.5    12.9
```

There are two special cases of stepwise regression: forward selection and backwards elimination.

Forward Selection. In forward selection, variables are entered into the equation as in Stepwise, but are never permitted to be removed. The procedure terminates when no variable that is not in the equation has a *p*-value less than the Alpha to Enter. To do forward selection using Minitab's Stepwise command, click the Methods button and select the Forward Selection option. Then no variables will ever be removed. (You may choose to use your own alpha or *F*-value if you wish.)

Backwards Elimination. Backwards elimination starts with all variables in the equation, then removes them, one by one, using the same rule that Stepwise uses. No variables are ever added back in. The procedure terminates when no variable in the equation has a *p*-value that is greater than Alpha to Remove. To do backwards elimination using Stepwise, click the Methods button, and click the Backward Selection option. Then no variable that is removed can reenter the equation. (Again you may choose to use your own alpha or *F*-value if you wish.)

Note that since backwards elimination starts with all variables in the model, Minitab must be able to fit this model. If you attempt to fit a model with highly correlated predictor variables (substantial multicolinearity) Minitab will inform you that it cannot fit a model.

Best Subsets Regression

The Best Subsets Regression command does **best subsets regression**, a method also called **all possible regressions.** In addition to specifying the response variable, and possible predictors, you can also specify predictors that are to be included in all models. You can click the Options button in the Best Subsets Regression dialog box to set a minimum and maximum number of predictors, how many "best" models to print, and whether or not to fit an intercept term. To perform Best Subsets Regression with the Peru data,

1. Choose **Stat ▸ Regression ▸ Best Subsets**.
2. Enter **Systol** as the response variable and **C1-C8** and **Fraction** as the predictors.
3. Click **OK**. Minitab displays the output shown in Figure 15.3.

Best Subsets Regression first looked at all one-predictor models and selected the model with the largest value, the model with Weight. Information on this model and on the next-best one-predictor model was printed. Then Best Subsets Regression looked at all two-predictor models, found the one with the largest R^2, and printed information on it and on the next-best one, and so on, until all nine predictors were used. Our task now is to determine which of these models are "good," and are thus candidates for further study.

Figure 15.3   Best subsets regression for systolic blood pressure

```
Best Subsets Regression: Systol versus Age, Years, ...

Response is Systol

                                                    F
                                              F     r
                                        W  H  o     a
                                  Y  e  e  r     P  c
                               e  i  i  C  e  C  u  t
                       Mallows  A  a  g  g  h  a  a  l  i
                               g  r  h  h  i  r  l  s  o
Vars  R-Sq  R-Sq(adj)    Cp      S    e  s  t  t  n  m  f  e  n
  1   27.2     25.2     28.5  11.338  X
  1    7.6      5.1     45.5  12.770                          X
  2   47.3     44.4     12.9   9.7772       X                 X
  2   42.1     38.9     17.5  10.251  X  X
  3   50.3     46.1     12.3   9.6273       X     X           X
  3   49.0     44.7     13.4   9.7509  X  X                   X
  4   59.7     55.0      6.1   8.7946  X  X  X                X
  4   52.5     46.9     12.4   9.5502  X  X        X          X
  5   63.9     58.4      4.5   8.4571  X  X  X        X       X
  5   63.1     57.6      5.1   8.5417  X  X  X           X    X
  6   64.9     58.3      5.6   8.4663  X  X  X        X  X    X
  6   64.9     58.3      5.6   8.4681  X  X  X        X     X X
  7   66.1     58.4      6.6   8.4556  X  X  X  X  X  X       X
  7   65.5     57.7      7.1   8.5220  X  X  X        X  X  X X
  8   66.6     57.7      8.1   8.5228  X  X  X  X  X  X     X X
  8   66.2     57.2      8.5   8.5760  X  X  X  X  X  X  X    X
  9   66.7     56.4     10.0   8.6554  X  X  X  X  X  X  X  X X
```

To aid us, the Best Subsets Regression command prints four statistics for each model: R^2, adjusted R^2, Cp, and s. On Minitab's output, R^2 and adjusted R^2 are both expressed as percentages.

R^2 was discussed in Chapter 11. It is calculated as

$$R^2 = \frac{\text{Regression SS}}{\text{Total SS}} = 1 - \frac{\text{Error SS}}{\text{Total SS}}$$

First, notice that maximizing R^2 is the same as minimizing SSE. Also notice that since SSE always decreases as the number of predictors increase, R^2 always increases as the number of predictors increases. Thus R^2 is not especially useful when you are comparing models with different numbers of predictors. In general we look for increasing values of R^2, until the rate of increase tapers off.

Then we stop adding predictors into our model. Here the number of useful predictors is probably about four or five.

Adjusted R^2 takes into account the number of predictors in the model. Suppose there are n observations in the data set and p parameters. Thus $p =$ (number of predictors) $+1$ if the equation has a constant term. The definition of R^2-adjusted is

$$R^2\text{-adjusted} = 1 - \frac{(\text{Error SS})/(n-p)}{(\text{Total SS})/(n-1)}$$

R^2-adjusted is directly related to MSE. As R^2-adjusted increases, MSE decreases. The fourth statistic, s, is the square root of MSE, and thus choosing a model based on small MSE or s is equivalent to choosing a model based on large R^2-adjusted.

In Figure 15.3 both R^2 and R^2-adjusted increased greatly when we went from a one-predictor model to a two-predictor model. Then they each increased more slowly. Once we reached five predictors, R^2-adjusted no longer increased at all, and in fact decreased a bit. Note that if you fit models without a constant, Best Subsets Regression does not print R^2 or R^2-adjusted.

Let's look at the remaining statistic, Cp. Again p is the number of parameters in the candidate model and n is the number of observations in the data set. MSE_m is calculated from the full model, the model with all the predictors in it. Then Cp is defined by

$$Cp = \frac{\text{SSE}_p}{\text{MSE}_m} - (n - 2p)$$

This statistic attempts to measure both random error and any bias from fitting a model with too few of the predictors. In general we look for models in which Cp is small and is also close to p. If a model has little or no bias, Cp should be close to p. If a model has considerable bias, Cp will tend to be larger than p. If Cp is small, the model is relatively precise (has a small variance) in estimating the true regression coefficients. Adding more predictors will not improve this precision much.

In Figure 15.3 the models with four and five predictors have small values of Cp (6.1 and 4.5, respectively), and values that are close to the number of parameters (5 and 6, respectively). Any models selected by automatic procedures should be tested further by looking carefully for outliers and influential observations. We investigate these issues in the next Section.

15.2 Diagnostics in Regression

Diagnostics for Individual Observations

The Regression-Storage dialog box shown on page 351 lists a number of values you can select to include as diagnostics for detecting unusual properties or problems with individual observations. You already saw how to store residuals or standardized residuals; the other diagnostics can be stored in the same way, by clicking their check boxes. Minitab then stores the values in the next available columns, assigning intuitive column names.

We assume the data set has n observations and the regression model has p parameters, including the constant term if it is in the equation. We use the following notation: Y_i is the response value for the ith observation; $\hat{Y}_i$ is the fitted value for the ith observation; $\mathbf{X}$ is the $n{\times}p$ design matrix.

Diagnostics of note include the following:

Hi (leverages). Select this option to store the **leverages**, the diagonal elements of the **hat matrix**. The $n{\times}n$ projection, or hat matrix, is defined as $\mathbf{H} = \mathbf{X}(\mathbf{X'X})^{-1}\mathbf{X'}$. It has the property that $\mathbf{H}Y = \hat{Y}$. The leverage of the ith observation is the ith diagonal, h_i (also denoted h_{ii}) of $\mathbf{H}$. Note that h_i depends only on the predictors; it does not involve the response Y. If h_i is large, the ith observation has unusual predictors. That is, it has predictor values that are far from the center of the data, using Mahalanobis distance, a metric that adjusts for the fact that the predictors may have different variances. This observation will have a large influence in determining the regression coefficients.

Many people consider h_i to be large enough to merit checking if it is larger than $2p/n$ or $3p/n$. If an observation has $h_i > 3p/n$, Minitab lists it as an unusual value and tags it with an X.

Deleted t Residuals. These are called **Studentized Residuals** or **Studentized deleted residuals**. To calculate the Studentized residual for the ith observation, first remove the ith observation from the data set. Then calculate the regression equation, using this smaller data set. Let $\hat{Y}_{(i)}$ = the fitted or predicted value for the deleted observation, and let its residual $e_{(i)} = Y_i - \hat{Y}_{(i)}$. Then the Studentized residual for the ith observation is $e_{(i)}/\text{stdev}(e_{(i)})$. Thus deleted t residuals are similar to standardized residuals, but the ith observation is omitted. Because deleted t residuals estimate the equation, with the ith observation deleted, the ith observation cannot influence the estimates. Therefore unusual Y values stand out more clearly. If the regression model is correct, each Studentized residual has a Student's t distribution with $(n-1-p)$ degrees of freedom.

Cook's Distance. Recall that leverages, stored by HI (leverages), tell us if an observation has unusual predictors. Standardized residuals tell us if an observation has an unusual response. **Cook's distance** combines these two into one overall measure of how unusual an observation is. Specifically Cook's distance for the ith observation is

$$\text{COOKD}_i = \frac{1}{p}\frac{h_i}{1-h_i}(\text{standardized residual}_i)^2$$

Generally it is a good idea to check observations where Cook's distance > $F(0.5, p, n-p)$. This is the median of an F-distribution which can be obtained with Minitab's Cumulative probability command.

DFITS. These are also called DFFITS. They combine leverage and Studentized residuals into one overall measure of how unusual an observation is. Specifically DFITS for the *i*th observation is

$$\text{DFITS}_i = \sqrt{h_i/(1-h_i)}(\text{Studentized residual}_i)^2$$

Generally you should check observations where $\text{DFITS}_i > 2\sqrt{p/n}$.

Regression Options

You can further customize a regression by using the options in the Regression-Options dialog box shown on p. 340.

Lack-of-Fit Tests. Use the Pure error lack-of-fit test to assess how well your model fits the data. To use it, you must have replicates. A **replicate** is one setting of the predictors for which you have two or more responses. In this case it is possible to calculate an estimate of σ^2, the variance of Y_i, without assuming any form for the model. The estimate is MSPE, defined below.

Suppose X_j is one unique value of a predictor. Suppose there are n_j responses for this value. Let Y_{ji} be the *i*th response corresponding to X_j and $\bar{Y}_j$ = the average of the responses at X_j. Then the **sum of squares for pure error** is

$$\text{SSPE} = \sum_j \sum_i (Y_{ji} - \bar{Y}_j)^2$$

If SSE is the sum of squares for fitting a specific model, then the sum of squares for lack-of-fit is

SSLOF = SSE − SSPE

As usual, each sum of squares has degrees of freedom associated with it. The degrees of freedom for SSPE is $\Sigma(n_j - 1)$. The degrees of freedom for SSLOF is df(SSLOF) = df(SSE) − df(SSPE). We then use these to form mean squares and an F-statistic as follows:

$$F = \frac{\text{MSLOF}}{\text{MSPE}} = \frac{\text{SSLOF}/\text{df(SSLOF)}}{\text{SSPE}/\text{df(SSPE)}}$$

If you click the Pure error check box, Minitab prints this F-statistic and its p-value. If the p-value is small, your model has significant lack-of-fit. For example if you fit a line and find significant lack-of-fit, a line is not an adequate model. Perhaps you need a quadratic curve or a transformation of one or more variables.

Durbin-Watson Statistic. This subcommand prints the Durbin-Watson statistic, used to test for autocorrelation in the data. Regression assumes that observations are independent. However, observations collected over time are often correlated, usually positively.

The Durbin-Watson statistic, D, is calculated from the residuals, e_i, as follows:

$$D = \frac{\sum_i (e_i - e_{i-1})^2}{\sum_i e_i^2}$$

Minitab prints the value of D. You need to use special tables to check the significance of D.[†]

There are also graphical ways to check for autocorrelation. Suppose you've done a regression and stored the residuals in "Resid." You can plot each residual, e_i, versus its lagged value, e_{i-1}. Minitab's command Stat ▶ Time Series ▶ Lag will shift a column down one cell to give you the lagged values. You enter the column containing the residuals as the series, store the lags in a separate column, and then plot the residuals versus the lagged values. If this plot shows a linear trend, the observations are autocorrelated.

15.3 Additional Regression Features

Storage of Some Results

You can store MSE and $(\mathbf{X'X})^{-1}$, using the Storage dialog box. The product $\text{MSE} \times (\mathbf{X'X})^{-1}$ is the estimated variance-covariance matrix of the coefficients in the regression equation.

Minitab stores the diagonal of the hat matrix (produced by the Hi leverages check box), but it does not store the full matrix. The full hat matrix is given by $\mathbf{H} = \mathbf{X}(\mathbf{X'X})^{-1}\mathbf{X'}$. If you wish, you can calculate it by using the matrix stored

---

[†] See almost any book on regression: for example, J. Neter, W. Wasserman, and M. Kutner, *Applied Linear Regression Models*. Homewood, IL: Irwin, 1989.

by the **X'X** inverse check box along with the matrix commands Calc ▸ Matrices ▸ Transpose and Arithmetic (which allows you to do multiplication).

Weighted Regression

The Regression command assumes that for each x, the amount of variation in the population of y is approximately the same. This variance is usually called the variance of y about the regression line, and is denoted by σ^2. Suppose the variances are not equal. Suppose the variance associated with the ith observation is denoted σ_i^2. If this is the case, we might be able detect it when we plot residuals versus a predictor or versus the fitted values. When variances are not equal, we often use a **weighted regression analysis**. This is what the Weights option in the Regression-Options dialog box does.

The weight associated with the ith observation is $w_i = 1/\sigma_i^2$. Thus observations with a high variance are given a low weight. Actually you can use $w_i = k/\sigma_i^2$ where k is any constant of proportionality. Of course in practice you do not know σ_i, so these values must be estimated from the data. There is no exact way to do this.

In many cases the variances are related to the level of a predictor, X, or to the level of the response, Y. Plotting the residuals versus each predictor and versus Y can often give you an idea as to what that relationship is. Here are some common situations and appropriate weights to use for each:

If $\sigma_i^2 = \sigma^2 X_i$, use $w_i = 1/X_i$ If $\sigma_i^2 = \sigma^2 Y_i$, use $w_i = 1/Y_i$

If $\sigma_i^2 = \sigma^2 X_i^2$, use $w_i = 1/X_i^2$ If $\sigma_i^2 = \sigma^2 Y_i^2$, use $w_i = 1/Y_i^2$

If $\sigma_i^2 = \sigma^2 \sqrt{X_i}$, use $w_i = 1/\sqrt{X_i}$ If $\sigma_i^2 = \sigma^2 \sqrt{Y_i}$, use $w_i = 1/\sqrt{Y_i}$

Ill-Conditioned Data

Predictor variables can have several types of problems. These can cause both statistical and computational difficulties.

Multicollinearity. If one predictor is highly correlated with other predictors, we say the predictors are multicollinear. If this correlation is moderately high, Minitab prints a warning message. If this correlation is very high, Minitab removes the predictor from the equation, prints a message, and fits the smaller model. In all cases calculations are done with a high degree of numerical accuracy.

Minitab uses the following rules: A predictor X_i is regressed on the remaining predictors. If R^2 for this regression is greater than 99%, a warning is printed. If R^2 is greater than 99.99%, X_i is removed from the regression model.

Multicollinearity does not affect the fits and residuals, but it does affect the coefficients. Their estimated standard deviations can be very large. Thus the fitted coefficients can vary greatly from one sample to another. There are cases in which no coefficient in the equation is statistically significant, but R^2 for the equation as a whole is very high. You can investigate the correlation structure of the predictors by using the Correlation command, and by regressing each predictor on the remaining predictors.

Some possible solutions to the problem of multicollinearity are

1. Eliminate predictors from the equations, especially if this has little effect on R^2.

2. Change predictors by taking linear combinations of them.

3. If you are fitting polynomials, subtract a value near the mean of the predictor before squaring it.

Small Coefficient of Variation. The coefficient of variation of a predictor, X_j, is

(standard deviation of X_j)/(mean X_j)

If a predictor has a small coefficient of variation, the variability in the predictor is very small relative to the magnitude of the predictor. This means the predictor is essentially constant. For example the variable Year with values from 1970 to 1975 has a small coefficient of variation. All the information in Year is in the fourth digit. This can cause numerical problems.

If the coefficient of variation is moderately small, Minitab prints a warning that tells you the predictor is nearly constant. If the coefficient of variation is very small, Minitab removes it from the equation. The solution is simple: Subtract a constant from the data. For example replace Year by Year $- 1969$, which has values from 1 to 6.

Exercises

15-1   Use Stepwise with the Peru data, but now go to the methods dialog box and use F-values as the criterion. Set both Enter and Remove F-values to 2 instead of the default 4. Any changes in the selected predictors?

15-2   We will look at two of the predictors in the Peru data.
   (a) Use regression to fit the equation with just the predictor Fraction = Years/ Age. Does fraction of life since migration appear to be a useful predictor of systolic blood pressure?

(b) Add Weight to the model in part (a). Compare this two-predictor model to the one in part (a). Is Fraction a useful predictor? Try to explain what is going on.

15-3   Use Stepwise to fit models to the Peru data. Use the original predictors plus derived variables that may be useful. For example, we might try Chin + Forearm + Calf as a composite measure of body fat, or Weight/Height as a measure of body mass. Can you think of others? Can you improve the model in Figure 15.1, p. 454, in a meaningful and useful way?

15-4   Use backwards elimination to fit models to the Peru data. Any changes from the results in Figure 15.3, p. 456?

15-5   Use regression to further investigate the 4- and 5-predictor models suggested for the Peru data by Best Subsets Regression.

15-6   In Figure 15.2, p. 454, Stepwise stopped with a 2-predictor model. Why didn't it go further when Best Subsets Regression, in Figure 15.3, p. 456, seems to indicate that a 4- or 5-predictor model fits the data better?

16

Additional Topics in Analysis of Variance

Minitab has a fairly extensive analysis of variance capability. It includes analysis of balanced and unbalanced designs, crossed and nested factors, analysis of covariance, random and mixed models for balanced designs, multiple comparisons for one-way designs, and MANOVA.

This chapter gives a brief introduction to a few features. To learn more, consult Minitab's extensive online Help and books on analysis of variance.

16.1 Multiple Comparisons with One-Way

If a one-way analysis of variance results in a significant F-test, we conclude that the population means are not all equal, but we do not know what the differences are. The procedures discussed in this section help you study these differences.

There are many procedures for making multiple comparisons among means. We will discuss three of the more common ones: Fisher's, Tukey's, and Dunnett's. These methods are usually presented as a set of tests for pairwise comparisons of means. Minitab, however, presents the results as a set of confidence intervals for pairwise differences, $\mu_i - \mu_j$. This allows you to assess the practical significance of differences between means, as well as the statistical significance. You can easily convert each confidence interval to a test if you wish. Reject H_0: $\mu_i = \mu_j$ if and only if the confidence interval for $\mu_i - \mu_j$ does not contain zero.

All three methods are available by clicking the Comparisons button in the One-way dialog box. In all three methods, the confidence intervals for $\mu_i - \mu_j$ have the same general form. Suppose group i has n_i observations with sample mean $\bar{x}_i$, and group j has n_j observations with sample mean $\bar{x}_j$. Then the confidence interval for $\mu_i - \mu_j$ is

$$\bar{x}_i - \bar{x}_j \pm Ds \sqrt{\frac{1}{n_i} + \frac{1}{n_j}}$$

where $s = \sqrt{MSE}$ and has ν degrees of freedom. Each method determines the value of the constant D in a different way. We discuss these below.

Minitab prints two error rates for each procedure: the individual and the familywise error rate. The **individual error rate** is the probability that a given confidence interval will not contain the true difference in group means. Multiple-comparison procedures, however, calculate not just one confidence interval but a set, or family, of confidence intervals. The **familywise error rate** is the probability that this family will contain at least one confidence interval that does not contain the true difference in group means.

Fisher's Least Significant Difference (LSD)

Fisher's method gives confidence intervals for all pairwise differences, $\mu_i - \mu_j$. You specify the individual error rate, α, by clicking the Fisher's check box in the One-way Multiple Comparisons dialog box and then entering it in the adjoining box. The value of D is the upper $\alpha/2$ point of the t distribution with v degrees of freedom. Minitab prints the critical value, D, in the output.

Minitab calculates the familywise error rate. This error rate is exact if all groups contain the same number of observations. If the group sizes are different, Minitab uses the Tukey-Kramer method to calculate an approximate familywise error rate. The true error rate will be slightly smaller, resulting in conservative (slightly larger) confidence intervals.

Tukey's Honestly Significant Difference (HSD)

Tukey's method also provides confidence intervals for all pairwise difference between group means.

You specify the familywise error rate, α, in the adjoining box to the Tukey's check box in the One-way Multiple Comparisons dialog box. The value of $D = Q/\sqrt{2}$, where Q is the upper α point of the Studentized range distribution with parameters $r =$ number of groups and $v =$ degrees of freedom for MSE. Minitab prints the critical value, Q, in the output.

Minitab calculates the individual error rate. This error rate is always exact. The familywise error rate is exact if all groups contain the same number of observations. If the group sizes are different, Minitab uses the Tukey-Kramer method. The true error rate will be slightly smaller, resulting in conservative (slightly larger) confidence intervals. The methods used by Fisher and Tukey are closely linked: The results are identical if the specified error rate for Fisher is equal to the individual error rate for Tukey, or, equivalently, if the specified error rate for Tukey is equal to the familywise error rate for Fisher. Which method you use depends on which error rate you want to specify.

Dunnett's Procedure

Dunnett's procedure is designed for the case in which you want to compare several experimental treatments with one control treatment. You specify the familywise error rate and the control group in the subcommand Dunnett. The value of D is calculated from special tables.[†] Minitab prints the critical value, D, on the output.

Minitab calculates the individual error rate. Both individual and familywise error rates are exact in all cases.

Example Using Fabric Data

Let's do an analysis of variance, using the Fabric data from Chapter 10. To make multiple comparisons using Tukey's method in the One-way command,

1. Restart Minitab to return all dialog box settings to their defaults.
2. Open the **Fabric1** worksheet.
3. Choose **Stat ▸ ANOVA ▸ One-Way**.
4. Enter **Charred** as the response and **Lab** as the factor.
5. Click the **Comparisons** button and then click the **Tukey's** check box.
6. Click **OK** twice. Figure 16.1 shows a portion of the extensive output.

The analysis of variance test says the means are significantly different. We did all three multiple comparison methods, using the default error rate of 0.05 for each.

Tukey. The output from Tukey prints 10 confidence intervals, controlling the familywise error rate. Here we took the default, 0.05. Therefore we are 95% confident that all 10 intervals cover their true differences. Each individual interval has an error rate of 0.00671, much smaller than the 0.05 in Fisher. This means each interval printed by Tukey is wider than the corresponding interval that will be found using Fisher's method.

Dunnett. There is no natural control in the fabric data set. Therefore in order to illustrate Dunnett's procedure, we will pretend that lab 5 uses special methods and is the standard of accuracy. We then test each of the other four labs against this control. This gives four confidence intervals. Minitab prints and plots these intervals. You will find that the interval for lab 4 does not contain zero. The other three do. Thus only lab 4 is significantly different from the control. The confidence interval for $\mu_5 - \mu_4$ is $(-1.0819, -0.2091)$. This tells us that lab 4 is below the control by at least 0.2091 and perhaps by as much as 1.0819.

[†] See, for example, B. J. Winer, *Statistical Principles in Experimental Design*, New York: McGraw-Hill, 1971.

Figure 16.1 Tukey's comparisons in one-way ANOVA

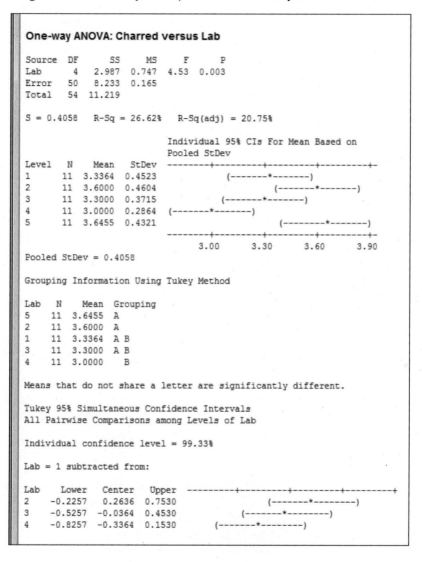

Figure 16.1   Tukey's comparisons in one-way ANOVA (continued)

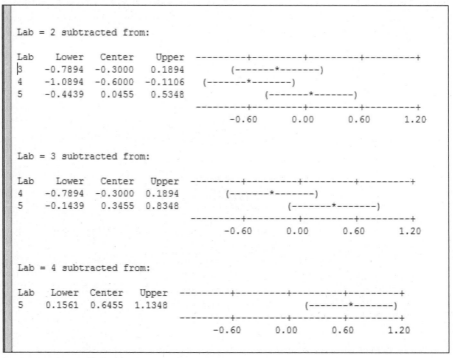

```
Lab = 2 subtracted from:

Lab    Lower    Center    Upper   ---------+---------+---------+---------+
 3    -0.7894  -0.3000   0.1894        (-------*-------)
 4    -1.0894  -0.6000  -0.1106   (-------*-------)
 5    -0.4439   0.0455   0.5348            (-------*-------)
                                  ---------+---------+---------+---------+
                                        -0.60     0.00      0.60      1.20


Lab = 3 subtracted from:

Lab    Lower    Center    Upper   ---------+---------+---------+---------+
 4    -0.7894  -0.3000   0.1894        (-------*-------)
 5    -0.1439   0.3455   0.8348                (-------*-------)
                                  ---------+---------+---------+---------+
                                        -0.60     0.00      0.60      1.20


Lab = 4 subtracted from:

Lab    Lower   Center    Upper    ---------+---------+---------+---------+
 5    0.1561  0.6455    1.1348                   (-------*-------)
                                  ---------+---------+---------+---------+
                                        -0.60     0.00      0.60      1.20
```

Fisher. Repeat the analysis using Fisher's method. The output from Fisher also prints 10 confidence intervals. Each has an error rate of 0.05. Look at the first one. This is a confidence interval for $\mu_2 - \mu_1$ and goes from −0.6112 to 0.0840. Because the interval contains zero, the corresponding null hypothesis, H_0: $\mu_2 = \mu_1$, cannot be rejected. There are just two intervals that exclude zero: the intervals for $\mu_4 - \mu_2$ and $\mu_5 - \mu_4$. These are the only cases in which there is evidence that the means are significantly different.

The family error rate is 0.277. This means that when we use Fisher (with this number of groups, this number of observations, and $\alpha = 0.05$), we have a 0.277 probability that at least one interval will not contain the corresponding true difference in means.

16.2 Multifactor Balanced Designs

In Chapter 10 we used the Balanced ANOVA command to analyze balanced two-factor designs. You can use the Balanced ANOVA command to fit balanced designs with up to nine factors. In this section we will discuss designs with three factors using this command.

You specify the model for the design in the Model box in the Balanced Analysis of Variance dialog box. Enter the response variable in the Response box and then enter the factors and interactions in your model in the Model box. The asterisk symbol, *, indicates an interaction term. Parentheses, ( ), indicate nesting.

Here are some examples using three factors:

```
A B C
A B C A*C B*C
A B C A*B A*C B*C A*B*C
```

The first model contains just main effects. The second model contains main effects and two interactions. The third is the full model, and contains all interactions.

Several special rules apply to specifying a model. You may omit the quotes around variable names. Because of this, any variable name used in the Model box must start with a letter and must contain only letters and numbers. If you want to use special symbols in a variable name, you must enclose the name in single quotes, as in other Minitab commands. You can always use column numbers such as C1, C2,... instead of column names. You may not put any extra text in the Model box, except after the symbol #.

You can fit reduced models. For example Y = A B C A*B is a three-factor model with just one two-way interaction. Models, however, must be hierarchical. For example if the term A*B*C is in the model, then all of its subterms (A, B, C, A*B, A*C, and B*C) must also be in the model.

Because models can be quite long and tedious to type, Minitab provides two shortcuts. A vertical bar (or an exclamation point) indicates crossed factors, and a minus sign removes terms. Here are some examples:

```
ANOVA Y = A|B|C
```

is equivalent to

```
ANOVA Y = A B C A*B A*C B*C A*B*C
```

and

```
ANOVA Y = A|B|C - A*B*C
```

is equivalent to

```
ANOVA Y = A B C A*B A*C B*C
```

Example Using the Potato Rot Data

The Potato Rot data was described in Exercise 10-11, p. 323, which analyzed the data by using two-way analysis of variance and plots. Recall that this data set has three factors: (1) amount of bacteria injected into the potato (1 = low amount, 2 = medium amount, 3 = high amount); (2) temperature during storage (10°C, 16°C); (3) amount of oxygen during storage (2%, 6%, 10%). The response variable is the amount of rot in the potato after five days of storage.

Let's use the Balanced ANOVA command to do a three-way analysis. There are seven terms in this model. To do a three-way analysis,

1. Open the **Potato** worksheet.
2. Choose **Stat ▸ ANOVA ▸ Balanced ANOVA**.
3. Enter **Rot** as the response variable.
4. Enter **Bacteria|Temp|Oxygen** as the model. Click **OK**. Figure 16.2 shows the results.

Figure 16.2   Three-way ANOVA for potato rot

```
ANOVA: Rot versus Bacteria, Temp, Oxygen

Factor      Type    Levels   Values
Bacteria    fixed      3     1, 2, 3
Temp        fixed      2     1, 2
Oxygen      fixed      3     1, 2, 3


Analysis of Variance for Rot

Source                DF       SS       MS       F       P
Bacteria               2   651.81   325.91   13.91   0.000
Temp                   1   848.07   848.07   36.20   0.000
Oxygen                 2    97.81    48.91    2.09   0.139
Bacteria*Temp          2   152.93    76.46    3.26   0.050
Bacteria*Oxygen        4    30.07     7.52    0.32   0.862
Temp*Oxygen            2     1.59     0.80    0.03   0.967
Bacteria*Temp*Oxygen   4    81.41    20.35    0.87   0.492
Error                 36   843.33    23.43
Total                 53  2707.04


S = 4.84003   R-Sq = 68.85%   R-Sq(adj) = 54.14%
```

As always, we start by checking the interactions. The three-factor interaction, Bacteria*Temp*Oxygen, is not significant, nor are the two-factor interactions, Bacteria*Oxygen and Temp*Oxygen. That leaves the three main effects and one two-way interaction. Now try fitting a model with just those terms. To fit the model with three main effects and one two-way interaction,

1. Click the **Edit Last Dialog** button.
2. Replace the model with **Bacteria Temp Oxygen Bacteria*Temp** and then click **OK**. Figure 16.3 shows the output.

In Figure 16.3 we fit a model with just these terms. Because this model is balanced, the sums of squares for terms in the models do not change. The sums of squares for the terms we dropped go into SSE. Thus SSE increases, as does its degrees of freedom. MSE, however, decreases. Now two factors, bacteria and temperature, and the interaction between them are all significant. The remaining factor, oxygen, is only marginally significant.

Figure 16.3   Three-way ANOVA using a reduced model

```
ANOVA: Rot versus Bacteria, Temp, Oxygen

Factor    Type   Levels  Values
Bacteria  fixed       3  1, 2, 3
Temp      fixed       2  1, 2
Oxygen    fixed       3  1, 2, 3


Analysis of Variance for Rot

Source          DF       SS      MS      F      P
Bacteria         2   651.81  325.91  15.68  0.000
Temp             1   848.07  848.07  40.79  0.000
Oxygen           2    97.81   48.91   2.35  0.106
Bacteria*Temp    2   152.93   76.46   3.68  0.033
Error           46   956.41   20.79
Total           53  2707.04


S = 4.55977   R-Sq = 64.67%   R-Sq(adj) = 59.29%
```

16.3 Unbalanced Designs

The command General Linear Model will fit what is called the **general linear model**. This includes balanced and unbalanced analysis of variance, analysis of covariance, and more. In this section we will use General Linear Model in a simple form to do an unbalanced two-way analysis. Output contains both the sequential sums of squares and the adjusted sums of squares (i.e., each term is fitted after all other terms are in the model). Automatic tests are done using the adjusted sums of squares. Observations that are considered unusual are flagged and printed out. The General Linear Model command does its calculations using a regression approach. First, a design matrix is formed from the factors and interactions. The design matrix will have as many columns as there are degrees of freedom in the model. The columns of the design matrix are then used as predictors.

As in Balanced ANOVA, you specify the model in the Model box of the General Linear Model dialog box. You use the same rules for specifying the model as you do in Balanced ANOVA. Unfortunately the analysis of unbalanced design is not as simple as in Balanced ANOVA.

In Chapter 10 we analyzed data from an experiment to study driving performance. The design was balanced, with four observations in each cell. Suppose the experiment did not go as planned. Suppose that on two of the drives the car broke down, so no result was obtained. Our carefully designed experiment is no longer balanced. Table 16.1 shows the data.

Table 16.1 Driving performance data with two observations omitted

| | Road type | | |
	First-class	Second-class	Dirt
Inexperienced Driver	4	23	16
	18	15	27
	8	21	23
	10	13	14
Experienced Driver	6	6	50
	13	8	15
	7	12	8
			17

To analyze an unbalanced design,

1. Open the **Drive** worksheet.
2. Choose **Data ▸ Delete Rows**. Enter **6 13 25:48** in the **Rows to delete** text box.
3. Drag over the four variables in **C1-C4** then press **Select** to enter them in the **Columns from which to delete these rows** box. See Figure 16.5.
4. Click **OK**.

Figure 16.4   Delete Rows dialog box

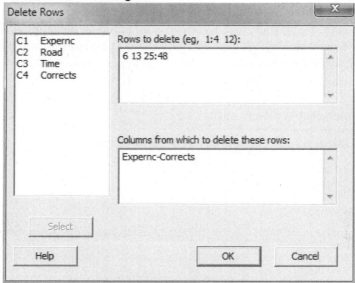

1. Choose **Stat ▸ ANOVA ▸ General Linear Model**.
2. Enter **Corrects** as the response and **Expernc | Road** as the model. See Figure 16.5.
3. Click **OK**. Figure 16.6 shows the output.

Figure 16.5   General linear model dialog box

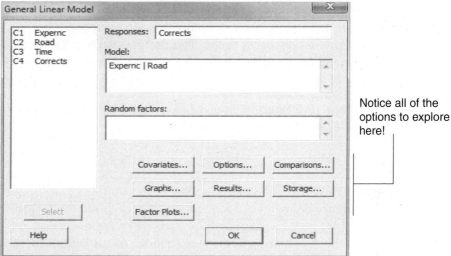

Notice all of the options to explore here!

In Figure 16.6 Minitab displays two types of sums of squares—sequential sums of squares (Seq SS) and adjusted sums of squares (Adj SS). The mean squares, F-statistics, and p-values are all based on the adjusted sums of squares. In Balanced ANOVA the sums of squares for the terms in the model add up to the total sum of squares. In General Linear Model the sequential sums of squares add up to the total sum of squares, but the adjusted ones do not.

Figure 16.6   General linear model results

```
General Linear Model: Corrects versus Expernc, Road

Factor   Type    Levels  Values
Expernc  fixed      2    1, 2
Road     fixed      3    1, 2, 3


Analysis of Variance for Corrects, using Adjusted SS for Tests

Source        DF  Seq SS  Adj SS  Adj MS    F      P
Expernc        1  125.67  147.27  147.27  5.78  0.029
Road           2  265.27  247.00  123.50  4.85  0.023
Expernc*Road   2   55.00   55.00   27.50  1.08  0.363
Error         16  407.33  407.33   25.46
Total         21  853.27


S = 5.04563   R-Sq = 52.26%   R-Sq(adj) = 37.34%
```

There are many ways to explain what this analysis means. Computationally it is based on doing regression, and it is often explained by using terminology from regression. We will discuss that approach later. Now we will look at what hypotheses are being tested. This design has six cells. Each cell represents one population. Each population has a mean. The notation for these are shown in Table 16.2.

Table 16.2 Notation for population means in the driving experiment

	Road type		
	First-class	*Second-class*	*Dirt*
Inexperienced driver	μ_{11}	μ_{12}	μ_{13}
Experienced driver	μ_{21}	μ_{22}	μ_{23}

For example μ_{11} is the mean number of corrections for the population of all inexperienced drivers driving on first-class roads. Similarly μ_{23} is the mean number of corrections for the population of all experienced drivers driving on dirt roads.

Suppose we want to see whether experience makes a difference. If we use the adjusted sum of squares to do the test, as GLM does, then the specific hypothesis we are testing is

$$H_0: \frac{\mu_{11} + \mu_{12} + \mu_{13}}{3} = \frac{\mu_{21} + \mu_{22} + \mu_{23}}{3}$$

Thus we are weighting the cell means in each row equally. In most cases this is a reasonable thing to do. Similarly if we use the adjusted sum of squares to see whether road type makes a difference, we are testing the hypothesis

$$H_0: \frac{\mu_{11} + \mu_{12}}{2} = \frac{\mu_{12} + \mu_{22}}{2} = \frac{\mu_{13} + \mu_{23}}{2}$$

Models in GLM

The sequential sums of squares can be used to do custom tests. We will not discuss these here. Note that sequential sums of squares depend on the order in which terms are listed in the Model box in the General Linear Model dialog box. Thus the sequential sums of squares for A B and B A will, in general, be different. The adjusted sums of squares, however, do not depend on the order of the terms and will always be the same.

Although models can be unbalanced in General Linear Model, they must be "full rank." Thus there must be enough data to estimate all the terms in your model. For example suppose you have a two-factor model with one empty cell. Then you can fit the model A B, but not A|B. Don't worry about figuring out whether or not your model is of full rank. Minitab will tell you whether it is or not. In most cases eliminating some of the high-order interactions (assuming, of course, that they are not important) will solve your problem.

Regression Approach to Analysis of Variance

When we discussed analysis of variance in Chapter 10, we did the calculations using a sum-of-squares approach. This works when the design is balanced. In unbalanced designs we must use regression to do the calculations.

If a factor B has b levels, then it has $(b-1)$ degrees of freedom. We form $(b-1)$ predictors, called **dummy variables**. The first dummy variable, B_1, is 1 when B is at its lowest level, -1 when B is at its highest level, and 0 otherwise. The second dummy variable, B_2, is 1 when B is at its second level, 0 when B is at its highest level, and 0 otherwise. The third dummy variable is 1 when B is at its third level, 0 when B is at its highest level, and 0 otherwise. We continue in this way until we have $(b-1)$ predictors.

We form one block of dummy variables for each factor. Suppose there is an interaction, say $A*B$, in the model, where A has a degrees of freedom and B has b degrees of freedom. Then $A*B$ has $(a-1)(b-1)$ degrees of freedom, so we need $(a-1)(b-1)$ dummy variables for $A*B$. To get these, we multiply each dummy variable for A by each dummy variable for B.

Table 16.3 shows the Driving Performance data used in Figure 16.5. To save space, we use E for experience and R for road type. E has two levels and thus just one dummy variable, called E1. R has three levels and thus two dummy variables, R1 and R2. The interaction has two degrees of freedom and thus two dummy variables. ER11 is the product of columns E1 and R1; ER12 is the product of columns E1 and R2.

To see the dummy variables that correspond to a given model, click the Storage button in the General Linear Model dialog box (shown on p. 473) and click the Design matrix check box. The first column of this matrix is the dummy variable for the constant term. It is a column containing only 1s. The remaining columns are for the factors and interactions in your model. This matrix is often called a **design matrix**.

First, we must have Minitab create the design matrix and copy it into the columns of the worksheet so that you can use it in the regressions. Then, to get the adjusted sums of squares that the General Linear Model command prints, we must do one regression for each term in the model. In each case we list the predictors for the term we want last on the regression line. The information we need is in the table of sequential sums of squares printed by the Regression command.

Table 16.3 Dummy variables for data from Table 16.1

Original Data				Dummy Variables				
Corrects	E	R		E1	R1	R2	ER11	ER12
4	0	1		1	1	0	1	0
18	0	1		1	1	0	1	0
8	0	1		1	1	0	1	0
10	0	1		1	1	0	1	0
23	0	2		1	0	1	0	1
15	0	2		1	0	1	0	1
21	0	2		1	0	1	0	1
13	0	2		1	0	1	0	1
16	0	3		1	−1	−1	−1	−1
27	0	3		1	−1	−1	−1	−1
23	0	3		1	−1	−1	−1	−1
14	0	3		1	−1	−1	−1	−1
6	1	1		−1	1	0	−1	0
13	1	1		−1	1	0	−1	0
7	1	1		−1	1	0	−1	0
6	1	2		−1	0	1	0	−1
8	1	2		−1	0	1	0	−1
12	1	2		−1	0	1	0	−1
20	1	3		−1	−1	−1	1	1
15	1	3		−1	−1	−1	1	1
8	1	3		−1	−1	−1	1	1
17	1	3		−1	−1	−1	1	1

To create the design matrix and run the regressions,

1. Choose **Stat ▸ ANOVA ▸ General Linear Model**.
2. Enter **Corrects** as the response and **Expernc|Road** as the model.
3. Click the **Storage** button and then select the **Design matrix** check box.
4. Click **OK** twice. Minitab does all of the calculations and stores the dummy variables in the matrix named XMAT1.
5. Choose **Data ▸ Copy ▸ Matrices to Columns**.
6. Enter **XMAT1** in the "Copy from matrix" box and enter **C5-C10** in the "In current worksheet, in columns" text box.
7. **Deselect** Name the columns containing the copied data
8. Click **OK**.

9. Name C6 **E1** (remember C5 contains the dummy variable for the constant term, which we won't be using), C7 **R1**, C8 **R2**, C9 **ER11**, and C10 **ER12**.

10. Choose **Stat ▸ Regression ▸ Regression**. Enter **Corrects** as the response and **R1 R2 ER11 ER12 E1** as the predictors. Click **OK**. Minitab produces the regression output.

11. Click the **Edit Last Dialog** button and replace the responses with **E1 ER11 ER12 R1 R2**. Click **OK**.

12. Click the **Edit Last Dialog** button and replace the responses with **E1 R1 R2 ER11 ER12**. Click **OK**.

Figure 16.7 shows the edited output from the three regressions corresponding to the three terms in the Driving Performance model.

The first regression is for the main effect experience. There is just one dummy variable, E1. Its SEQ SS is 147.27, which is the Adj SS listed on the General Linear Model output. This sum of squares is the additional sum of squares for fitting the model with all five predictors in it over the model with just the predictors R1, R2, ER11, and ER12—that is, the model without E1.

The second regression is for road. There are two dummy variables, R1 with SEQ SS = 246.96, and R2 with SEQ SS = 0.03. The sum, 246.96 + 0.03 = 246.99, is the Adj SS for road (except for rounding error). This sum of squares is the additional sum of squares for fitting the model with all five predictors in it over the model with just the predictors E1, ER11, and ER 12—that is, the model without R1 and R2.

The last regression is for the interaction. There are two dummy variables: ER11, with SEQ SS = 10.48, and ER12, with SEQ SS = 44.52. The sum, 10.48 + 44.52 = 55.00, is the Adj SS for the interaction. This sum of squares is the additional sum of squares for fitting the model with all five predictors in it over the model with just the predictors E1, R1, and R2.

Figure 16.7   Edited Regression output from various GLM calculations

---

General Linear Model: Corrects versus Expernc, Road

```
Analysis of Variance for Corrects, using Adjusted SS for Tests

Source        DF   Seq SS   Adj SS   Adj MS     F      P
Expernc        1   125.67   147.27   147.27   5.78   0.029
Road           2   265.27   247.00   123.50   4.85   0.023
Expernc*Road   2    55.00    55.00    27.50   1.08   0.363
Error         16   407.33   407.33    25.46
Total         21   853.27
```

Corrects versus R1, R2, ER11, ER12, E1

```
The regression equation is
Corrects = 13.4 - 4.06R1 - 0.06R2 - 1.94ER11 + 2.06ER12 + 2.61E1

Source  DF  Seq SS
R1       1  242.19
R2       1    1.37
ER11     1   16.02
ER12     1   39.09
E1       1  147.27
```

Corrects versus E1, ER11, ER12, R1, R2

```
The regression equation is
Corrects = 13.4 + 2.61E1 - 1.94ER11 + 2.06ER12 - 4.06R1 - 0.06R2

Source  DF  Seq SS
E1       1  125.67
ER11     1   19.86
ER12     1   53.41
R1       1  246.96
R2       1    0.03
```

Corrects versus E1, R1, R2, ER11, ER12

```
The regression equation is
Corrects = 13.4 + 2.61E1 - 4.06R1 - 0.06R2 - 1.94ER11 + 2.06ER12

Source  DF  Seq SS
E1       1  125.67
R1       1  264.69
R2       1    0.58
ER11     1   10.48
ER12     1   44.52
```

16.4 Analysis of Covariance

In analysis of variance all variables in the model are categorical. Often we want to use models that contain continuous variables, called **covariates**, as well. the General Linear Model include models that have factors (categorical variables) as well as covariates (continuous variables). We will look at a simple case: one factor and one covariate.

First let's analyze data for the preprofessionals (ED = 1) in the Cartoon data (described in the Appendix, p. 524). We'll use one-way analysis of variance to see whether the factor Color makes a difference. To perform the one-way analysis of variance,

1. Open the **Cartoon** worksheet. We'll first subset it so we are using only the data for ED = 1.
2. Choose **Data ▸ Subset Worksheet**. Click the **Specify which rows to include** option button. Click the **Condition** button and then enter **ED=1** as the expression. Click **OK** twice. Minitab subsets the data into a new worksheet. Use this new worksheet in subsequent analyses.
3. Choose **Stat ▸ ANOVA ▸ General Linear Model**.
4. Enter **Cartoon1** as the response and **Color** as the model. Click **OK**. Minitab produces the output shown in Figure 16.8.

The *F*-test shows no evidence that the factor Color makes a difference. If we look at the Otis scores for the preprofessionals, we see that they vary greatly, from 78 to 129. This variation in ability might be masking a small but significant Color effect. Analysis of covariance allows us to "correct" for this variation.

Figure 16.8   One-way ANOVA of cartoon data

```
General Linear Model: Cartoon1 versus Color

Factor  Type   Levels  Values
Color   fixed       2   0, 1


Analysis of Variance for Cartoon1, using Adjusted SS for Tests

Source  DF   Seq SS   Adj SS  Adj MS    F      P
Color    1    7.664    7.664   7.664  1.00  0.322
Error   51  391.128  391.128   7.669
Total   52  398.792


S = 2.76933   R-Sq = 1.92%   R-Sq(adj) = 0.00%


Unusual Observations for Cartoon1

Obs  Cartoon1     Fit   SE Fit  Residual  St Resid
 19   1.00000  6.53846  0.54311  -5.53846    -2.04 R
 25   1.00000  6.53846  0.54311  -5.53846    -2.04 R
 30   0.00000  5.77778  0.53296  -5.77778    -2.13 R

R denotes an observation with a large standardized residual.
```

Now use General Linear Model to do an analysis of covariance. The model is the same as the model for a one-way analysis with one extra term, the covariate. We will use the Covariates button to tell Minitab which variables in our model are covariates. We will also print out information on the coefficients, information we will use later. To do an analysis of covariance,

1. Click the **Edit Last Dialog** button and this time click the **Covariates** button. Enter **Otis** in the Covariates box. Click **OK**. You return to the main General Linear Model dialog box.
2. Click the **Results** button and click the last option button which says **In addition, coefficients for all terms**. Click **OK** twice. Minitab produces the output shown in Figure 16.9.

The analysis of variance table does a test for the covariate, Otis. The result is very significant, as we would expect. A person's performance on the cartoon test depends on the person's ability, which is what the Otis score measures. The color factor is also significant. Thus, once we account for a person's ability, we see a statistically significant color effect.

Figure 16.9   Results of using a covariate in ANOVA of cartoon data

```
General Linear Model: Cartoon1 versus Color

Factor  Type   Levels  Values
Color   fixed       2  0, 1


Analysis of Variance for Cartoon1, using Adjusted SS for Tests

Source  DF   Seq SS   Adj SS   Adj MS      F      P
Otis     1  171.855  183.930  183.930  44.39  0.000
Color    1   19.740   19.740   19.740   4.76  0.034
Error   50  207.198  207.198    4.144
Total   52  398.792


S = 2.03567   R-Sq = 48.04%   R-Sq(adj) = 45.97%


Term         Coef  SE Coef      T      P
Constant   -8.168    2.169  -3.77  0.000
Otis      0.13661  0.02050   6.66  0.000
Color
0          0.6152   0.2819   2.18  0.034


Unusual Observations for Cartoon1

Obs  Cartoon1      Fit   SE Fit  Residual  St Resid
 28   1.00000  5.56022  0.39312  -4.56022    -2.28 R
 29   1.00000  6.24326  0.39795  -5.24326    -2.63 R

R denotes an observation with a large standardized residual.
```

Regression Approach to Analysis of Covariance

Suppose we use Regression to do the calculations that General Linear Model did in Figure 16.9. First, we form the dummy variables for the factors and their interactions, as we do in analysis of variance. Then we add the covariates as additional predictors. This example has one factor, Color, with one degree of freedom, so it has just one dummy variable. We will call it C. C = 1 if Color is at its low level of 0; and C = −1 if Color is at its high level of 1. The second predictor is Otis, the covariate. To generate the dummy variable and then run the regression,

1. Choose **Stat ▸ ANOVA ▸ General Linear Model**. Click the **Storage** button, make sure the Design matrix check box is selected, and click OK twice.

2. Choose **Data ▸ Copy ▸ Matrix to Columns**. Copy from the newest matrix, **XMAT3** to **C10-C12**.[†]

3. Deselect **Name the columns containing the copied data** and click **OK**.

4. In the subsetted worksheet, name C12 **C**.

5. Choose **Stat ▸ Regression ▸ Regression**. Enter **Cartoon1** as the response variable and **C** and **Otis** as the predictors. Click **OK**. Minitab displays the output shown in Figure 16.10.

Figure 16.10 Analysis of cartoon data using regression

```
Regression Analysis: Cartoon1 versus C, Otis

The regression equation is
Cartoon1 = - 8.17 + 0.615 C + 0.137 Otis


Predictor      Coef  SE Coef      T      P
Constant     -8.168    2.169  -3.77  0.000
C            0.6152   0.2819   2.18  0.034
Otis        0.13661  0.02050   6.66  0.000


S = 2.03567   R-Sq = 48.0%   R-Sq(adj) = 46.0%


Analysis of Variance

Source          DF       SS      MS      F      P
Regression       2  191.594  95.797  23.12  0.000
Residual Error  50  207.198   4.144
Total           52  398.792


Source  DF   Seq SS
C        1    7.664
Otis     1  183.930


Unusual Observations

Obs      C  Cartoon1    Fit  SE Fit  Residual  St Resid
 28  -1.00     1.000  5.560   0.393    -4.560     -2.28R
 29  -1.00     1.000  6.243   0.398    -5.243     -2.63R

R denotes an observation with a large standardized residual.
```

[†] To learn the Minitab name for the stored Design matrix you can choose **Data ▸ Display Data...** Note the name(s) of the stored matrices and click **Cancel.** (It will be the last matrix name shown: either XMAT1, XMAT2 or XMAT3. For this illustration we will assume it is XMAT3.)

Figure 16.10 shows output from the regression. One regression equation is printed, but it represents two separate lines, one for each level of the factor. The lines have the same slope, 0.137, the coefficient of Otis, but they have different constant terms.

When Color is at its low level, C = 1. Then the line is

CARTOON1 = −8.17 + 0.615(1) + 0.137(Otis) = −7.555 + 0.137(Otis)

When Color is at its high level, C = −1. Then the line is

CARTOON1 = −8.17 + 0.615(−1) + 0.137(Otis) = −8.785 + 0.137(Otis)

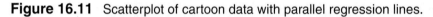

Recall that the low level of Color is for the people who saw black-and-white slides. Thus the line above with open delta symbols is for these people who saw black-and-white slides. The other level of Color, plotted as open diamonds is for the people who saw color slides. Thus the second line is for these people who saw color slides. Figure 16.11 shows a working plot of the data and the two regression lines. In Minitab the data and the fitted values will be plotted in four different colors in addition to the four different symbols for easy identification.

Figure 16.11   Scatterplot of cartoon data with parallel regression lines.

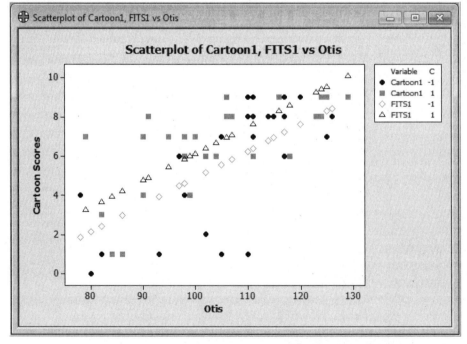

To produce this working plot, first save the fitted values from the regression. When plotted against the predictor, Otis, they will lie on the two parallel regression lines. Now use Graph ▸ Scatterplot ▸ With Groups to plot Cartoon1 versus Otis and FITS1 versus Otis with C (Color dummy variable) as the grouping variable. Be sure to click the Multiple Graphs button and select the option to place both graphs on the same page. Notice that the "line" for black-and-white slides (C = −1) is above the line for color slides. Black-and-white slides seem to be a better teaching device than color slides, at least in this experiment. There is, however, quite a lot of variability, even if we look only at people with about the same Otis scores. Note that Graph ▸ Scatterplot (With Regression and Groups) would *not* produce the desired plot since that command effectively includes an "interaction" term in the model and the resulting lines are not forced to be parallel.

With a little more effort we could have Minitab produce a presentation quality plot of the regression lines. You need to proceed somewhat as we did to graph a regression plane on p. 362. Any "calculated line" may be added to a scatterplot. Calculation with the regression equation shows that the two lines run from endpoints at $(78, 3.131)$ to $(129, 10.118)$ for C=1 and $(78, 1.901)$ to $(129, 8.888)$ for C=−1.

Look at the output from General Linear Model in Figure 16.9, p. 482. The General Linear Model command printed the same three coefficients that Regression printed. Therefore we could have calculated the values required for the two lines using information from General Linear Model.

16.5 Random Effects and Mixed Models

Sometimes the levels of a factor are not of intrinsic interest in themselves, but are a sample from a larger population. Then the factor is called **random**.

A common example is subject. For example suppose a large data-entry company is interested in comparing three keyboard designs for use with computers. The company selects 20 employees at random to test the keyboards. Each employee, or subject, tests each keyboard once. Keyboard design is called a fixed factor because we are interested in these three designs. Subject is called a random factor because we are not interested in these 20 subjects but want to generalize to the population of all subjects, all current and future employees who will use keyboards.

We have already discussed an example with a random factor, the randomized block design in Chapter 10. In a randomized block design the blocking factor is usually random. Randomized block designs, however, are a very simple case. The analysis is the same whether the blocking factor is fixed or random, so we did not need to discuss random factors in Chapter 10. With more complicated designs, we must know which factors are random and which are fixed in order to do the correct analysis.

Minitab's Balanced ANOVA command will help you analyze a model with both fixed and random effects (called a mixed model). The sums of squares are the same in all cases. What changes is the way you do the tests; specifically what term you use in the denominator of each F-test. In a model in which all factors are fixed, MSE is used as the denominator for all F-tests. In mixed models different terms may use different denominators for F-tests.

We will use a formal example to illustrate ANOVA. Table 16.4 shows data for a two-factor design.

In doing the analysis of variance, you'll use the Results button to display a table of expected mean squares. This is the information used to do the proper tests. The literature contains two methods for analyzing mixed models. Most books use what is called the restricted model. Minitab's default is the unrestricted model, which we will not discuss here.[†] You'll select the restricted model option. To analyze the model when both factors are fixed,

1. Enter the data in Table 16.4 into a new Minitab worksheet. Name C1 **Y**, C2 **A** and C3 **B**. Enter all 24 data points into the first 24 rows of C1. Then, in C2, enter a **1** or **2**, identifying whether the data points come from A1 or A2. In C3 enter a **1**, **2**, or **3**, identifying whether the data points come from B1, B2, or B3. (Using Autofill, p. 84, and Generating Patterned Data, p. 82, are good ways to do this.)

2. Choose **Stat ▸ ANOVA ▸ Balanced ANOVA**. Enter **Y** as the response and **A|B** as the model.

3. Click the **Results** button and then select the **Display expected mean squares and variance components** check box. Click **OK** to return to the main dialog box.

4. Click the **Options** button and then select the **Use the restricted form of the model** check box. Click **OK** to return to the main dialog box.

5. Click **OK**. Figure 16.12 shows the output.

[†] For example, see S. R. Searle, *Linear Models*: New York: John Wiley, 1971. Chapter 9 includes a discussion of these two models.

Figure 16.12   Both factors are fixed

```
ANOVA: Y versus A, B

Factor  Type     Levels  Values
A       random      2     1, 2
B       random      3     1, 2, 3


Analysis of Variance for Y

Source  DF       SS        MS        F      P
A        1   1488.38   1488.38   11.66   0.076
B        2    915.58    457.79    3.59   0.218
A*B      2    255.25    127.63    4.88   0.020
Error   18    470.75     26.15
Total   23   3129.96


S = 5.11398    R-Sq = 84.96%    R-Sq(adj) = 80.78%


                                    Expected Mean Square
                 Variance   Error   for Each Term (using
       Source    component   term   restricted model)
1  A              113.40      3     (4) + 4 (3) + 12 (1)
2  B               41.27      3     (4) + 4 (3) + 8 (2)
3  A*B             25.37      4     (4) + 4 (3)
4  Error           26.15            (4)
```

To analyze the model when A is fixed and B is random,

1. Click the **Edit Last Dialog** button.
2. Leave the model as it is, but enter **B** into the Random factors box.
3. Click **OK**. Figure 16.13 shows the output.

Figure 16.13   Factor A is fixed and factor B is random

```
ANOVA: Y versus A, B

Factor  Type     Levels  Values
A       fixed       2    1, 2
B       random      3    1, 2, 3


Analysis of Variance for Y

Source  DF       SS        MS       F      P
A        1   1488.38   1488.38   11.66  0.076
B        2    915.58    457.79   17.50  0.000
A*B      2    255.25    127.63    4.88  0.020
Error   18    470.75     26.15
Total   23   3129.96


S = 5.11398   R-Sq = 84.96%   R-Sq(adj) = 80.78%


                                   Expected Mean Square
                   Variance  Error  for Each Term (using
       Source     component  term   restricted model)
1  A                           3    (4) + 4 (3) + 12 Q[1]
2  B                53.95      4    (4) + 8 (2)
3  A*B             25.37       4    (4) + 4 (3)
4  Error           26.15           (4)
```

To analyze the model when both factors are random,

1. Click the **Edit Last Dialog** button.
2. Leave the model as it is, but enter **A** and **B** into the Random factors box.
3. Click **OK**. Figure 16.14 shows the output.

Figure 16.14   Both factors are random

```
ANOVA: Y versus A, B

Factor  Type     Levels  Values
A       random      2    1, 2
B       random      3    1, 2, 3


Analysis of Variance for Y

Source  DF      SS        MS        F      P
A        1  1488.38   1488.38   11.66  0.076
B        2   915.58    457.79    3.59  0.218
A*B      2   255.25    127.63    4.88  0.020
Error   18   470.75     26.15
Total   23  3129.96


S = 5.11398    R-Sq = 84.96%    R-Sq(adj) = 80.78%


                                   Expected Mean Square
                 Variance  Error   for Each Term (using
        Source   component  term   restricted model)
1  A              113.40     3      (4) + 4 (3) + 12 (1)
2  B               41.27     3      (4) + 4 (3) + 8 (2)
3  A*B             25.37     4      (4) + 4 (3)
4  Error           26.15            (4)
```

In the first analysis of variance in Figure 16.12, p. 487, both factors are fixed. In this case each term is tested, using MSE for the denominator of the F-statistic. This denominator is often called the error term for the test. The Expected Mean Square table actually tells you the error for each term. Here it is term 4, Error. This is how the F-statistics shown in the Analysis of Variance table are calculated.

In the second analysis of variance, A is fixed and B is random. Terms B and A*B are tested, using MSE for error, but A is not. According to the Expected Mean Square table, the proper error term is 3, the mean square for A*B or MSAB. This changes the F-statistic for A in the Analysis of Variance table. Now F = MSA/MSAB.

In the third analysis of variance A and B are both random. Terms A and B are both tested, using term 3, the MS for A*B. Thus the F-statistic for A is MSA/MSAB and the F-statistic for B is MSB/MSAB. The interaction term is still tested by using MSE.

Mixed Models, Expected Mean Squares, and Variance Components

When you use the Balanced ANOVA command, you write a "computer" version of the analysis of variance model. Let's look at the formal version of that model for the two-factor case. Suppose factor A has a levels, factor B has b levels, and each cell has n observations. In texts this model is usually written as

$$Y = \mu + a_i + \beta_j + a\beta_{ij} + \varepsilon_{ijk}, \ i = 1,\ldots, a; j = 1,\ldots, b; k = 1,\ldots, n$$

The a_i are the effects of factor A, the β_j are the effects of factor B, the $a\beta_{ij}$ are the effects of the interaction, and the ε_{ijk} are the random error. In all cases are independent random variables, with each $\varepsilon_{ijk} \sim N(0, \sigma_\varepsilon^2)$.

There are additional conditions, depending on which terms are fixed and which are random. Here are the four cases:

1. If A and B are both fixed, then

$$\sum_i a_i = 0 \qquad \sum_j \beta_j = 0 \qquad \sum_i \sum_j a\beta_{ij} = 0$$

2. If A is fixed and B is random, then

$$\sum_i a_i = 0 \qquad \sum_i a\beta_{ij} = 0 \text{ for each } j$$

β_j and $a\beta_{ij}$ are random variables with each $\beta_j \sim N(0, \sigma_\beta^2)$ and each $a\beta_{ij} \sim N(0, \sigma_{\alpha\beta}^2)$.

3. If A is random and B is fixed, then

$$\sum_j \beta_j = 0 \qquad \sum_j a\beta_{ij} = 0 \text{ for each } i$$

a_i and $a\beta_{ij}$ are random variables with each $a_i \sim N(0, \sigma_\beta^2)$ and each $a\beta_{ij} \sim N(0, \sigma_{\alpha\beta}^2)$.

4. If A and B are both random, then

a_i, β_j, and $a\beta_{ij}$ are random variables with each $a_i \sim N(0, \sigma_\beta^2)$, each $\beta_j \sim N(0, \sigma_\beta^2)$, and each $a\beta_{ij} \sim N(0, \sigma_{\alpha\beta}^2)$.

The terms σ_a^2, σ_β^2, and $\sigma_{\alpha\beta}^2$ are called **variance components**. Minitab gives estimates of these in the EMS table.

If you are familiar with the formulas for expected mean squares, you may recognize them in the output from the three different models. When we do a test, we want the expected mean square for the numerator and that for the denominator of the F-statistic to differ by just one term, the term being tested. Let's look at the second example in Figure 16.13, in which A is fixed and B is random. Shown below is Minitab's output, along with the expected mean squares as they are usually written in texts.

Source	Variance component	Error term	Expected Mean Square (using restricted model)
1 A		3	$(4) + 4(3) + 12Q[1]$ $\sigma_\varepsilon^2 + 4\sigma_{\alpha\beta}^2 + 12\sum_i a_i^2/(a-1)$
2 B	53.95	4	$(4) + 8(2)$ $\sigma_\varepsilon^2 + 8\sigma_\beta^2$
3 A*B	25.37	4	$(4) + 4(3)$ $\sigma_\varepsilon^2 + 4\sigma_{\alpha\beta}^2$
4 Error	26.15		(4) σ_ε^2

Minitab uses parentheses to represent a variance component. For example, (3) represents the third variance component, $\sigma_{\alpha\beta}^2$. Minitab uses square brackets to represent a quadratic expression involving the constants for a fixed term. For example, Q[1] represents $\sum_i a_i^2/(a-1)$.

The hypotheses for the three tests can be expressed as follows:

Test for A H_0: $a_1 + a_2 + \ldots + a_a = 0$ or, equivalently, $\sum_i a_i^2 = 0$

Test for B H_0: $\sigma_\beta^2 = 0$

Test for A*B H_0: $\sigma_{\alpha\beta}^2 = 0$

If H_0 for A is true, the EMS for A and for A*B are the same. Therefore the F-statistic for A is MSA/MSAB. If H_0 for B is true, the EMS for B and Error are the same. Therefore the F-statistic for B is MSB/MSE. If H_0 for A*B is true, the EMS for A*B and Error are the same. Therefore the F-statistic for A is MSAB/MSE. The variance component, 53.95, listed next to factor B on the output, is an estimate of σ_β^2. Similarly 25.37 is an estimate of $\sigma_{\alpha\beta}^2$ and 26.15 is an estimate of σ_ε^2.

The estimates of the variance components are the usual unbiased analysis of variance estimates. They are obtained by setting each calculated MS on the output equal to the formula for its expected mean square. This gives a system of linear equations in the unknown variance components. This system is then solved. Unfortunately this method can result in negative estimates, which should be set to zero. Minitab, however, prints the negative estimates because they sometimes indicate that the model being fitted is inappropriate for the data. Variance components are not estimated for terms that are fixed.

Three-Factor Mixed Model

When you have more than two factors, there may be no exact F-test; that is, you may not be able to find a term to use for the denominator of the F-test. We will look at a simple example.

A company ran an experiment to see how several conditions would affect the thickness of a coating substance it manufactures. The manufacturing process was run at three settings: 35, 44, and 52. Three operators were chosen from a large pool of operators employed by the company. The experiment was run at two different times, selected at random. At each time each operator made two determinations of thickness at each setting. Two factors, Operator and Time, are random; the other factor, Setting, is fixed. The data are shown in Table 16.4.

Table 16.4 Data for a three-factor Analysis of Variance

	Time					
	1			*2*		
	Operator			*Operator*		
Setting	*1*	*2*	*3*	*1*	*2*	*3*
35	38	39	45	40	39	41
	40	42	40	40	43	40
44	63	72	78	68	77	85
	59	70	79	66	76	84
53	76	95	103	86	86	101
	78	96	106	82	85	98

We can write the model as

$$Y_{ijkl} = \mu + T_i + O_j + S_k + TO_{ij} + TS_{ik} + OS_{jk} + TOS_{ijk} + \varepsilon_{ijkl}$$

where T = Time, 0 = Operator, S = Setting, and Y is the response, Thickness. Because Operator and Time are random, and because an interaction involving a random factor is random, and because error is always random, the following terms are random: T_i, O_j, TO_{ij}, TS_{ik}, OS_{jk}, TOS_{ijk}, and ε_{ijkl}

These terms are all assumed to be normally distributed random variables with mean zero and variances given by σ_T^2, σ_O^2, σ_{TO}^2, σ_{TS}^2, σ_{OS}^2, σ_{TOS}^2, and σ_ε^2.

These variances are called variance components. The expected mean square output contains estimates of these variances.

In the restricted model any term that contains one or more subscripts corresponding to fixed factors is required to sum to zero over each fixed subscript. In the example there is one fixed factor, namely, Setting. Thus

$$\sum_k S_k = 0 \qquad \sum_k TS_{ik} = 0 \qquad \sum_k OS_{jk} = 0 \qquad \sum_k TOS_{ijk} = 0$$

Now let's produce an analysis of variance using the restricted model. To do the analysis of variance,

1. Enter the data in Table 16.4 into a new worksheet. (Again Autofill, p. 84, and Generating Patterned Data, p. 82, are good ways to enter the factor levels.
2. Name C1 **Thickness**, C2 **Time**, C3 **Operator**, and C4 **Setting**.
3. Choose **Stat ▸ ANOVA ▸ Balanced ANOVA**.
4. Enter **Thickness** as the response and enter **Time|Operator|Setting** as the model.
5. Enter **Time Operator** as the random factors. The restricted Setting should still be enforced, as should the storage of the expected mean square.
6. Click **OK**. Figure 16.15 shows the output.

The *F*-tests for all terms are given. In each case the appropriate denominator is shown in the expected mean square table. Note that the *F*-test for the factor Setting is inexact, as Minitab tells you in the footnote. How did Minitab get this answer?

Let's look at the main effect for Setting. The expected mean square for Setting is

$$(8) + 2(7) + 4(6) + 6(5) + 12Q[3] = \sigma_\varepsilon^2 + 2\sigma_{TOS}^2 + 4\sigma_{OS}^2 + 6\sigma_{TS}^2 + 12[\sum_k S_k^2]/(a-1)$$

We remove the term for the Setting factor, $12\sum_k S_k^2/(a-1)$, to get the denominator for the test. This gives

$$(8) + 2(7) + 4(6) + 6(5) = \sigma_\varepsilon^2 + 2\sigma_{TOS}^2 + 4\sigma_{OS}^2 + 6\sigma_{TS}^2$$

Figure 16.15 Analysis of variance when there is no exact *F*-test

```
ANOVA: Thickness versus Time, Operator, Setting

Factor      Type     Levels  Values
Time        random       2   1, 2
Operator    random       3   1, 2, 3
Setting     fixed        3   35, 44, 53


Analysis of Variance for Thickness

Source                  DF      SS       MS      F       P
Time                     1     9.0      9.0   0.29   0.644
Operator                 2  1120.9    560.4  18.08   0.052
Setting                  2 15676.4   7838.2  55.84   0.001 x
Time*Operator            2    62.0     31.0   9.15   0.002
Time*Setting             2   114.5     57.3   2.39   0.208
Operator*Setting         4   428.4    107.1   4.46   0.088
Time*Operator*Setting    4    96.0     24.0   7.08   0.001
Error                   18    61.0      3.4
Total                   35 17568.2

x Not an exact F-test.


S = 1.84089   R-Sq = 99.65%   R-Sq(adj) = 99.32%


                            Variance Error  Expected Mean Square for Each
     Source                 component term  Term (using restricted model)
1  Time                       -1.222    4   (8) + 6 (4) + 18 (1)
2  Operator                   44.120    4   (8) + 6 (4) + 12 (2)
3  Setting                              *   (8) + 2 (7) + 4 (6) + 6 (5) + 12
                                            Q[3]
4  Time*Operator               4.602    8   (8) + 6 (4)
5  Time*Setting                5.542    7   (8) + 2 (7) + 6 (5)
6  Operator*Setting           20.778    7   (8) + 2 (7) + 4 (6)
7  Time*Operator*Setting      10.306    8   (8) + 2 (7)
8  Error                       3.389        (8)


* Synthesized Test.


Error Terms for Synthesized Tests

                              Synthesis of
Source     Error DF  Error MS  Error MS
3 Setting      4.24    140.4   (5) + (6) - (7)
```

There is no term in the model that has this as its expected mean square. Therefore Setting has no exact *F*-test. Minitab "synthesizes" the test by finding a linear combination of terms in the model that gives this denominator. The combination that does it is

$$\text{MSTS} + \text{MSOS} - \text{MSTOS} = [(8) + 2(7) + 6(5)] + [(8) + 2(7) + 4(6)] - [(8) + 2(7)]$$
$$= (8) + 2(7) + 6(5) + 4(6)$$

Thus

$$F = \text{MSS}/[\text{MSTS} + \text{MSOS} - \text{MSTOS}]$$
$$= 7838.2/[57.3 + 107.1 - 24.0]$$
$$= 140.4$$

This gives an approximate F-test. Now we need the degrees of freedom for the denominator. For a linear combination of the form $\sum_i a_i \text{MS}_i$ we use the approximation

$$\text{df(denominator)} = \frac{[\sum_i a_i \text{MS}_i]^2}{\sum_i [(a_i \text{MS}_i)^2 / \text{df}_i]}$$

Using this formula to calculate the degrees of freedom for the test on Setting, Minitab calculates

$$\frac{[\text{MSTS} + \text{MSOS} - \text{MSTOS}]^2}{\left[\frac{(\text{MSTS})^2}{\text{df}_{TS}}\right] + \left[\frac{(\text{MSOS})^2}{\text{df}_{OS}}\right] + \left[\frac{(\text{MSTOS})^2}{\text{df}_{TOS}}\right]} = \frac{[57.3 + 107.1 - 24]^2}{\left[\frac{57.3^2}{2}\right] + \left[\frac{107.1^2}{4}\right] + \left[\frac{24^2}{4}\right]}$$

Minitab reports a value of 4.24. This small discrepancy is due to the fact that we are using rounded numbers in our calculations, and Minitab is using numbers whose accuracy extends to many more decimal places. Thus to be conservative, we should truncate our value to 4.

Exercises

Exercises for Section 16.1

16-1   Use the One-way command with the Tukey comparison selected to investigate differences due to temperature, using the Plywood data in Exercise 10-2, p. 319. Compare the analysis of variance tables, the display of individual confidence intervals, and the results from Tukey's test.

16-2   Exercise 10-9, p. 322, shows data from an experiment to study crop damage due to a beetle. There are two factors, attractant and disease, each with two levels. You can use One-way to study the individual cells in a two-way design if you make each cell a separate level. Create a new factor, called treatment, which is 1 if attractant = yes and disease = yes, 2 if attractant = yes and disease = no, 3 if attractant = no and disease = yes, and 4 if attractant = no and disease = no.

(a) Use One-way to analyze the beetle data in Exercise 10-9, p. 322, if you have not already done so.

(b) Use One-way, with the Tukey comparison selected, to analyze the beetle data. Use the factor treatment. Compare the results to those in part (a).

(c) Level 4 of treatment could be considered the control since neither attractant nor disease pellets were used. Analyze the data, using Dunnett's procedure. How do the results compare to those in parts (a) and (b)?

16-3   Table 10.3, p. 305, shows data from a two-way design used to study driving a car under different conditions.

(a) Analyze the data as a two-way design if you have not already done so.

(b) Use the technique described in Exercise 16-2 to convert the data to a one-way analysis. Use One-way with the Tukey procedure. Interpret the results and compare them to those in part (a).

Exercises for Section 16.2

16-4   In this exercise we will do further analyses using the Potato data, described in Exercise 10-11, p. 323.

(a) Use Balanced ANOVA to store the fits and residuals so you can analyze the residuals. Some possible displays are (1) a histogram of the residuals, (2) a plot of the residuals versus the fitted values, and (3) plots of the residuals versus each factor. Do you see any problems?

(b) Figure 16.2, p. 470, showed that there was one significant interaction. Investigate this further. Use the Cross Tabulation command to get cell means. Use appropriate plots (either hand-drawn or done with Minitab) to display your results.

16-5   Plywood is made by cutting a thin, continuous layer of wood off logs as they are spun on their axis. Several of these thin layers then are glued together to make plywood sheets. Considerable force is required to turn a log hard enough for a sharp blade to cut off a layer. Chucks are inserted into the centers of both ends of the log to apply the torque necessary to turn the log.

A study was done to determine how various factors affect the amount of torque that can be applied to a log. There are three factors: the diameter of the test logs; the distance the chucks were inserted into the logs; and the temperature of the logs at the time they were tested. For each treatment combination, 10 trials were done. On each trial the average torque that could be applied to the log before the chuck spun out was recorded. Then the torque for the 10 trials was averaged. This average is the response variable. The data are shown Table 16.5 and are stored in the file Plywood.

Table 16.5 Plywood data

Diameter (inches)	Penetration (inches)	Temperature (°F)	Torque	Diameter	Penetration	Temperature	Torque
4.5	1.00	60	17.30	7.5	1.00	60	29.55
4.5	1.50	60	18.05	7.5	1.50	60	31.50
4.5	2.25	60	17.40	7.5	2.25	60	36.75
4.5	3.25	60	17.40	7.5	3.25	60	41.20
4.5	1.00	120	16.70	7.5	1.00	120	23.20
4.5	1.50	120	17.95	7.5	1.50	120	25.90
4.5	2.25	120	18.60	7.5	2.25	120	35.65
4.5	3.25	120	18.55	7.5	3.25	120	37.60
4.5	1.00	150	15.75	7.5	1.00	150	22.55
4.5	1.50	150	16.65	7.5	1.50	150	22.90
4.5	2.25	150	15.25	7.5	2.25	150	28.90
4.5	3.25	150	15.85	7.5	3.25	150	35.20

(a) Analyze the data, using a three-way analysis of variance. Can you fit the full model?

(b) Several interactions are significant. Investigate these further.

(c) Sometimes it is easier to look at subsets of the data separately to see what is going on, especially when interactions are present. Analyze the data separately for 4.5-inch logs and then for 7.5-inch logs. How do the results compare?

(d) Use appropriate displays to present your conclusions.

16-6   The driving corrections data presented in Table 10-3, p. 319, are part of a three-way design. The third factor was time of day. A total of 48 drivers were used; 24 drove during the day and 24 drove at night. The number of driving corrections made by each driver was recorded. The results are shown in Table 16.6.

Analyze these data. Use appropriate displays (graphs and tables) and tests.

Table 16.6 Driving correction data

	Day			Night		
	First-class road	Second-class road	Dirt road	First-class road	Second-class road	Dirt road
	4	23	16	21	25	32
Inexperienced	18	15	27	14	33	42
Driver	8	21	23	19	30	46
	10	13	14	26	20	40
	6	2	20	11	23	17
Experienced	4	6	15	7	14	16
Driver	13	8	8	6	13	25
	7	12	7	16	12	12

Exercises for Section 16.3

16-7   We will investigate two factors, Color and ED, from the Cartoon data (described in the Appendix, p. 524).

(a) Use the score on the cartoon test given immediately after presentation. Is there a difference between using color and black-and-white slides? Is the educational level of the participants important?

(b) Repeat part (a), using the score on the realistic test given immediately after presentation. Compare your results to those in part (a).

(c) Repeat part (a), using the delayed cartoon score. How do these results compare to those of part (a)?

(d) Repeat part (a), using the delayed realistic score. How do these results compare to those of part (b)?

(e) Many people did not take the delayed test. Do you think this affects your analysis and conclusions? Use appropriate tests and displays to investigate.

16-8   Researchers at Penn State University studied the feasibility of using a variety of fast-growing poplar trees as a renewable energy source. In an effort to determine how to maximize yield, the researchers designed an experiment to study three factors: site, irrigation, and fertilizer. Two sites were used: site 1 had rich, moist soil, and site 2 had dry, sandy soil. Each site was divided into four areas. Area 1 received no irrigation and no fertilizer; area 2 received only fertilizer; area 3 received only irrigation; and area 4 received both fertilizer and irrigation. Trees were grown under each treatment combination and harvested. The dry weight of the wood in each tree was measured in kilograms. Data are shown in Table 16.7

and stored in the Poplars worksheet.

Analyze the data as a three-way design, using General Linear Model. What conditions maximize yield?

Table 16.7 Poplar data

Site	No irrigation		Irrigation	
	No fertilizer	Fertilizer	No fertilizer	Fertilizer
Rich and Moist	2.59	3.56	2.83	4.28
	1.89	5.69	1.70	2.36
	1.61	1.54	2.63	2.85
	1.92	2.86	2.48	3.83
	1.09	2.99	2.34	4.79
	1.32	2.11	0.99	4.33
	0.64	2.73	0.39	2.12
	0.16	1.85	3.24	3.83
		0.66		
		1.93		
Dry and Sandy	1.24	3.82	2.26	3.37
	3.24	1.10	2.17	4.34
	2.51	4.16	2.84	2.90
	2.68	4.92	1.68	5.21
	1.71	4.38	0.35	5.12
	2.61	1.10	2.71	6.19
	0.25	4.49	0.99	4.80
	0.82	1.80	1.14	
	2.27		2.36	
			4.05	

Exercises for Section 16.4

16-9   In this exercise we will repeat the Cartoon analysis done in the text, using only data for the professionals (ED = 1).

(a) Use a one-way analysis of variance to see whether there is a difference between color and black-and-white cartoon slides. Use only the Cartoon data for the professionals, using Subset Worksheet as necessary.

(b) Use the Otis score as a covariate. Compare the results to those in part (a).

(c) Calculate the regression line for each level of the factor Color.

16-10   (a) Repeat the analysis of Exercise 16-9, using only the data for students.

(b) Compare the results to those of Exercise 16-9. Investigate any differences.

16-11   The Furnace1 worksheet contains data from the study of an energy-saving device for furnaces. Use analysis of covariance to see whether the device made a difference.

17

Creating a Statistical Report

Once you have analyzed your data with both graphical and numerical methods, you would like to share your results with your colleagues or instructor. Minitab provides a simple word processing tool, the ReportPad, into which you can easily import Minitab graphs and numerical results. You can then add your own comments, change fonts, and print the report. Alternatively, you can export the report in RTF format (Rich Text Format) to a more sophisticated word processor, such as Microsoft Word® or Adobe FrameMaker®, or in HTML format (HyperText Markup Language) destined for a Web page. The ReportPad is part of Minitab's Project Manager. You may want to review the material on the Project Manager on p. 36 before continuing here.

We will illustrate some of the possibilities using the Pulse data set described in the Appendix, p. 529.

17.1 Adding Session Output to the ReportPad

First we will reset Minitab to its defaults and open the Pulse data file.

1. Either restart Minitab or else choose **File ▸ New ▸ Minitab Project** and click **OK**.
2. Open the Pulse Worksheet by choosing **File ▸ Open Worksheet** and select **Pulse** (*not* Pulse1). (Review p. 35, if needed.)

Now let's produce the descriptive statistics for the Height variable breaking it down by Sex. Here are the steps:

1. Choose **Stat ▸ Basic Statistics ▸ Display Descriptive Statistics**.
2. Enter **Height** as the variable and **Sex** as the By variable.
3. Click **OK**.

Minitab displays the descriptive statistics for the Height variable in the Session Window. We would like to include these in our report. To add these results to the ReportPad,

1. First make sure the blinking text cursor is somewhere within the Section of Session Window results that you wish to copy.

2. **Right-click** anywhere within the descriptive statistics text in the Session Window. A menu of choices appears as shown to the right.

3. Click the **Append Section to Report** choice and Minitab adds the results to the ReportPad.

       💡 *TIP Make sure you have not highlighted any text before you right-click. If so, you will have the choice of adding **just** the selected lines to the ReportPad.*

To see what is in your report you will need to open the ReportPad. There are several ways to do this. One way is to first open Minitab's Project Manager. Recall that to open the Project Manager you may use any one of the following:

- Choose Window ➤ Project Manager
- Click the Project Manager button 🔲 on the separate Project Manager toolbar
- Click the restore button on the Project Manager minimized title bar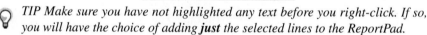
- Use the keyboard shortcut, Ctrl+I (I for Information)

Once the Project Manager is open you will see the ReportPad folder listed in the left pane. Click on that folder to the see the contents of the ReportPad.

       💡 *TIP If the Project Manager Toolbar has been enabled, you can click on the ReportPad button, 🔲, to go directly to the ReportPad. Recall that you may enable any of the toolbars by selecting them from Tools ➤ Toolbars.*

At this point your report should appear as in Figure 17.1.

Figure 17.1   Minitab ReportPad with descriptive statistics

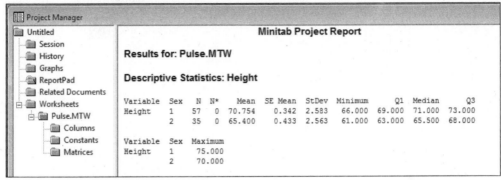

Sometimes you want to copy only selected lines of output from the Session Window. For example, you may want your report to contain basic regression results, but not all of the diagnostics Minitab often displays. To append selected lines from the Session Window to the ReportPad,

1. Drag with the mouse to **Highlight** the lines that you want to append.
2. **Right-click** anywhere in the Session Window. A menu of choices appears.
3. Click the **Append Selected Lines to Report** choice and Minitab adds the selected lines to the end of the report in the ReportPad.

17.2 Adding a Graph to the ReportPad

Let's create dotplots of the Height variable separately for males and females and add the graph to our report. Here is review from Chapter 4, p. 105, for creating the graph.

1. Choose **Graph ▸ Dotplot ▸ One Y, With Groups** Click **OK**. See Figure 4.1, p. 103.
2. Select **Height** as the Graph variable.
3. Select **Sex** as the categorical variable for grouping and click **OK**.

The two dotplots appear in their own Graph Window. To add the graph to your report,

1. **Right-click** anywhere within the Graph Window.
2. Choose **Append Graph to Report** from the menu that appears.

The dotplot graph is added at the end of the report as shown in Figure 17.2.

Figure 17.2   Minitab ReportPad with descriptive statistics and dotplots

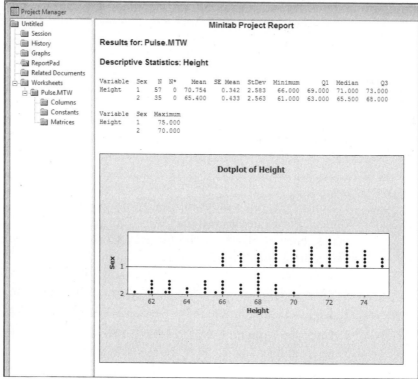

Graph Size

Graphs either printed from Minitab or appended to the ReportPad are, by default, six inches wide by four inches tall. To change this size,

1. Choose **Tools ▸ Options**.
2. Click on the **plus sign (+)** beside Graphics in the left pane.
3. Click on **Graph Size**.
4. Enter your desired sizes in the True Size (inches) text boxes for Width and Height. For example, **3** and **2** to obtain smaller graphs for your report but with the same aspect ratio, height to width.

 *TIP This will change **all** future graph sizes until you change it back.*

TIP If you make graphs smaller, Minitab will also make symbols and labels proportionately smaller. You may want to increase their sizes either before or after making the graph smaller.

You may also change the sizes of graphs using the Embedded Graph Editor after they are appended to a report in the ReportPad or transferred to a word processor. More information is given in the next Section.

17.3 Editing in ReportPad

Text Editing

As with any editor or simple word processor, the ReportPad allows you to

- insert, delete, copy, and paste text
- change fonts, font sizes, and colors
- change font styles, such as italic and bold
- alter alignment such as left, right, and center

By default, Minitab enters a report title of **Minitab Project Report**. As we have done here, Minitab uses an 11 point, bold, Arial typeface and centers the text. For our report, let's change this text to Analysis of Students' Height, keep it in the bold Arial font, but increase the size to 14 points and change the color to blue. To do so,

1. Highlight the default title, **Minitab Project Report**, by dragging across it or by triple-clicking it.
2. Type **Analysis of Students' Height**. This text replaces the original text.
3. **Highlight** your new title. (It will still be 11 point, bold, black, Arial.)
4. **Right-click** the report and select **Font** from the menu that appears.
5. Select **14** from the Size list and **Blue** from the Color drop-down list. Leave the Font and Font Style as is. See Figure 17.3.[†]
6. Click **OK** and Minitab makes the changes in the report title.

 TIP Minitab results that are displayed in table format such as descriptive statistics, cross tabulations, and regression and ANOVA output, should be kept in a fixed width font such as Courier. Otherwise, the columns in the table will not line up properly. Alternatively, you could use your word processor to convert it to a fully formatted table with any fonts you chose.

[†] The font choices available will depend on the fonts installed on your system.

Figure 17.3   Font dialog box

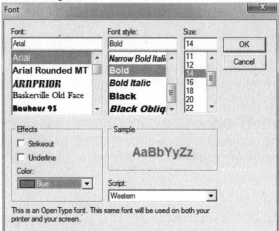

Graph Editing

Minitab's Embedded Graph Editor can be used to alter Minitab graphs after they have been appended to the ReportPad or copied to another word processor. This editor is a separate application from the Minitab data analysis software.

Let's change the background color of the graph to light blue. To do so,

1. Make sure the ReportPad is active.
2. **Double-click** on the graph in the ReportPad. Minitab opens the Embedded Graph Editor with the dotplot graph displayed.
3. **Double-click** the default grey background of the graph.
4. Make sure the **Graph Attributes** tab is selected.
5. Under Fill patterns, select **Custom**.
6. Use the Background color drop-down list to select **Light Blue** from the many choices available. See Figure 17.4.
7. Click **OK** and Minitab makes the change to the graph.
8. Select **File ► Exit and Return**. The edited graph appears in the ReportPad.

 *TIP When you use the Embedded Graph Editor to edit a graph in ReportPad or in an external word processor, the original graph is **not changed**.*

Double-clicking on other graph elements brings up more dialog boxes for editing those elements. In our example it would be good to edit the Sex labels and convert them from 1 to Male and 2 to Female.

Figure 17.5 shows our report after editing the graph background color, size, and labels. Consult the online Minitab Help for more information on graph editing and the Embedded Graph Editor.

Figure 17.4   Edit Graph and Figure Regions dialog box

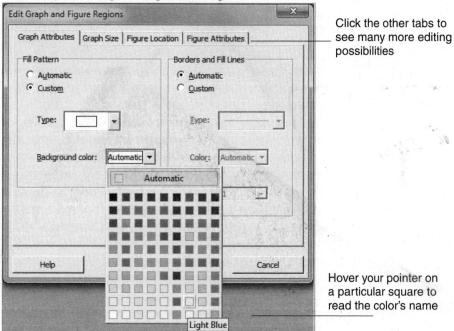

Click the other tabs to see many more editing possibilities

Hover your pointer on a particular square to read the color's name

Figure 17.5   ReportPad after substantial graph editing

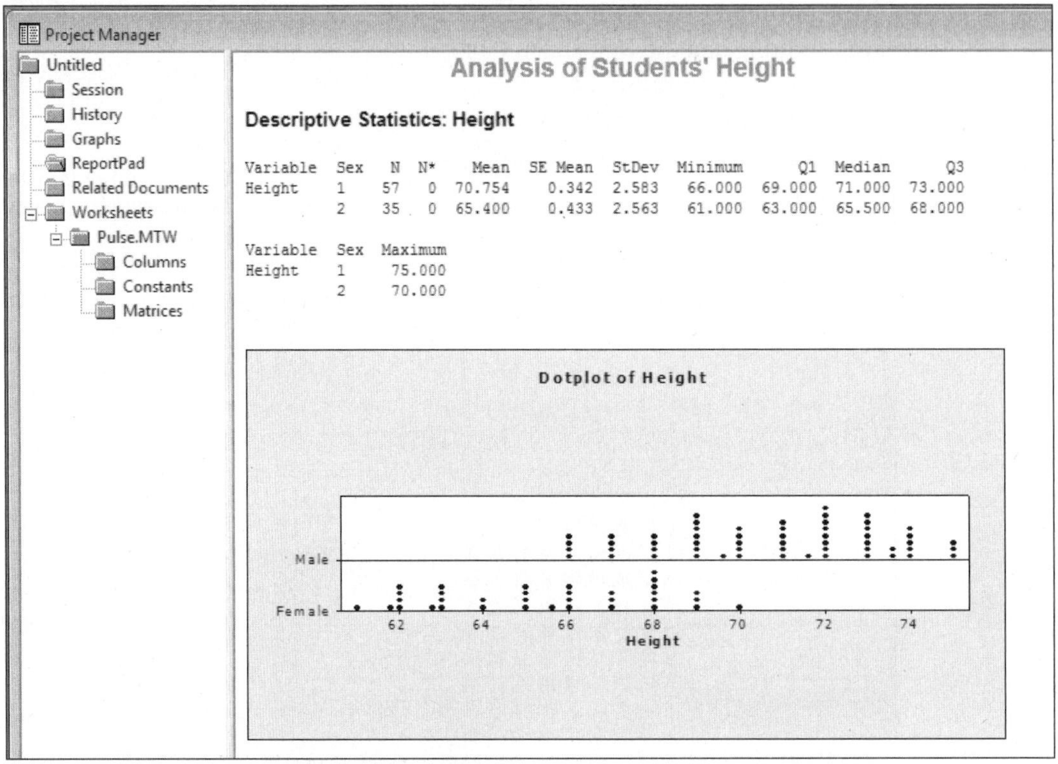

17.4 Saving a Report

The contents of the ReportPad can be saved as either RTF format (Rich Text Format) for use in a more sophisticated word processor or in HTML format (HyperText Markup Language) for use on a Web page. Let's save our report for use on a Web page.

1. Make the ReportPad active by clicking on it or by clicking the ReportPad button, ⊞ .
2. Choose **File ▸ Save Report As**.
3. Select **Web Page** from the Save As Type drop-down list.
4. Minitab puts .htm into the filename text box. Type a name of your choice into File name box retaining the .htm extension.
5. **Navigate** to a suitable place to save the file and click **Save**.

You may view the file with any Web browser and add it to other Web pages with a text editor or Web page editor.

 TIP Any graphs in your report will be saved as jpeg files in the same directory. They will have cryptic names such as TM00000700.JPG. If you move your HTML file, you must move these files to the same place.

17.5 Copying a Report to a Word Processor

With a true word processing application, you have many more formatting options for your report. For example, you can convert Minitab tables to better looking tables using any available font and you can place graphs side by side.

There are three ways to transfer the ReportPad contents to a word processing program while keeping its current formatting intact.

- Save the ReportPad contents in RTF format and then open the RTF document in the word processor
- Copy the ReportPad contents and then Paste into the word processor
- Move the ReportPad contents to the word processor deleting it from ReportPad

To copy the contents to a word processor,

1. **Right-click** the ReportPad folder in the left pane of the Project Manager. (Be sure to click the ReportPad *folder*, not the Report itself.)
2. Select **Copy to Word Processor** from the choices given.
3. Select a place to **Save** your report file. The default name of the file will be Untitled1.rtf. You may choose another name if you wish. Click **OK**.
4. Your default word processing program opens with the ReportPad contents displayed.

The word processor file will have a default filename of Untitled1.rtf (or Untitled2.rtf if Untitled1.rtf already exists). You can now edit the document further and save it with a more appropriate file name and in any format permitted.

 TIP Your default word processor is the one that Windows associates with RTF file types.

Graphs displayed in your word processor can still be edited with the Embedded Graph Editor. Double-click the graph and the Minitab graph editing tools and menus replace the menus of the word processor. Edit the graph as in the previous section. When you have finished editing the graph, click outside the graph and the word processor menus reappear.

Exercises

17-1   Use the Pulse data set.
    (a) Compute descriptive statistics of both first and second pulse rates for all the students.
    (b) Transfer the descriptive statistics to the ReportPad and write a short report to interpret the results.
    (c) Print the report.

17-2   Use the Pulse data set.
    (a) Display dotplots of both first and second pulse rates for all the students in the same graph.
    (b) Transfer the plots to the ReportPad and write a short report to interpret the dotplots.
    (c) Print the report.

17-3   Use the Pulse data set.
    (a) Compute descriptive statistics of both first and second pulse rates for all the students.
    (b) Display dotplots of both first and second pulse rates for all the students in the same graph.
    (c) Transfer the descriptive statistics and dotplots to a word processor and write a short report to interpret the results.
    (d) Print the report.

17-4   Use the ReportPad and Embedded Graph Editor to edit the default graph shown in Figure 17.2, p. 504, to get the graph shown in Figure 17.5, p. 508. You will need to change the size of the graph and edit the default title and several of the graph labels.

18

Minitab Macros

Minitab is normally used interactively—each command is carried out as soon as you click OK in a dialog box or type the command in the Session Window and press the Enter key. But you may extend Minitab's functionality through Minitab macros to carry out tasks such as displaying a special graph, or to automate repetitive tasks.

Three types of macros are available in Minitab:

- Global macros are a form of macro which uses your current Worksheet for work space.
- Local macros are a more advanced form of macro which can use their own local Worksheet in addition to the current Worksheet.
- Execs are text files containing a series of Minitab commands. They are an older form of Minitab macro.

Both global and local macros allow you to create a program of Minitab commands and to use control statements such as loops and IF statements. Both types also allow you to invoke other macros from within a macro. Since macros are based on Session commands you may want to review the command syntax given in Chapter 2 before proceeding further.

A Minitab macro is a text file that you create with a simple text processor, such as Notepad.[†] Once you create a macro file, you invoke it in the Session Window or in the Command Line Editor by typing the percent sign, %, followed by the macro file name, as in %Monthly. Global macros can be designed to request additional information from the user after they are invoked. Local macros require more memorization as the user needs to know exactly what information they require *before* they are invoked. Local macros operate very much like built in Minitab commands.

In this *Handbook* we will concentrate on studying Global macros. See the Minitab Help system for information on Local macros and Execs.

[†] Notepad™ is a simple text editor that comes with Microsoft Windows. You can open it by clicking Minitab's Tools menu. Any word processor could be used but you must remember to save the file as a text (ASCII) file.

18.1 Global Macros

Every global macro begins with the statement GMacro and ends with the statement EndMacro. GMacro must be the first line of your macro because it labels the macro type as global, not local. The basic structure of a GMacro file is:

- GMacro statement
- Template (usually just the macro filename, such as Monthly)
- Body of the macro
- EndMacro statement

The body of the macro can contain

- Minitab commands
- Special Macro statements (such as IF, THEN, PAUSE, CALL and GOTO)
- Invocation of other global macros

Figure 18.1 shows an example of a very simple global macro. When it is run, Minitab will generate 50 samples, each of size 9, from a normal distribution with mean 80 and standard deviation 5. Then it will compute the sample means of the samples and graph the histogram of the sample means. All of the statements following the pound sign, #, are ignored by Minitab. They serve to document the macro so that you or someone else can understand what you are trying to do.

Figure 18.1 A simple global macro

```
GMacro              # Mandatory first statement
Means               # Template that is only used in local macros
Random 50 c1-c9;    # Generate 50 normal samples in C1-C9
 Normal 80 5.       # Use a mean of 80 and sigma of 5
RMean c1-c9 c10     # Compute sample means and store in C10
Histogram c10       # Display the histogram of the sample means
EndMacro            # Mandatory last statement
```

If we save these lines into a text file named Means.mac, for example, we could execute all of these commands by simply typing %Means at the Minitab prompt, MTB >, in the Session Window.

18.2 Writing a Macro

When you write a macro, you store Minitab commands in a text file. An instructor might store commands to do a complicated exercise; then students would not have to type them. A researcher may have an analysis that she does each week, when she gets a new data set. If she stores the commands to do the analysis, she will not have to retype them each week.

Writing a macro will usually involve some "trial-and-error." You will try some commands in interactive Minitab to make sure they will do what you want. If you have Enabled Commands, Minitab will display the commands in the Session Window even if you are using the menus to submit the commands. This is a good way to learn how the commands work. Once you have a better idea of what commands you will need, you can copy-and-paste from either the Session Window, or, better still, from the History folder, into your text editor. The History folder lists all of the commands and subcommands that you have used but none of the output is listed there. You will nearly always start with a very simple macro, add features, clean up output, enhance graphics, and so forth, as you develop the macro.

To illustrate macro writing, suppose we want to develop a Minitab macro that simulates bivariate data with a specified correlation coefficient. We use the statistical theory that says that if X and Y are independent standard normal variables, then X and $Z = \rho X + \sqrt{1 - \rho^2} Y$ have a bivariate normal distribution with zero means, unit variances, and correlation coefficient ρ.

To start we will suppose that we want 30 pairs with correlation coefficient $\rho = 0.7$. (However, we know that we will eventually want to generalize to arbitrary sample sizes and correlation coefficients.)

We can do this with two commands:

```
Random 30 C1 C2
Let C2 = 0.7*C1+Sqrt(1-0.7*0.7)*C2
```

Suppose now we want to specify the value for the correlation coefficient in a Worksheet constant before we run the macro. We can replace 0.7 by a constant, K1, as in

```
Random 30 C1 C2
Let C2 = K1*C1+Sqrt(1-K1*K1)*C2
```

If we specify the value of K1 before we execute the macro we will generate pairs with correlation coefficient K1. Minitab also has a way to make the macro much more interactive. The following commands will request the user to enter a value for ρ when the macro runs. We indented the subcommand just for increased readability. (a good practice to follow.)

```
Set C1;
 File "terminal".
End
```

At this point Minitab will display the Data > prompt. Whatever the user types will be stored in column C1. To ensure that just one number is entered, it is good practice to include another subcommand, NObs 1, just after the File "terminal" subcommand.[†] The four lines would then read:

```
Set C1;
 File "terminal";
 NObs 1.
End
```

Notice that now a semicolon follows "terminal" to indicate that there are more subcommands to follow. A period follows NObs 1 to indicate that it is the last subcommand.

However, when these commands execute, all the user will see displayed in the Session Window will be Data > and nothing will happen until something is entered and the Enter key is pressed.

To inform the user as to what data is wanted, we need to precede the Set command by a command which displays an informative message. The Note command serves this purpose. Whatever text follows the Note command will be displayed on the screen when the Note command is encountered. Here we could use

```
Note Enter the required correlation coefficient (between -1 and
+1) and press the Enter key.
```

We should also add a note at the end to tell the user where the results are stored. Now our workable macro reads

```
GMacro
Corr
Note Enter the required correlation coefficient (between -1 and
+1) and press the Enter key.
Set C1;
 File "terminal";
 NObs 1.
End
Let K1 = C1
Random 30 C1 C2
Let C2 = K1*C1+Sqrt(1-K1*K1)*C2
Note The Pairs are in C1 and C2.
EndMacro
```

[†] NObs stands for Number of Observations.)

It would also be nice to have the sample size input by the user, to have the sample correlation coefficient calculated, and have the scatterplot produced. Figure 18.2 displays a listing of a complete macro that will carry out these tasks. Use the Minitab Help system to discover the use of any of the commands which we have not discussed in this *Handbook*. An excellent way to learn about macros is to study macros written by others line-by-line.

Figure 18.2 A Minitab macro to simulate data with a given correlation coefficient

```
GMacro # Macro name: CORR.MAC for Release 14 and later
Corr    # Written by Jon Cryer    Jon-Cryer@uiowa.edu
NoTitle
Note This macro will simulate data pairs with your specified
Note correlation coefficient. The variables will each have zero mean
Note and unit standard deviation.
Note (Uses C1, C2, K1, and K2)
Note
Note Enter the required correlation coefficient (between -1 and +1)
Note and press the Enter key.
Set C1;
 File "terminal";
 NObs 1.
End
Let K2=C1
Note
Note Enter the number of data pairs desired and press Enter.
Set C1;
 File "terminal";
 NObs 1.
End
Let K1=C1
Random K1 C1 C2
Let C2=K2*C1+Sqrt(1-K2*K2)*C2
Center C1 C2 C1 C2
Name C1 'X' C2 'Y'
Sort C1 C2 C1 C2
Corr C1 C2;
 NoPValues.
Note
Note The pairs are placed in C1 and C2 sorted by C1.
Plot 'Y'*'X'
Erase K1 K2
EndMacro
```

18.3 Invoking a Macro

To invoke, or run, a global macro from Minitab, enter the percent symbol % followed by the macro filename after the Minitab prompt, MTB >, in the Session Window. For example, to invoke a macro file named Corr.mac, enter the command: %Corr.

Notes on invoking macros

- When you invoke a macro, by default Minitab looks for the macro file first in the current directory, then in the Macros subdirectory. If the macro is not in one of those default directories, you must specify the device and directory by including a full directory path when you invoke the macro. For example, %E:\MyMacros\Corr.

- If the macro filename or path includes any spaces, you must put the path and name in single or double quotes. For example, %'C:\My Documents\My Macros\Corr'

- The default filename extension for macros is mac. When you invoke a macro that has an extension of mac, you only need to type the file name, as in %Corr. If the extension is not mac, you must type both the file name and the extension, as in %Corr.txt.

18.4 Variable Column and Row Numbers

Column Numbers

When you write a macro that operates on data in the active Worksheet, you need to know where that data is located. Sometimes it might be in column C1, other times in C9. Fortunately, Minitab macros have a special feature, called the **CK capability**, that allows you to use a variable column number.

Any place where a column name is used in a command, such as in `Mean C3`, you may use CK1 (or CK2, CK3, or...) to name the column. The constant K1 (or K2, K3, or...) serves as a placeholder. When Minitab executes the command, K1 is replaced by its current value. If K1 is equal to 3, `Mean CK1` calculates the mean of column C3.

As a very simple example, suppose the macro is supposed to display a dotplot of the data in some column. (Obviously, we don't need a macro to do this! This is only an illustration.) Here are the commands that would accomplish that task:

```
Gmacro
MyDotplot
Note Enter the number for the data column.
Note(For example, enter 1 for C1, 2 for C2,...)
Set C100
 File "terminal";
 NObs 1.
End
Let K100=C100
Dotplot CK100
EndMacro
```

 TIP Since this macro operates on the user's data, it is good practice to use columns and constants within the macro that would not likely infringe on the user's data, for example C100 and K100. It is also good practice to provide a **Note** *at the beginning of the macro alerting the user that the macro will use these columns and constants. If necessary, they could abandon the macro if it were to mess up some of their data. Figure 18.2 shows an example of this kind of note.*

TIP If Minitab is displaying the Data > prompt but you wish to abandon the macro, type **End** *and press the Enter key.*

Row Numbers

Individual elements of rows in a column may be accessed using the **Let** command as the following examples show.

```
Let C2(3) = 5 + K1
Let C1 = C2(3)*C5
```

In the first example, the sum, 5 + K1, is put into row 3 of C2. The rest of C2 remains the same. In the second example, all the rows of C5 are multiplied by the number in the third row of C2.

Here is a fragment of a macro showing how this feature might be used.

```
Note Enter the sample size and number of replications.
Set C100;
 File "terminal";
 NObs 2.
Let K100=C100(1)    # sample size
Let K101=C100(2)    # number of replications
```

18.5 Control Statements

Control statements allow you to control the sequence in which commands in a macro are executed. For example, if you want the macro to

- perform some action a set number of times, use a `Do-EndDo` loop
- perform some action only if some condition is true or false, use an `If` statement
- repeat a block of commands as long as some condition is true, use a `While-EndWhile` loop.
- run another macro from within your macro, use `Call` and `Return`

The following commands are available to let you control macros:

```
Do Enddo
If ElseIf Else EndIf
While Endwhile
Next
Break
Goto Mlabel
Call Return
YesNo
Exit
```

Here is a fragment of a macro that calculates the running means and variances of the data in column C1. It illustrates a `Do-EndDo` loop and uses a varying constant, K1, to reference row numbers in C2 and C3.

```
Let K2=N(C1)      # get total sample size
Let C2(1)=C1(1)  # initial values for mean
Let C3(1)=0       # and variance
Do K1=2:K2
  Let C2(K1)=(1-1/K1)*C2(K1-1)+C1(K1)/K1           # update mean
  Let C3(K1)=(1-1/(K1-1))*C3(K1-1)+(C1(K1)-C2(K1-1))**2/K1
                                                    # update variance
EndDo
```

When this macro finishes running, column C2(1) contains the mean of the first element of C1, C2(2) contains the mean of the first two elements of C1 and so forth. Similarly, the rows of C3 contain running variances of the consecutive elements of C1.

> 💡 *TIP If a macro seems to be running quite long, possibly in an endless loop, you can press **Ctrl+Break** at any time to abandon the macro without finishing.*

18.6 Available Macros

Here is a list of the Minitab global macros that are used in this *Handbook* and are available from the Data Library at www.CengageBrain.com associated with this book. All of the macros are interactive and request the user enter parameters, sample sizes, number of repetitions, or other required information.

Macro Name	*Description*
AreaSampler	Selects repeated random samples without replacement.
Corr	Simulates bivariate normal data with a specified correlation coefficient.
BinoNorm	Displays a binomial histogram with approximating normal curve.
GMeanCI	Displays a sequence of Student *t* confidence intervals for the mean.
MeanCI	Demonstrates the robustness of Student *t* confidence intervals.
Monthly	Displays a time series plot with special monthly plotting symbols.
PoissonNorm	Displays a Poisson histogram with approximating normal curve.
Stats	Calculates running means and standard deviations of a variable.
ZMeanCI	Demonstrates the lack of robustness of Z confidence intervals.

Exercises

18-1   Figure 18.2, p. 515, lists a complete Minitab macro. Use the Minitab Help system to discover the purpose of the `NoTitle` command used in that macro.

18-2   Figure 18.2, p. 515, lists a complete Minitab macro. Use the Minitab Help system to discover the purpose of the `Center C1 C2 C1 C2` command shown near the end of that macro.

18-3   Use the Minitab Help system to discover the function of the `YesNo` macro control command.

18-4   Use the Minitab Help system to discover the function of the `MTitle` macro command.

18-5   Use the Minitab Help system to discover the function of the `WTitle` macro command.

18-6   Write a Minitab macro that asks the user which column contains a set of data values and then "prints" those data values in the Session Window.

18-7   Write a Minitab macro that asks the user which column contains the X values and which column contains the Y values and then plots the scatterplot of the (X,Y) pairs.

18-8   Use a Do-EndDo loop that runs K1 from 1 to K2 in a macro that calculates the mean of data in C1, storing the mean in C2. Here are some commands you might use:

```
Let K2 = Count(C1)
Let C2(1) = C2(1)+C1(K1)
```

18-9   Use a Do-EndDo loop that runs K1 from 1 to K2 in a macro that calculates a moving average of length 2 for data in C1, storing the moving averages in C2. Here are some commands you can use:

```
Let K2 = Count(C1)-1
Let C2(K1) = (C1(K1)+C1(K1+1))/2
```

Save the file with name MoveAvg.mac. To test your macro, enter the following data into C1: 4 2 3 6 5 4. When Minitab runs your commands, your Worksheet should contain the values shown in Table 18.1.

Table 18.1 Moving averages

C1	C2
4	3.0
2	2.5
3	4.5
6	5.5
5	4.5
4	

Appendix

Data Sets Used in this Handbook

Many data sets are distributed with the Minitab software. Table A.1 gives a complete list. They can be opened with the File ▸ Open Worksheet command. Press the button next to **Look in Minitab Sample Data folder** near the bottom of the dialog box. Then scroll down until you find the one you are looking for.

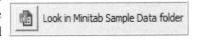

Many additional new data sets are listed in Table A.2. These can be downloaded from www.CengageBrain.com. Search for *Minitab Handbook*, sixth edition, and click on the button in the lower left labeled **Access Now**.

Longer data sets that are used throughout this book are described in this appendix. These data sets are available with the Minitab software and their contents are not displayed in this book.

Table A.1 Data sets available with Minitab as saved Worksheets

Name	Description	Text citation
ACID	Acid data set	Table 13.4, p. 420, Table 13.5, p. 421
ALFALFA	Alfalfa data	Table 10.17, p. 325
AUTO	Automobile accident data	Table 5.10, p. 179
AZALEA	Azalea data	Table 13.8, p. 425
BEARS	Data on wild bears	Exercise 1-7
BEETLE	Beetle data	Table 10.15, p. 323
BILLIARD	Billiard ball data	Table 10.6, p. 314
CAP	Cap torque data	Table 14.2, p. 447
CARTOON	Cartoon data	This appendix
CHOLEST	Cholesterol data	Table 9.1, p. 276
CITIES	Temperatures for various cities	Table 5.7, p. 177
DOCS	Documents data	Table 14.4, p. 449
DRIVE	Driving performance data	Table 10.3, p. 305, Table 16.6, p. 498
EMPLOY	Employment	This appendix, Table 5.4, p. 169

Table A.1 Data sets available with Minitab as saved Worksheets (continued)

Name	Description	Text citation
Fa	Regression data from Frank Anscombe	Table 11.12, p. 380
FABRIC	Fabric data, organized as in Table 10.1	Table 10.1, p. 300
FABRIC1	Fabric data, organized as used in Table 13.3	Table 13.3, p. 418
FALLS	Accidental deaths from falls data	Table 12.6, p. 396
FURNACE	Furnace data	This appendix
FURNACE1	Furnace data	Table 5.1, p. 146, Section 5.2, p. 155
GRADES	Full grades data set	This appendix
Ga	Sample A, first 100 observations	
Gb	Sample B, second 100 observations	
HCC	Health care center data	Table 4.5, p. 143
KEY	Ignition key data	Table 14.1, p. 428
LAKE	Lake data set	This appendix, Section 2.1, p. 43, Section 4.2, p. 101
LEAF	Leaf-surface resistance data	Table 5.11, p. 181
LIVER	Manganese in cow liver data	Table 10.13, p. 321
MAPLE	Maple trees data	Table 11.8, p. 376
MARRIAGE	Marriage rate data	Table 5.5, p. 175
MEATLOAF	Meat loaf data	Table 10.18, p. 326
OXYGEN	Oxygen in steel data	Table 10.14, p. 322
PANCAKE	Pancake data	Table 10.4, p. 309
PARKINSN	Parkinson's disease data	Table 9.3, p. 282, Table 13.2, p. 416
PENDULUM	Pendulum data, as a two-way design	Table 10.12, p. 320
PERU	Peru data	This appendix, Chapter 15
PLATING	Plating data	Section 14.3, p. 437
PLYWOOD	Plywood data (full set)	Table 16.5, p. 497
PLYWOOD1	Plywood data (short version)	Table 10.11, p. 319
POPLARS	Poplar trees data	Table 16.7, p. 499
POTATO	Potato rot data	Table 10.16, p. 324
PRES	Presidents data set	Table 13.6, p. 422
PULSE	Pulse data	This appendix, Table 2.2, p. 52, Section 4.4, p. 111, Section 4.5, p. 121
PULSE1	Pulse data (portion of full data set)	Table 2.2, p. 52, Section 4.3, p. 132
REACTOR	Penn State University nuclear reactor data	Table 6.4, p. 213

Table A.1 Data sets available with Minitab as saved Worksheets (continued)

Name	Description	Text citation
RESTRNT	Restaurant data	This appendix, Section 4.7, p. 127
SAT	SAT scores data	Table 11.4, p. 373
SCORES	Test scores data	Table 11.1, p. 327
STREAM	Stream data	Section 11.5, p. 346, Table 11.9, p. 378, Table 11.10, p. 378
TRACK 15	Olympic track data, 1500 m	Table 11.6, p. 374
TRACKM	Men's Olympic track data	Table 5.2, p. 157
TRACKMW	Track records data for men and women	Table 5.6, p. 176
TREES	Trees data	This appendix, Exercises 2-12, p. 67, 2-13, p. 67, 11-2, p. 371, 11-12, 375
TUMOR	Tumor data	Table 11.13, p. 381
TWAIN	Mark Twain data	Table 12.14, p. 401, Table 12.15, p. 402
WIND	Wind chill data	Table 5.8, p. 177
WINE	Wine data	Table 13.9, p. 426

Table A.2 Additional data sets available from the Data Library associated with this book at www.CengageBrain.com

Name	Description	Text citation
Absent	Company absenteeism data	Table 14.5, p. 450
AllSamples	All possible samples from a small population	Exercise 7-14, p. 239
Areas	The areas of 100 properties	Table 7.1, p. 227
BreadLoaves	Weights of bread loaves from a bakery	Exercise 14-3, p. 448
CapTorque	Procter & Gamble study of caps on containers	Exercise 14-1, p. 447
ChestSize	Chest sizes on 5738 Scottish soldiers	Table 6.3, p. 208
Counties	Population sizes of the 100 Iowa counties	Figure 4.21, p. 120, Figure 4.22, p. 121
EdwinMoses	Edwin Moses' winning times in 122 races	Table 5.12, p. 181
Forbes 2011 Top 100	100 richest Americans as listed by *FORBES* magazine in September 2011	Brushing, p. 107
Forbes 2011 Top 400	400 richest Americans as listed by *FORBES* magazine in September 2011 in spreadsheet format	Excel spreadsheet p. 94
Helium Football	Will a football filled with helium kick farther than one filled with ordinary air?	Exercise 9-8, p. 294, Exercise 9-18, p. 298
Maze Times	Times for males and females to complete maze	Exercise 9-16, p. 298

Name	Description (continued)	Text citation
Moving	Man-hours and volume for a moving company	Exercise 11-42, p. 384
Property	Various variables on real properties	Table 5.3, p. 163
Redwoods	Height and diameter of 21 California redwoods	Exercise 11-41, p. 383
Richest2002.xls	*Excel* worksheet file of richest Americans in 2002	Section 3.3, p. 92
SearchSites	Data on hits on internet search sites	Exercise 4-20, p. 141
Temperatures.dat	A *text file* of monthly average temperatures over several years in Dubuque, Iowa	Table 2.3, p. 53
UIAges	Ages of University of Iowa students	Figure 4.18, p. 117, Figure 4.20, p. 119
USRichest2002	Forbes magazine data of richest Americans	Exercise 4-11, p. 140, Section 4.3, p. 102
WindChill	Wind chill data in an alternative layout	Table 5.9, p. 178

The datasets below are provided as Minitab Worksheets for your convenience.

Name	Description	Text citation
Battery	Battery temperature and voltage measurements	Table 11.7, p. 374
BirdFat	Migratory bird data sampled on two dates	Table 13.7, p. 424
BookPrices	Book length and price data	Table 11.3, p. 371
Boys Shoes	Amount of wear on boys' shoes	Table 9.2, p. 277
ChestGirth	Chest girth for 1516 soldiers, 1885 Potomac army	Table 4.2, p. 115, Table 6.1, p. 183
Gravity	Force of gravity measurements	Exercise 9-13, p. 296
Magnify	Apparent size and distance of an object	Table 11.14, p. 382
Samples	Ten samples each of size 5	Table 14.3, p. 449
TaskTime	Times to complete a task	Table 9.8, p. 297
Thickness	Coating thickness experiment with three factors	Table 16.4, p. 492

Cartoon

When educators make an instructional film, they have two objectives: Will the people who watch the film learn the material as efficiently as possible? Will they retain what they have learned?

To help answer these questions, an experiment was conducted to evaluate the relative effectiveness of cartoon sketches and realistic photographs, using both color and black-and-white visual materials.

A short instructional slide presentation was developed. The topic chosen for the presentation was the behavior of people in a group situation and, in particular, the various roles or character types that group members often assume. The presentation consisted of a five-minute lecture on tape, accompanied by eighteen slides. Each role was identified as an animal. Each animal was shown on two slides: once in a cartoon sketch and once in a realistic picture. All 179 participants saw all of the eighteen slides, but a randomly selected half of the participants saw them in black-and-white while the other half saw them in color.

After they had seen the slides, the participants took a test (immediate test) on the material. The eighteen slides were presented in a random order, and the participants wrote down the character type represented by each slide. They received two scores: one for the number of cartoon characters they correctly identified and one for the number of realistic characters they correctly identified. Each score could range from 0 to 9, since there were nine characters.

Four weeks later, the participants were given another test (delayed test) and their scores were computed again. Some participants did not show up for this delayed test, so their scores were given the missing-value code,*.

The primary participants in this study were preprofessional and professional personnel at three hospitals in Pennsylvania involved in an in-service training program. A group of Penn State undergraduate students were also given the test as a comparison. All participants were given the OTIS Quick Scoring Mental Ability Text, which yielded a rough estimate of their natural ability.

Some questions that are of interest here are as follows: Is there a difference between color and black-and-white visual aids? Between cartoon and realistic depictions? Is there any difference in retention? Does any difference depend on educational level or location? Does adjusting for OTIS scores make any difference? The data are stored in the Minitab worksheet called Cartoon.

Description of Cartoon Data Variables

Variable	Description
1 ID	Identification number
2 COLOR	0 = black-and-white, 1 = color (no participant saw both)
3 ED	Education: 0 = preprofessional, 1 = professional, 2 = college student
4 LOCATION	Location: 1 = hospital A, 2 = hospital B, 3 = hospital C, 4 = Penn State student
5 OTIS	OTIS score: from about 70 to about 130
6 CARTOON1	Score on cartoon test given immediately after presentation (possible scores are 0, 1,..., 9)
7 REAL1	Score on realistic test given immediately after presentation (possible scores are 0, 1,..., 9)
8 CARTOON2	Score on cartoon test given four weeks (delayed) after presentation (possible scores are 0, 1,..., 9. * is use for missing observations.)
9 REAL2	Score on realistic test given four weeks (delayed) after presentation (possible scores are 0, 1,..., 9. * is use for missing observations.)

Furnace

Wisconsin Power and Light studied the effectiveness of two devices for improving the efficiency of gas home-heating systems. The electric vent damper (EVD) reduces heat loss through the chimney when the furnace is in its off cycle by closing off the vent. It is controlled electrically. The thermally activated vent damper (TVD) is the same as the EVD, except that it is controlled by the thermal properties of a set of bimetal fins set in the vent. Ninety test houses were used, 40 with TVDs and 50 with EVDs. For each house, energy consumption was measured for a period of several weeks with the vent damper active and for a period with the damper not active. This should help show how effective the vent damper is in each house.

Both overall weather conditions and the size of a house can greatly affect energy consumption. A simple formula was used to try to adjust for this. Average energy consumed by the house during one period was recorded as (consumption)/[(weather)(house area)], where consumption is total energy consumption for the period, measured in BTUs, weather is measured in number of degree days, and house area is measured in square feet. In addition, various characteristics of the house, chimney, and furnace were recorded for each house. A few observations were missing and recorded as *, Minitab's missing-data code.

The data are stored in the Minitab worksheet called Furnace.

Description of Furnace Data Variables

Variable	Description
1 TYPE	Type of furnace: 1 = forced air, 2 = gravity, 3 = forced water
2 CH.AREA	Chimney area
3 CH.SHAPE	Chimney shape: 1 = round, 2 = square, 3 = rectangular
4 CH.HT	Chimney height (in feet)
5 CH.LINER	Type of chimney liner: 0 = unlined, 1 = tile, 2 = metal
6 HOUSE	Type of house: 1 = ranch, 2 = two-story, 3 = tri-level, 4 = bi-level, 5 = one and one-half stories
7 AGE	House age in years (99 means 99 or more years)
8 BTU.IN	Average energy consumption with vent damper in
9 BTU.OUT	Average energy consumption with vent damper out
10 DAMPER	Type of damper: 1 = EVD, 2 = TVD

Grades (Also Ga, Gb)

Scholastic Aptitude Tests (SAT) are often used as criteria for admission to college, as predictors of college performance, or as indicators for placement in courses. The data below are a sample of SAT scores and freshman-year grade-point averages (GPAs) from a northeastern university. (The university wishes to remain anonymous.) The sample of 200 students was broken down randomly into two samples of size 100 for ease of use in this handbook. These two samples are stored as Ga and Gb. The combined sample is stored as Grades.

Description of Grades Data

Variable	Description
1 VERB	Score on verbal aptitude test
2 MATH	Score on mathematical aptitude test
3 GPA	Grade-point average (0 to 4, with 4 the best grade)

Lake

These lakes are all in the Vilas and Oneida counties of northern Wisconsin. Measurements were made in 1959–1963. The data are stored in the Minitab worksheet called Lake.

Description of Lake Data Variables

Variable	Description
1 AREA	Area of lake in acres
2 DEPTH	Maximum depth of lake in feet
3 PH	pH, a measure of acidity (a lower pH is more acidic; a pH of 7 is neutral; a higher pH is more alkaline)
4 WSHED	Watershed area in square miles
5 HIONS	Concentration of hydrogen ions

Peru

A study was conducted by some anthropologists to determine the long-term effects of a change in environment on blood pressure. In this study they measured the blood pressure of a number of Indians who had migrated from a very primitive environment, high in the Andes mountains of Peru, into the mainstream of Peruvian society, at a much lower altitude.

A previous study in Africa had suggested that migration from a primitive society to a modern one might increase blood pressure at first, but that blood pressure would tend to decrease back to normal over time.

The anthropologists also measured the height, weight, and a number of other characteristics of each subject. A portion of their data is given below. All these data are for males over 21 who were born at a high altitude and whose parents were born at a high altitude. The skin-fold measurements were taken as a general measure of obesity. Systolic and diastolic blood pressure are usually studied separately. Systolic blood pressure is often a more sensitive indicator.

The data are stored in the Minitab worksheet called Peru.

Description of Peru Data Variables

Variable	Description
1 AGE	Age in years
2 YEARS	Years since migration
3 WEIGHT	Weight in kilograms (1kg = 2.2 lb)
4 HEIGHT	Height in millimeters (1mm = 0.039 in.)
5 CHIN	Chin skin fold in millimeters
6 FOREARM	Forearm skin fold in millimeters
7 CALF	Calf skin fold in millimeters
8 PULSE	Pulse rate in beats per minute
9 SYSTOL	Systolic blood pressure
10 DIASTOL	Diastolic blood pressure

Pulse

Students in an introductory statistics course participated in a simple experiment. The students took their own pulse rates (which is easiest to do by holding the thumb and forefinger of one hand on the pair of arteries on the side of the neck). They were then asked to flip a coin. If their coin came up heads, they were to run in place for one minute. Then all the students took their own pulses again. The pulse rates and some other data are given below. The data are stored as PULSE.

Description of Pulse Data Variables

Variable	Description
1 PULSE1	First pulse rate
2 PULSE2	Second pulse rate
3 RAN	1 = ran in place, 2 = did not run in place
4 SMOKES	1 = smokes regularly, 2 = does not smoke regularly
5 SEX	1 = male, 2 = female
6 HEIGHT	Height in inches
7 WEIGHT	Weight in pounds
8 ACTIVITY	Usual level of physical activity: 1 = slight, 2 = moderate, 3 = considerable

Restaurant

The survey is described in Chapter 4. The data are stored in the worksheet called Restrnt.

Description of Restaurant Data Variables

Variable	Description
1 ID	Identification number
2 OUTLOOK	Values 1, 2, 3, 4, 5, 6, 7, denoting outlook, from very unfavorable to very favorable
3 SALES	Gross 1979 sales, in $1000s
4 NEWCAP	New capital invested in 1979, in $1000s
5 VALUE	Estimated market value of the business, in $1000s
6 COSTGOOD	Cost of goods sold as a percentage of sales
7 WAGES	Wages as a percentage of sales
8 ADS	Advertising as a percentage of sales
9 TYPEFOOD	1 = fast food, 2 = supper club, 3 = other
10 SEATS	Number of seats in dining area
11 OWNER	1 = sole proprietorship, 2 = partnership, 3 = corporation
12 FT.EMPL	Number of full-time employees
13 PT.EMPL	Number of part-time employees
14 SIZE	Size of restaurant: 1 = 1 to 9.5 full-time equivalent employees, 2 = 10 to 20 full-time equivalent employees, 3 = over 20 full-time equivalent employees, where full-time equivalent employees equals (number of full-time) + (1/2)(number of part-time)

Trees

Foresters need to be able to estimate the amount of timber in a given area of a forest. Therefore, they need a quick and easy way to determine the volume of any given tree. Of course, it is difficult to measure the volume of a tree directly. But it is not too difficult to measure the height, and even easier to measure the diameter. Thus the forester would like to develop an equation or table that makes it easy to estimate the volume of a tree from its diameter and/or height. A sample of trees of various diameters and heights was cut, and the diameter, height, and volume of each tree was recorded. Below are the results of one such sample. This sample is for black cherry trees in the Allegheny National Forest, Pennsylvania. (Of course, different varieties of trees and different locations will yield different results, so separate tables are prepared for each variety and each location.) The data are stored in the worksheet called Trees.

Description of Tree Data Variables

Variable	Description
1 DIAMETER	Diameter in inches at 4.5 feet above ground level
2 HEIGHT	Height of tree in feet
3 VOLUME	Volume of tree in cubic feet

Credits

Chapter 2

52: Table 2.2, Pulse data from a classroom experiment by Brian L. Joiner. **67: Exercise 2-12**, Trees data from H. Arthur Meyer, *Forest Mensuration* (State College, PA: Penns Valley Publishers, Inc., 1953).

Chapter 4

94: http://www.forbes.com/forbes-400/

Chapter 4

527: Lake data collected by the Wisconsin Department of Natural Resources and provided by Alison K. Pollack and Joseph M. Eilers.

Chapter 5

146: Table 5.1, Furnace data from a study conducted by Tim LaHann and Dick Mabbott of the Wisconsin Power and Light Company. **153: Figure 5.8**, Cartoon data from Stephen Kauffman, "An Experimental Evaluation of the Relative Effectiveness of Cartoons and Realistic Photographs in Both Color and Black and White Visuals in In-Service Training Programs," Master's thesis, The Pennsylvania State University, 1973. **176: Table 5.6**, Men's Olympic track data from *The World Almanac and Book of Facts*, 1993 (New York: Pharos Books, 1993). **169: Table 5.4**, Employment data from R. B. Miller and Dean W. Wichern, *Intermediate Business Statistics* (New York: Holt, Rinehart & Winston, 1977). Copyright © 1977 by Holt, Rinehart & Winston. **174: Figure 5.33**, Radiation data from the Las Vegas Radiation Facility, Las Vegas, Nevada. **175: Table 5.5**, Marriage rate data from *The World Almanac and Book of Facts*, 1993 (New York: Pharos Books, 1993). **176: Table 5.6**, Track records data from *The World Almanac and Book of Facts*, 1993 (New York: Pharos Books, 1993). **177: Table 5.7**, Temperatures for various cities from *The World Almanac and Book of Facts*, 1993 (New York: Pharos Books, 1993). **177: Table 5.8**, Wind chill data from the U.S. Army. **179: Exercise 5-13**, *Automobile accident data from Paul M. Hurst, "Blood Test Legislation in New Zealand," Accident Analysis and Prevention 10* (Pergamon Press, Ltd. 1978), pp. 287–296. Copyright © 1978, Pergamon Press, Ltd. **181: Exercise 5.11**, Leaf surface resistance data provided by Thomas Starkey of The Pennsylvania State University.

Chapter 6

183: Table 6.1, Chest girth data from Roger Carlson, "Normal Probability Distributions," in *Statistics by Example: Detecting Patterns*, F. Mosteller, W. H. Kruskal, R. F. Link, R. S. Pieters, and G. R. Rising, editors (Reading, MA: Addison-Wesley, 1973). Copyright © 1973 Addison-Wesley Publishing Company, Inc. **208: Table 6.3**, Chest size data from S. M. Stigler, *The History of Statistics*, (Belknap Press of Harvard University, 1986), p. 207. **213: Table 6.4**,

Penn State nuclear reactor data from testimony presented to the Atomic Energy Commission, 1971.

Chapter 9

276: Table 9.1, Cholesterol data for heart attack victims collected by a Pennsylvania medical center, 1971. **277: Table 9.2**, J. S. Hunter, *Statistics for Experimentation* (New York: Wiley & Sons, 1978). Copyright © 1978 by John Wiley & Sons, Inc. Reprinted by permission. **282: Table 9.3**, Parkinson's disease data obtained by Gastone Celesia in a study at the University of Wisconsin. **294: Exercise 9-8** and **298: Exercise 9-18**, DASL Data Library http://lib.stat.cmu.edu/DASL/Datafiles/Heliumfootball.html. **298: Exercise 9-16**, Times to complete a maze, http://www.amstat.org/education/stew/pdfs/AnAmazingComparison.pdf.

Chapter 10

300: Table 10.1, Fabric flammability test conducted by the American Society for Testing Materials. Related analyses are "ASTM Studies DOC Standard FF-30-71 on Flammability of Children's Sleepwear," Materials Research Standards, May 1972; and John Mandel, Mary N. Steel, and L. James Sharman, "National Bureau of Standards Analysis of the ASTM Interlaboratory Study of DOC/FF 3-71 Flammability of Children's Sleepwear," ASTM Standardization News, May 1973. **305: Table 10.3**, Driving performance data from David C. Howell, *Statistical Methods for Psychology*, Third Edition (Boston: PWS-Kent, 1992), p. 418. **309: Table 10.4**, Pancake data provided by Gregory Mack and John Skilling. **314: Table 10.6**, Billiard ball data from Albert Romano, *Applied Statistics for Science and Industry* (Boston: Allyn and Bacon, Inc., 1977), p. 300. **319: Table 10.11**, Plywood data from a study directed by Frank J. Fonszak of the U.S. Forest Products Research Laboratory. **320: Table 10.12**, Pendulum data from John Mandel, "The Acceleration of Gravity," in *Statistics by Example: Detecting Patterns*, F. Mosteller, W. H. Kruskal, R. F. Link, R. S. Pieters, and G. R. Rising, editors (Reading, MA: Addison-Wesley, 1973.) Copyright © 1973 Addison-Wesley Publishing Company, Inc. **321: Table 10.13**, Manganese in cow liver data from the National Institute of Standards and Technology. **324: Table 10.16**, Potato rot data collected at the University of Wisconsin. **325: Table 10.17**, Alfalfa data from an experiment conducted by the University of Wisconsin. **326: Table 10.18**, Meat loaf data from a study at the University of Wisconsin conducted by Barbara J. Bobeng and Beatrice David.

Chapter 11

327: Table 11.1, Test score data collected in a statistics class at The Pennsylvania State University. **378: Figure 11.9**, Stream data gathered as part of an environmental impact study. **373: Table 11.4**, SAT scores data set from The Digest of Educational Statistics: 1983–1984 Volume, U.S. Government Publication, and *The World Almanac and Book of Facts*, 1993 (New York: Pharos Books, 1993). **373: Table 11.5**, SAT scores data set from *The Digest of Educational Statistics: 1983–1984 Volume, U.S. Government Publication, and The World Almanac and Book of Facts*, 1993 (New York: Pharos Books, 1993); **374: Table 11.6**, Track data from *The World Almanac and Book of Facts*, 1993 (New York: Pharos Books, 1993). **376: Table 11.8**, Maple tree data provided by Erik V. Nordheim of the University of Wisconsin. **378: Table 11.9**, Stream data gathered as part of an environmental impact study. **378: Table 11.10**, Stream data gathered as part of an environmental impact study. **380: Table 11.12**, Regression data from F. J. Anscombe, "Graphs in Statistical Analysis," *The American Statistician* 27(1), February 1973, pp. 17–21. Reprinted by permission of the American Statistical Association. **382: Table 11.14**, Magnifying glass data collected by a student in an introductory physics class. **383: Exercise 11-41**, from *Even You Can Learn Statistics* by David Levine and David Stephan, Kindle edition. **384: Exercise 11-42**, from *Even You Can Learn Statistics* by David Levine and David Stephan, Kindle edition.

Chapter 12

385: Table 12.1, Penny tossing data described in W. J. Youden, *Risk, Choice and Prediction* (North Scituate, MA: Duxbury Press, 1974). **389: Table 12.2**, Further discussions of the ESP data are given in Marvin Lee Moon, "Extrasensory Perception and Art Experience," Ph.D. thesis, Pennsylvania State University, 1973. An additional article by Moon appeared in *American Journal of Research*, April 1975. **396: Table 12.6**, Accidental deaths data from *The World Almanac and Book of Facts*, 1984 (New York: Newspaper Enterprise Association, Inc., 1984). **397: Table 12.7**, Precipitation data collected in State College, Pennsylvania; Table 12.8, Migratory geese data from R. K. Tsutakawa, "Chi-Square Distribution by Computer Simulation," in *Statistics by Example: Detecting Patterns*, F. Mosteller, W. H. Kruskal, R. F. Link, R. S. Pieters, and G. R. Rising, editors (Reading, MA: Addison-Wesley, 1973.) Copyright © 1973 Addison-Wesley Publishing Company, Inc. **398: Table 12.9**, Harrisburg, Pennsylvania infant mortality data from Vilma Hunt and William Cross, "Infant Mortality and the Environment of a Lesser Metropolitan County: A Study Based on Births in One Calendar Year," *Environmental Research* 9, 1975, pp. 135–151; **398: Table 12.10**, Harrisburg, Pennsylvania infant mortality data from Vilma Hunt and William Cross, "Infant Mortality and the Environment of a Lesser Metropolitan County: A Study Based on Births in One Calendar Year," *Environmental Research* 9, 1975, pp. 135–151. **399: Table 12.11**, Harrisburg, Pennsylvania infant mortality data from Vilma Hunt and William Cross, "Infant Mortality and the Environment of a Lesser Metropolitan County: A Study Based on Births in One Calendar Year," *Environmental Research* 9, 1975, pp. 135–151; **399: Table 12.12**, Car defects data from L. A. Klimko and Camil Fuchs, "An Analysis of the Data from Wisconsin's Dealer Based Demonstration Motor Vehicle Inspection Program," Statistical Laboratory Report 77/7, Department of Statistics, University of Wisconsin, Madison, 1977. **400: Table 12.13**, Car defects data from L. A. Klimko and Camil Fuchs, "An Analysis of the Data from Wisconsin's Dealer Based Demonstration Motor Vehicle Inspection Program," Statistical Laboratory Report 77/7, Department of Statistics, University of Wisconsin, Madison, 1977. **401: Table 12.14**, Mark Twain data from Claude Brinegar, "Mark Twain and the Quintus Curtius Snodgrass Letters: A Statistical Test of Authorship," *Journal of the American Statistical Association* 58, 1963, pp. 85–96.

Chapter 13

420: Table 13.4 and **421: Table 13.5**, Acid data was collected in a chemistry class. **422: Table 13.6**, Presidents data from *The World Almanac and Book of Facts*, 1984 (New York: Newspaper Enterprise Association, Inc., 1984). **424: Table 13.7**, Bird fat data from Jack Hailman, "Notes on Quantitative Treatments of Subcutaneous Lipid Data," *Bird Banding* 36, 1965, pp. 14–20. **425: Table 13.8**, Azalea data from a study done at The Pennsylvania State University by Professor Stanley Pennypacker, Department of Plant Pathology. **426: Table 13.9**, Wine data from W. Kwan, B. R. Kowalski, and R. K. Skogerboe, *Journal of Agriculture and Food Chemistry* 27, 1979, p. 1321.

Chapter 14

428: Table 14.1, Ignition key data from N. Farnum, *Modern Statistical Quality Control and Improvement* (North Scituate, MA: Duxbury Press, 1994), p. 185. **439: Figure 14.8**, Plating data from *Statistics for Business: Data Analysis and Modeling, 2 edition*, J. D. Cryer and R. B. Miller (Belmont, CA: Duxbury Press, 1994), p. 317. **447: Table 14.2**, Cap torque data from *Statistics for Business: Data Analysis and Modeling*, J. D. Cryer and R. B. Miller (Belmont, CA: Duxbury Press, 1993), p. 292. **449: Table 14.4**, Documents data from N. Farnum, Modern Statistical Quality Control and Improvement (North Scituate, MA: Duxbury Press, 1994), p. 245. **448: Exercise 14-3**, from *Statistical Quality Control*, J. Ledolter and C. W. Burrill (New York: John Wiley & Sons, Inc., 1999), p. 321. **449: Exercise 14-6**, from *The Practice of Business Statistics*, D. S. Moore, G. P. McCabe, W. M. Duckworth, and S. L. Sclove (New York: W. H. Freeman, 2003). p. 12-62.

Chapter 16

497: Table 16.5, Plywood data from a study directed by Frank J. Fonszak of the U.S. Forest Products Research Laboratory. **498: Table 16.6**, Driving performance data from David C. Howell, *Statistical Methods for Psychology*, Third Edition (Boston: PWS-Kent, 1992), p. 418. **499: Table 16.7**, Poplar data from P. R. Blankenhorn, T. W. Bowersox, C. H. Strauss, and others, "Net energy and economic analyses for producing Populus hybrid under four management strategies," final report to U.S. Department of Energy, Washington, DC, ORNL/Sub/79-07918/1, 1985.

Appendix

528: Peru data from P. T. Baker and C. M. Beall, "The Biology and Health of Andean Migrants: A Case Study in South Coastal Peru," *Mountain Research and Development* 2(1), 1982. **530**: Restaurant data from The Wisconsin Restaurant Survey by William A. Strang of the Small Business Development Center, University of Wisconsin, Madison, 1980. We also thank Robert Miller for his help with the analysis of these data and with the writing of Chapter 5 for earlier editions.

Index